Application of
Directional Blasting in Mining
and Civil Engineering

Application of Directional Blasting in Mining and Civil Engineering

Second revised and enlarged edition

A.A. Chernigovskii

1986
A.A. BALKEMA/ROTTERDAM

Translation of:

Primenenie Napravlennogo Vzryva v Gornom Dele i Stroitel'stve.
Nedra Publishers, Moscow, 1976

© *1986 Copyright reserved*

Translator: Subhash C. Dhamija
Editor: V. Pandit

ISBN 90 6191 573 2

Printed in India at Printsman Press, Faridabad

UDC 235.674.5

The author describes the mechanism of breaking mineral rocks through blasting and offers practical recommendations on the application of directional blasting in mining and civil engineering. He also introduces criteria for cost effectiveness of transporting rock mass from stope to pile by means of energy of explosion, and presents ballistic tables to be used in constructing the trajectory of movement of a rock fragment during blasting operations while taking into account the topography of the land. As an improvement upon the first edition, the author presents a more detailed theoretical treatment on the mechanism of fragmentation of a monolithic or fissured rock mass through blasting.

The book is intended for engineers and technicians planning blasting operations. It will act as a useful aid for researchers involved with the problem of improving blast efficiency.

Contents

viii

List of Symbols Used in the Text and Their Definitions

A_s width of the stope, m

a distance between concentrated charges, m

a_{sec} distance between consecutive sections, m

B_{ch} width of the slab charge, m

b_d drag factor, m^{-1}

c exponent for deep-hole charge

c_{inst} a constant accounting for the process of expansion of the gas chamber

c_{rock} velocity of sound in rock, m/s

c_x head drag coefficient

D distance between center of gravity of blasted rock before and after explosion along the horizontal, m

D_{cal} calculated distance of the boundary of the muck pile, m

D_{det} rate of detonation

$D_{g.c}$ diameter of gas chamber, m

D_m mean distance of muck-pile edge from contour of slab charge and height of rock, m

D_s slant distance, m

D_v range of fragment trajectory in vacuum, m

D_w velocity of propagation of shock-wave front in the medium, m/s

d hole diameter, m

E modulus of elasticity

E_{pot} potential energy, J

e specific energy required for the formation of a unit area of the re-formed surface, J/m^2

e_{exp} specific energy of explosion, J/kg

F_{adit} area of cross section of the charge adit, m^2

F_{cham} surface area of the charge chamber, m^2

F_{cone} area of the base of an elementary cone of rock throw, m^2

f_1 reduced cross-sectional area of charge adit, m^2

f_{acc} accuracy factor

f_{dir} coefficient of directivity

F_{fr} coefficient of static friction

$F_{fr.d}$ coefficient of dynamic friction

f_{ob} focal length of the objective, m

G_{sh} shear modulus

g acceleration due to gravity, m/s^2

H height of the gas chamber, m

H_0 height of gas chamber for a slab charge of infinite length and breadth, m

H_{ch} length of the charge, m

$H_{g.c}$ thickness of the gas chamber along the normal to the slab-charge system, m

H_h length of the hole, m

h overburden thickness, m

h_m mean thickness of rock layer as it moves in the air, m

i proportionality factor which depends on the physicomechanical properties of soil and explosives

i_e index of displacement by explosion

j_{air} retardation due to air drag, m/s^2

k edge effect

k_{adit} $(\eta/\eta_0)_F$

k_b size coefficient depending on $\hat{x}_1$

k_{ch} coefficient of loading

k_{dy} coefficient taking into account the dynamic head and elastic properties of the medium

k_e computed specific consumption of explosive charge for blasting, kg/m^2

k_{exp} computed specific consumption of explosives, kg/m^2

$k_{f.f}$ form factor coefficient

k_{fr} average coefficient of friction

k_g correction factor for gravity

k_{loc} ratio between the maximum and the actual average sizes of fragments

k_{loos} coefficient of loosening

k_{safe} coefficient depending on the physicomechanical properties of rock

k_U $(\eta/\eta_0)_U$

k_w correction for explosion depth

L charge length, m

L_{ch} total length of the charged areas of the hole in one section, m

L_g total length of the trajectory from the origin to the point O, m

L_h height of the photographic frame, m

L_{hole} depth of the blast hole, m

L_p pile width, m

L_{print}	print width, m
l	height gradient
l_{crack}	width of the fissure opening, m
l_{stem}	length of stemming, m
ΔM	mass of the rock, kg
$m_a = a/W$	coefficient of convergence for concentrated charges in a given row
m_b	critical coefficient of convergence for linear charges or rows of concentrated charges
N	number of holes or cracks
N_f	number of re-formed fragments of size x_f
N_{sec}	number of holes in one section
n	exponent depending on correlation between trajectory length and breadth of slab charge
n_e	blast-efficiency factor
P	mean mass per unit length of a linear charge, kg/m
P_0	mass per unit length of charge
P_{actual}	actual value of peripheral charge
P_h	mass of explosive in 1 m length hole, kg/m
P_{lat}	actual value of mass of lateral charges per unit length, kg/m
P_{sub}	mass per unit length of the additional charges, kg/m
Q	mass of one row of charge, kg
Q_{eff}	effective mass of concentrated charges, kg
Q_{sec}	mass of explosive charges in all the holes of one section, kg
Q_{sp}	mass of a spherical charge, kg
q	specific consumption of explosives, kg/m^3
q_{actual}	actual specific consumption of explosives, kg/m^2
q_{imp}	specific consumption of explosives in case of implosion, kg/m^3
q_{loos}	specific consumption of loosening
q_m	mean specific consumption of explosives, kg/m^3
R	radius of the sphere occupied by explosion products, m
R_1	distance between center of gas chamber and free face at any given moment of time, m
R_{cen}	the distance between the center of gravity of the volume of explosion products and the center of the gas chamber, m
R_{cham}	radius of gas chamber, m
R_{crat}	radius of crater, m
R_{first}	radius of gas chamber at the end of first phase of explosion, m
R_0	radius of the charge chamber at initial pressure, m
$\hat{r}_{def}$	relative distance of the fragment undergoing shear deformation from the charge
r_0	radius of the charge, m
S	distance in the vertical plane between the center of gravity of the blasted rock before and after explosion, m

S_{air} coefficient characterizing the retarding action of air

S_{ch} area of cross section of a charge, m^2

S_{crack} surface area of microfissures in a unit volume of crushed rock, m^2

S_f area of the middle section of a rock fragment, perpendicular to the velocity vector, m^2

S_p area of the muck pile, m^2

S_Σ total surface area of all fragments in a unit volume of crushed rock, m^2

T duration of flight of a rock fragment on blasting, s

Δt_b time required for breaking of rock due to shock stress, s

t_{def} time during which shear deformation takes place, s

t_{del} delay time, s

t_{eff} the time in which the rock fragment in the gas stream has attained half its velocity, s

t_{exp} time for expansion of explosion products, s

t_{sh} time in which an elementary layer of rock is broken under the effect of shock-shear deformation, s

t_w time of arrival of compressional wave, s

U volume of rock with fragments varying in size from 0 to x_f, m^3

U_{adit} adit volume, m^3

U_{ch} volume of charge, m^3

U_{cham} chamber volume, m^3

U_{def} potential energy of deformation, J

U_e^+ content of large separated blocks in the rock mass before blasting

U_f true volume of the fragment, m^3

$U_{g \cdot c}$ volume of gas chamber, m^3

U_l ratio between the volume of a large-size fragment and the total volume of the rock mass

U_{sec} volume of adit or hole section between blocking charges, m^3

U_{stem} volume of compressed stemming, m^3

u specific volume of explosion products, m^3

u_{cr} increase in the critical value of specific volume of explosion products, m^3

u_{det} specific volume of explosion products at the moment of detonation, m^3

u_{rel} relative velocity, m/s

v mass velocity at the shock-wave front, m/s

v_0 initial velocity of projection of blasted rock, m/s

v_{cr} critical velocity for breaking of rock, m/s

v_{crack} velocity of propagation of a crack of general breaking, m/s

v_{ful} direction of projection of exploded rock which coincides with the direction of fulmination of the charge system

v_{max} maximum velocity of the medium (rock) under the action of gravity and air drag, m/s

v_{pr} component of the initial velocity of projection on the free face, m/s

v_{φ} initial velocity of projection of blasted rock at an angle φ to the line of least resistance, m/s

W depth of laying of concentrated charges, m

W_{des} depth of design pit, m

$\hat{x}$ relative size of rock fragment, m

x_{crack} size of an individual rock determined by the network of fissures, m

x_f size of fragment, m

x_{fine} size of a fine fraction of rock fragment, m

$\hat{x}_l$ mean size of large fragment of rock, m

x_m mean fragment size, m

z_{rest} coefficient of restoration

α_{opt} optimum angle of inclination of the slab charge with the horizontal, deg

α_q correction coefficient of specific consumption of explosives

$\beta_{1\,max}$ maximum angle at which rock deposited in peripheral region of a slab charge is blown away, deg

β_b coefficient accounting for the symmetry of explosion, load, and the nature of brittle and elastoplastic breaking of rock

γ angle of rotation of the wave front of fragments to the horizontal, deg

γ_b breaking value for γ

γ_{lim} limiting angle of inclination of the charge, deg

γ_{thrust} coefficient of lateral thrust

ε relative deformation during compression of stemming due to shock wave

ζ_c coefficient depending on physicomechanical properties of rock to be blasted

η blast efficiency

η_0 blast efficiency of slab charge of infinite linear dimensions

η_{crush} portion of energy expended in crushing

η_{diss} portion of energy dissipated

η_{kin} portion of energy expended in providing kinetic energy to the fragments

η_{lim} limiting value of blast efficiency

η_{plast} energy utilized in plastic deformation

η_{rem} portion of energy remaining in explosion products

μ Poisson coefficient

μ_{cr} specific value of explosion products at pressure Σ_{cr}

μ_{diss} coefficient describing dissipation of explosion energy

ξ_i stiffness coefficient of impact

Π_{sp} springing index, m³/kg

ρ density of a medium in its natural state

ρ_{air} density of air

Σ pressure of explosion products

Σ_0 initial pressure of explosion products, N/m²

Σ_{at} atmospheric pressure

Σ_{cr} critical pressure of explosion products, N/m²

Σ_{det} pressure of charge detonation products, N/m²

Σ_y yield stress, N/m²

σ instantaneous breaking strength, N/m²

σ_1 tensile stress, N/m²

σ_2 compression stress, N/m²

σ_r radial component of compression stress, N/m²

σ_{red} reduced stress, N/m²

σ_t tangential component of tensile stress, N/m²

τ_b tangential stress at which rock breaks, N/m²

τ_{fr} frictional force, N/m²

ψ_{cy} coefficient of energy losses for a cylindrical charge

ψ_e coefficient of energy losses

ψ_{sp} coefficient of energy losses for a spherical charge

Ω ratio between the dimensions of the slab charge and height of the gas hole

Introduction

Theoretical investigations and experience in blasting operations reveal that the slab-charge system is most effective for achieving a maximum throw efficiency. An ideal slab charge is that charge, the length and breadth of which are considerably greater than its thickness. An example of an *ideal slab charge* is a thin plate of explosive material. Calculations show that in a majority of cases involving the use of slab charges for the purpose of explosive casting of rock, the thickness of the charge varies from a fraction of a millimeter to several millimeters. Evidently, it is very difficult to lay a slab charge of such thickness under rock to be exploded by the usual techniques. Therefore, in practice, an ideal slab charge can be replaced by an equally effective system of chamber or deep-hole charges, these being placed in the same plane at distances not exceeding a certain limiting value. During firing of such slab charges the compressional waves from each charge merge at a certain distance. The resultant merged wave is then propagated parallel to the plane of charges and moves the medium in accordance with the same principles as are applicable in the case of firing of a charge in the form of an ideal slab. For the sake of brevity, a slab system of chamber or deep-hole charges, will be hereafter referred to as a slab charge.

A good illustration of the action of slab charges is the firing of multirow chamber charges used in excavations and for creating wide trenches. Another example may be the firing of blast-hole charges arranged in a row and laid parallel to the free face. At particular ratios of the line of least resistance to the charge diameter and the distance between rows, rock is separated from the massif along the plane of the blast-hole charges, and moves in a direction perpendicular to this plane.

The main advantage of the slab-charge method over other known techniques is that in the case of blasting using slab charge, rock fragments traverse approximately along the normal to the free face. As a result, the blast characteristics, such as directivity and accuracy of throw are considerably improved. Rock fragments can be projected in a desired direction by arranging the slab charge at a particular angle to the horizontal.

A high accuracy of throw, as a rule, is very useful in overburden removal operations in the construction of dams, cofferdams, canals and other hydraulic engineering constructions. In these instances of utilization of blast energy, the rock as a whole, or a considerable portion of it, blown away as a result of blasting, must fall within the boundaries marked on the free face. For example, the upstream apron of the rock-fill dam of the Baipazinsk hydroelectric complex was made by blasting a gravel-loam mixture which had earlier been placed over slab charges. According to the plan, 80–90% of the blasted earth should have fallen within the apron contour. It has been confirmed through experience that such rigid requirements can, in principle, be met by using slab charges.

Equally rigid requirements for explosive casting of overburden to a muck pile are laid down in open-pit mining for mineral deposits. In the Soviet Union the application of the method of open-pit mining for mineral resources is becoming increasingly popular. For example, during 1950 to 1970, the proportion of the open-pit method for coal and iron-ore mining in the overall mine production increased by 2.5 and 1.5 times respectively [1]. The reason for this preference lies in the possibility of using heavy duty machines in open-pit mining as they are highly effective for excavation and removing muck in large volumes. Such a possibility is, however, absent in the case of underground mining. Therefore, the bulk of commercial mineral mining has steadily increased during the last decade, mainly on account of the progress made in the open-pit mining method.

Mining operations are immediately preceded, as a rule, by very heavy (in terms of scale and bulk of technical equipment) mining-capital outlays and overburden removal operations connected with breaking, extraction, loading and transportation of large volumes of rock, both covering and bearing minerals, to the muck pile [2, 3]. Drilling and blasting methods are mainly used for loosening the rocks before they are extracted and removed. However, the application of blasting is not confined to loosening of hard rocks alone.

During the last decades, directional large-scale blasting methods have been widely used in overburden removal and mining operations in such open-pit mines as those at Altyn-Topkansk, Korkinsk, Cheremkhovsk, etc.

Overburden removal operations can be undertaken in two different ways: by the explosive excavation method, and with the help of directional explosive casting. In the latter case the blast not only breaks the massif but also moves a considerable part of the crushed rocks beyond the spoil contours. It should be borne in mind that blasting is carried out in the first case for breaking the overlying rocks before they are removed with excavators. However, blasting has only a secondary significance in these operations because in overburden removal operations (i.e. removal of the gangue from stope to bank), excavators and various means of transportation are being used.

Stripping of minerals covered with rocks involves several processes: blast loosening, excavation, transportation and dumping of rocks. Mass blasting used for overburden operations, combine these processes into a single operation.

Experience gathered by Soviet scientists over the last thirty years has revealed the following advantages of mass blasting [4–6]:

—it is a unique method of mechanization of labor-oriented digging and mining operations by virtue of which much work can be accomplished in a short time;

—large-scale blasting does not involve high-cost preparatory operations and highly skilled labor; hence, it is possible to considerably reduce the number of equipments, machinery and means of transportation;

—while carrying out directional large-scale blasting in overburden operations, the approach roads and a major part of the industrial and residential complexes can be simultaneously constructed along with the preparations for mass blasting. As a result, the open-pit mine can be ready for operation in a very short period.

Till recently, directional large-scale blasts used for various operations were conducted by the technique of chamber charges. It has been established through practice and theoretical studies that an increase in the mass of the chamber charge with a simultaneous increase in the depth of placing of the charge increases the effectiveness of large-scale blasts. However, the use of chamber charges for mining-capital outlays and overburden operations has a number of disadvantages, namely: difficulty in making a pit of the desired contour, impossibility of removing the entire blasted rock to the trench banks, high cost of rock blasted per unit volume, high consumption of explosives, etc. [1]. Besides, the directivity and accuracy of throw in the case of chamber charges is very poor because a considerable part of the blasted rock remains within the boundary of the designed pit or in its proximity. Most of these disadvantages have now been eliminated by devising new techniques of directional blasting, using cheaper explosives and high-efficiency drilling rigs.

Theoretical and experimental investigations recently carried out in the Soviet Union have revealed that slab charges, in addition to various other advantages, help in considerably improving such important blast characteristics as directivity and accuracy of throw [7–9] in explosive casting of overburden.

Research in the utilization of slab charges for directional rock blasting in the Soviet Union began for the first time, at the Siberian Division of the USSR Academy of Sciences, under the guidance of academician M.A. Lavrent'ev [7]. According to the newly proposed technique of directional blasting, a layer of explosives is placed around the rock to be displaced. The explosive layer, on detonation and subsequent transfer of a pulse through the rock, creates a kind of solid wall which prevents the rock fragments from flying in lateral directions. Theoretical study [10] has established that with such types of slab charges around encasing rock, a one hundred percent directional accuracy can be achieved.

Trial tests of this method [11] have confirmed theoretical conclusions.

However, the utilization of this technique entails an increased consumption of explosives and labor in comparison with the usual methods.

A one-sided directed blast is also possible with other techniques utilizing slab charges. In particular, investigations [8, 9, 12] have established that slab charges, including wedge-shaped charges, can be used for a fairly effective directional blasting provided the optimal conditions of blasting are observed. In such cases the positioning of deep-holes and chamber charges, which are substitutes for an ideal slab charge, and the sequence of explosions are highly significant.

The blast efficiency, which depends on parameters of stemming [13] assumes great practical importance in designing slab charges. The value of blast efficiency determines the economic advantages of using the slab-charge method. The effectiveness of directional blasting also depends on the selection of optimum blasting conditions (e.g. angle of inclination of the slab charge, thickness of the layer to be blasted and grades of explosives).

Theoretical and experimental investigations of slab charges using techniques for removing the blasted rock to long distances, were first begun in the Soviet Union. However, this method has not been widely accepted in construction and overburden removal operations in the Soviet Union mainly because adequately precise practical recommendations have not yet been evolved for determining the conditions under which the slab-charge method can be economically viable. Besides, there is no complete account of the basis for designing a slab-charge system for stripping minerals and for hydraulic-engineering installations.

An endeavor has been made in the present book to overcome the short-comings present in the existing text books on designing slab-charge systems under different mining and geological conditions.

Chapter 1

Principles of Designing
Slab-charge Systems

1. DISTINCTIVE FEATURES OF THE SLAB-CHARGE METHOD

The chief advantage of the slab-charge system over the chamber charge method is that the former affords an improvement in such practically important parameters of rock blasting as directivity and accuracy of throw. The directivity of throw can be qualitatively expressed as the directivity factor f_{dir}, which is numerically equal to the ratio between the volume of the muck projected in a given direction and the total volume removed to the free face by blasting. The term 'given direction' implies the direction toward that part of the area which is situated on one side of the vertical plane passing through the center-of-mass of the system of charges and lies perpendicular to the given plane of symmetry of the blast throw.

Another parameter of practical importance in directional blasting for removal of burden and construction operations is the accuracy of throw characterized by that part of blown-away volume of rock which is removed to a predetermined area on the free face. Numerically, this parameter is expressed as the *accuracy factor* f_{acc} which is equal to the ratio between the volume of the rock thrown away to a predetermined area and the total volume of rock moved to the free face through blasting. The accuracy factor for a slab-charge system can be higher than 90% under favorable conditions.

Let us study the peculiarities of rock movement when blasting is carried out with the help of chamber and slab-charge systems on stopes (Fig. 1). In the given instance the chamber charge has been prepared by charging the cavity already made in the rock massif and the slab charge has been made by parallel deep-hole charges arranged in one plane. In Fig. 1 the dotted lines indicate the trajectories of the rock blown away by firing a chamber charge and the solid lines show the trajectories of rock fragments blown away by a slab charge. The masses of chamber and slab charges are equal. As is evident, a chamber charge projects rock fragments in all directions. As a result, a considerable part of the rock remains within the limits of the open-pit mine.

2

In case of a free-face configuration, as shown in Fig. 1, the chamber charge blast moves only that volume of rock beyond the open-pit mine limits which is situated within the solid angle *BOC*. The blasted rock is dumped on the left side and has to be subsequently removed with excavators. Therefore, it is obvious that the chamber-charge technique under these conditions is uneconomic. In case of a slab charge the picture is different; the rock moves as a compact mass in a given direction. The reason for this phenomenon is that in this case the initial direction of trajectory of the rock fragments is perpendicular to the base of the slab charge. The blasted rock can be moved to the desired distance and in the required direction by varying the angle of inclination of the slab charge to the horizontal, and varying the mass of the charge as dictated by the specific conditions.

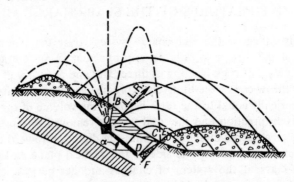

Fig. 1. Diagram of rock blasting along a slope on firing of a slab
or chamber charge of equal mass.

While plotting the trajectory of motion of a rock fragment on firing a chamber charge it was presumed that at the time of blasting the fragments move away radially from the center of the charge. In reality, the fragments do not strictly move in a radial direction. However, this feature does not greatly distort the general throw pattern.

Yet another advantage of the slab-charge method is that the blast efficiency is slightly greater than that of concentrated charges, other factors (such as equal mass of charges, similar rocks and equal depth of charge laying) being identical. The reason for this phenomenon is that a slab-charge blasting moves the entire volume of rock with equal speed except for a small quantity at the end areas (Fig. 2, a). Chamber-charge blasting under similar conditions (Fig. 2, b) differs significantly from slab-charge blasting in its high velocity gradient. As a result, forces of friction are developed which transform the energy of forward motion into heat and plastic deformation, thus reducing the blast efficiency of a chamber charge.

Since in a slab-charge blasting the fragments fly as a compact mass, air drag has much less effect on their movement as compared to that in chamber-

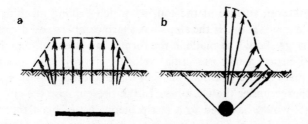

Fig. 2. Diagram of initial velocities of projection by firing of slab
charge or a chamber charge of equal mass.

charge blasting. Consequently, as proved through computations, slab charges
are helpful in moving a large mass of rock to comparatively longer distances
with a relatively low specific consumption of explosives. This is the major
difference between slab charges and chamber charges. During blasting of
chamber charges the rock moves in fragments and the range of its displacement is reduced sharply because of air drag.

In practice an ideal slab charge can be substituted by an equally effective
system of parallel linear (deep-hole, blast-hole, trench) charges (Fig. 3, a) or a
system of parallel rows of concentrated (chamber, sprung-hole) charges (Fig.
3, b), laid in one plane, at distances not exceeding certain limiting values.

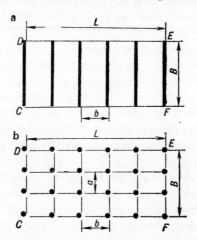

Fig. 3. Methods of preparing a system of slab charges:
a—system of linear charges, b—system of concentrated charges.

In the first case, a number of parallel holes are drilled, charges are laid in
them and the upper parts of the holes are filled with stemming. In the second
case, chambers are drilled in the massif to be blasted and they are then charged and stemmed. Chambers (tunnels) are also made by the springing method.
The parallel deep-hole charge method, as a rule, is used on slopes with a

more or less even surface at the top, on which drilling machines and other equipment can be set up. If the slope has sharp cambers and depressions then the slab charge should be made in the form of a system of chamber charges. In order to ensure a uniform initial velocity of projection along the entire slope, the mass of the chamber charges under the depression should be less than the charge under the chamber. In the present case the use of parallel-hole charges would have led to a sharp inequality of blast throw along the slope; at depressions the initial velocity of projection would have been excessively high, and at the chamber, very low and insufficient for moving the rock of this area to the predetermined distance.

Such a peculiarity of blasting of parallel-hole charges on uneven slopes arises due to the fact that the material to be blasted in the holes has the possibility of moving freely from a high-pressure zone to zones of lower pressure. Because the explosion products, while expanding in the direction of the line of least resistance at high velocity, create a low-pressure zone in the depression, the explosion products of other zones tend to move into this zone. As a result, the energy released on explosion is distributed extremely unevenly along the slope. This disadvantage is totally eliminated if instead of a number of deep-hole charges, we use a system of chamber charges, or if the holes are charged areawise and the intermediate space is filled with an inert material. In the given instance each chamber charge, or part of the deep-hole charge, exerts its energy only in the direction of that part of the rock which is located above these charges.

The system of concentrated (sprung-hole) charges is preferable if the ore, deposited under the horizontal free face, is required to be strip-mined.

It is necessary to emphasize the special significance of the tamping process in all those cases where rock blasted by slab charges moves to a predetermined distance. It has been established that if the explosion products burst out into the atmosphere because of weak stemming, they carry away with them a considerable portion of the explosive energy, thereby greatly weakening the blast effect.

The rock blasted by slab charges is dug out along the plane of charges, only if the distances between the charges do not exceed certain critical values, otherwise the rock deposited between the charges is not removed. As a result, a range of alternate humps and depressions is formed on the bed of the stope because of the explosion.

In Fig. 4 a, the horizontal shaded line shows the volume of burden that remains intact during blasting. This volume of rock remains in the zone of blasting because it does not experience a velocity component in the direction of the line of least resistance. If the distance between charges is reduced to such an extent that the gas chambers come in contact with each other, practically the entire overburden situated between the plane of charges and free face, will be thrown out beyond the boundary of the muck pile (Fig. 4, b).

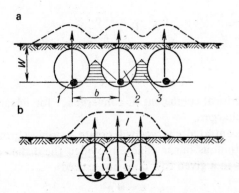

Fig. 4. Diagram of interaction of gas chambers and
rock contiguous to them during blasting of a) dispersed and b) concentrated charges:

1—charge, 2—gas chamber; 3—volume of undisturbed rock.

Experience in blasting practice [14] reveals that the bottom of the trench formed due to firing of several parallel rows of concentrated or linear charges, will be an even surface if the distance between consecutive rows of charges (see Fig. 4) does not exceed:

$$b = W n_e^{\frac{2}{3}}, \tag{1.1}$$

where n_e is the blast efficiency factor which is equal to the ratio between the radius of the crater R_{crat} and firing depth of the concentrated charge W.

The mean mass of the row of charges p related to a unit length of such a row is linked with the depth of charges W and the blast efficiency factor n_e by the equation

$$q = k_{exp} W^2 n_e^2, \tag{1.2}$$

where k_{exp} is the computed specific consumption of explosives for the blast [15]. Since the specific consumption of explosives q when blasting several linear charges with mass p per unit length or rows of concentrated charges, the mass of each of them being Q, is expressed by the following formulas:

$$q = \frac{p}{Wb}; \tag{1.3}$$

$$q = \frac{Q}{abW}, \tag{1.4}$$

therefore, by using the equations given above, let us determine the unknown critical distance between linear charges or rows of concentrated charges:

$$b = W \sqrt{\frac{q}{k_{exp}}} \; ; \tag{1.5}$$

$$m_b = \frac{m}{W} = \sqrt{\frac{q}{k_{exp}}} \; , \tag{1.6}$$

where m_b is the critical coefficient of convergence for linear charges or rows of concentrated charges.

A row of concentrated charges will create a stoping ground equal to that created by a continuous linear charge provided the distance a between concentrated charges in a given row does not exceed

$$a = W n^{2/3}. \tag{1.7}$$

Since the mass of a concentrated charge $Q = k_{exp} \, W^3 \, n^3 = q W^3 \, m_b m_a$; where $m_a = a/W$ is the coefficient of convergence for concentrated charges in a given row, therefore, taking into consideration the equation for coefficient m_b given above, we find

$$a = W \left[\frac{q}{k_{exp}} \right]^{3/7} \; ; \tag{1.8}$$

$$m_a = \frac{a}{W} = \left[\frac{q}{k_{exp}} \right]^{3/7} . \tag{1.9}$$

However, the value of a can be calculated by a formula identical to formula (1.5),

$$a = W \sqrt{\frac{q}{k_{exp}}} \; ; \tag{1.10}$$

$$m_a = \frac{a}{W} = \sqrt{\frac{q}{k_{exp}}} \; . \tag{1.11}$$

In this case, the error does not exceed 10% for a specific consumption of explosives $q < 5$ kg/m³.

In case of explosive ripping of hard rock by a row of deep-hole charges parallel to each other, the blasted rock in the plane of deep-hole charges is not broken. The rock pile comes directly in contact with the bench formed by explosion. This happens because in case of explosive ripping of hard rock, the coefficient of convergence of holes m_a is considerably greater than its critical value calculated by equation (1.11). Besides, it should be borne in mind that as a result of bursting of explosion products through the hole mouth, the blast efficiency of the explosion is sharply reduced after the escape of stemming. Therefore, the actual specific consumption of explosives q calculated by formula (1.11) will be considerably less than its calculated value. This can be taken into account by the correction coefficient α_q. In this case equation (1.11) will take the following form:

$$m_a = \sqrt{\frac{q\alpha_q}{k_{exp}}}.$$

The specific consumption of explosives for loosening the rocks of average hardness is $q = 0.5$ kg/m³. In this case the mean value of $k_{exp} = 2$ kg/m³ and $\alpha_q = 0.5$. In order to achieve a uniform blasting of the overburden along the plane of deep-hole charges, the coefficient of convergence of the charges should not exceed $m_a = 0.35$. In mining practice, the value of m_a generally lies in the range of 0.8 to 1.2.

2. EXTERIOR BALLISTICS

The movement of rock blasted by firing a slab charge depends considerably on the correlation between the size of slab charge, its mass, firing depth and angle of inclination of the charge to the horizontal.

If the mass of a slab charge is comparatively small and its length and breadth considerably exceed the thickness of the layer of rock to be blasted, the dividing line between the explosion products and moving rock runs parallel to the slab charge. The expanding gas chamber is in the shape of a thin parallelepiped with rounded edges (Fig. 5, a, shaded area). As a result, almost all points of the bulk of the blasted rock, with the exception of insignificant areas adjacent to the charge ends, move at velocities equal in both magnitude and direction. Here, it can be considered that the contours of the rock during the entire period of motion remain parallel to the initial position at the time of explosion.

Visual examination, theoretical analysis and filming of the blasting process during the firing of slab-charge systems reveal (see Sec. 6) that the rock blast can be plane-parallel (see Fig. 5, a), i.e. contours of blown-away rock are parallel to the initial contour of the free face, if the following conditions are fulfilled:

$$\xi = \frac{L}{quW} \geqslant 4; \qquad (1.12)$$

$$\chi = \frac{B}{quW} \geqslant 4, \qquad (1.13)$$

where L and B are the length and breadth of the slab charge respectively; q the specific consumption of explosives; u the specific volume of explosion products following their adiabatic expansion to atmospheric pressure; W the thickness of the rock layer to be blasted; and ξ and χ—the relative length and breadth of the slab charges respectively.

These conditions for a plane-parallel blast correspond to a firing depth of $W > 4$ m and specific consumption of explosives $q \leqslant 6$ kg/m³. For other

8

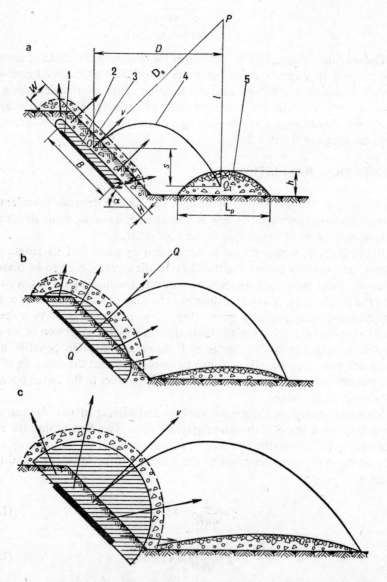

Fig. 5. Diagram showing relationship between scatter of rock by slab-charge
blasting and specific consumption of explosives:
a—low; b—average; c—high; 1—slab charge; 2—gas chamber; 3—free face
in the final stage of blasting; 4—trajectory of center of gravity;
5—muck-pile contour.

values of q and W the conditions, under which a plane-parallel blast is feasible,
are computed from the curves shown in Figs. 14 and 15.

The specific consumption of explosives in expression (1.12) henceforth implies the ratio between the total mass of a slab charge and the rock volume situated along the normal to the face on which the entire system of charges is placed. Strictly speaking, this specific consumption of explosives should be called estimated specific consumption because the actual volume of rock blown away to lateral areas due to blasting is slightly greater than the calculated. This anomaly will be removed by a special correction. However, in a majority of practical cases this difference between actually blasted and calculated volumes is negligible. Therefore, we will not make any distinction between the calculated and actual specific consumption of explosives.

Product qWu is numerically equal to the thickness of the gas chamber if the explosion products do not expand to the lateral areas or when such expansion can be ignored (for example, in case of a slab charge with very large linear dimensions). Therefore, the dimensionless quantities $\xi = L/quW$ and $\chi = B/quW$ represent the relative length and breadth of a slab charge, respectively. The physical significance of the product qWu becomes clear if we recall that the product qW expresses the mass of explosives required for 1 m² of a slab charge. If this mass is multiplied by the specific volume u of explosion products, we obtain the height of the gas chamber above the slab charge.

Let us study the movement of rock blasted by a slab charge of relatively small size and high specific consumption of explosives. In this case the gas chamber becomes dome-shaped at the end of the expansion stage (Fig. 5, b). As a result, rock fragments fly in radial directions at velocities which decrease from the center of the slab charge toward the periphery. Due to these factors, the blasted rock forms a pile on the free face which is almost elliptic in shape.

It has been experimentally proved that rock fragments fly in radial directions, as stated above, when the following conditions are fulfilled:

$$\left.\begin{aligned} 2 \leqslant \frac{B}{qWu} < 4 \,; \\ 2 \leqslant \frac{B}{qWu} < 4.^* \end{aligned}\right\} \qquad (1.14)$$

If the specific consumption of explosives is excessively high and the linear dimensions of the slab charges are small, then blasting by a slab charge explosion has the same characteristics as blasting by concentrated charges (see Fig. 5, c). Such an instance of explosion is not dealt with in this book.

3. DERIVATION OF BASIC DESIGN FORMULAS

Let us determine design dependence for a plane-parallel blast which is realized on fulfillment of conditions (1.12) and (1.13). Because the rock in this case

*The two conditions are identical in the original book—Translator.

flies as a compact mass and with a relatively low velocity, we will ignore the effect of air drag.

Let us denote the coordinates of the center of gravity of the muck pile with respect to the center of gravity of the rock to be blasted by D and S (see Fig. 5, a). This volume of rock simultaneously executes two motions: a) a uniform and rectilinear motion in the direction of the projection velocity vector; and b) a uniformly accelerated motion under the action of the gravitational force. Let us denote the time of movement of the center of gravity of the flyrock from the moment of firing to the moment of its falling on the horizontal surface by T. During time T the center of gravity traverses the path OP executing a uniform and rectilinear motion. In ballistics, this path is called the *slant distance* and is expressed by the formula $D_s = v_0 T$, where v_0 is the initial velocity of projection of the blasted rock.

In the case of a uniformly accelerated motion, the center of gravity moves vertically and traverses a path $pO_1 = l$, where l is the height gradient. From the known formula for the path traversed in a uniformly accelerated motion, we have

$$l = \frac{1}{2} gT^2,$$

where, g is the acceleration due to gravity.

By eliminating T from these equations, we obtain the formula for determining the initial velocity which is responsible for the movement of the flyrock to a predetermined location:

$$v_0 = D_s \sqrt{\frac{g}{2l}}. \tag{1.15}$$

From this, it is evident that it is necessary to know the slant distance and height gradient in order to calculate v_0. For determining these parameters in a given explosion layout, it is necessary to draw, through the center of gravity O of the blasted rock, a straight line perpendicular to the slab charge; thereafter, a line vertical to the horizontal should be drawn through the center of gravity of the same rock volume at the muck pile (point O_1). These straight lines intersect at P. The segments OP and PO_1 multiplied by the layout scale will determine the values D_s and l respectively.

For practical calculations, it is more convenient to use the range D and relative elevation S in place of slant distance and height gradient respectively. From Fig. 5, it is evident that D is the distance, along the horizontal, between the center of gravity of the blasted rock before and after explosion; S is a similar distance in the vertical plane. The relation between these parameters can be expressed as follows:

$$D_s = \frac{D}{\sin \alpha};$$

$$l = S + D \cot \alpha.$$

Substituting these expressions in formula (1.15) we have

$$v_0 = \sqrt{\frac{1}{2} gDf(\alpha, \zeta)}. \tag{1.16}$$

Here, the following notations have been used:

$$f(\alpha, \zeta) = \frac{1}{(\zeta + \cot \alpha) \sin^2 \alpha}, \tag{1.17}$$

$$\zeta = \frac{S}{D}, \tag{1.18}$$

where α is the angle of inclination between the slab charge and the horizontal.

The relation between v_0 and specific consumption q of explosives on the basis of the law of conservation of energy can be expressed as follows:

$$\frac{1}{2} \rho v_0^2 = q e_{\exp} \eta, \tag{1.19}$$

where, ρ is the density of the rock to be blasted; $e_{\exp}$ the specific energy of explosives and η the blast efficiency of a slab charge with finite dimensions. The left-hand side of this equation denotes the kinetic energy per cubic meter rock; the right-hand side of the equation represents that part of the blast energy expended on setting a cubic meter of rock in forward motion. From expressions (1.16) and (1.19), we obtain the following formulas:

$$q = f(\alpha, \zeta) \frac{\rho g}{4 \eta \, e_{\exp}} D; \tag{1.20}$$

$$v_0 = \sqrt{\frac{2 e_{\exp}}{\rho} \eta q}. \tag{1.21}$$

From equation (1.19) and formula (1.20), it is evident that the density ρ characterizes the property of inertia of the blasted rock: since for an identical range D, the required specific consumption q of explosives increases in direct proportion to density ρ (other explosion parameters remaining unchanged). The blast efficiency η, occurring in the derived expressions, represents the ratio between the kinetic energy of blasted rock and total blast energy of a slab charge.

Blast efficiency η depends, to a considerable extent, on the dimensions of the slab charge, the type of rock being blasted and the specific consumption and type of explosives. Its value varies within fairly wide limits: from a fraction of one percent to several tens of a percentage. It is higher for hard rock and lower for loose soil in which, in contrast with hard rock, a considerable dissipation of blast energy is observed. Due to this phenomenon the blast efficiency in sand not saturated with water, is less than 1% when the explosives are of average power and the specific consumption $q = 0.5$ kg/m³. In hard dense rock the blast efficiency may rise to several scores of a percent since both, the loss due to dissipation and work done in crushing in this case, are much less [8].

It should be remembered that the value of blast efficiency η for a throw action is not only determined by the rock density ρ. It is often observed that blast efficiencies differ considerably for rocks of similar density. Blast efficiency depends on the dissipation of blast energy within the rock itself. Therefore, it is obvious that the rock density ρ is not the determining parameter for η. It characterizes, as stated above, only the properties of inertia of the rock being blasted.

Blast efficiency, depending on the grade of explosives, cannot exceed a certain critical value [16]. Figure 6 shows the curves of relationship of blast efficiency and the adiabatic index χ and extent of expansion of the gas chamber characterized by the ratio between the chamber radius r_{cham} and charge radius r_0. These curves have been plotted on the basis of [16]. From the curves it is evident that at the moment of expansion of explosion products to atmospheric pressure, when $r_{cham}/r_0 = 9$ to 11, the maximum efficiency for adiabatic indexes χ of 1.3; 1.24 and 1.2 is equal to 0.85; 0.81 and 0.76 respectively.

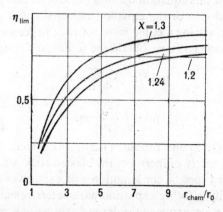

Fig. 6. Relationship of maximum possible value of blast efficiency η_{max} and the extent of expansion of gas chamber r_{cham}/r_0 for certain given values of adiabatic index χ.

The blast efficiency of an actual slab charge is considerably lower than the indicated maximum possible value and in case of a constant specific consumption of explosives, it decreases with a decrease in charge dimensions. Such a phenomenon occurs because of considerable losses of blast energy due to dissipation and, to a considerable extent, because of the high edge effect for an actual charge; i.e. scattering of explosion products to lateral areas which results in a loss of part of the blast energy in the form of plastic deformation of the rock adjacent to the ends of the slab charge. Therefore, in future calculations we will differentiate between the blast efficiency η of an actual slab charge and the blast efficiency η_0 of slab charges of infinite linear dimensions.

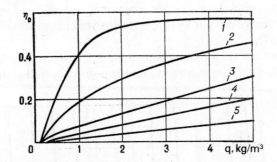

Fig. 7. Relationship of blast efficiency η_0 of slab charge and specific consumption of explosives:

1—hard rock (in the absence of a gas outburst); 2—semi-hard rock; 3—clay; 4—loam; and 5—sand.

Blast efficiency η_0 for a given type of rock can be evaluated on the basis of initial projection velocity of the concentrated charges. Curves of blast efficiency as a function of specific consumption q plotted in Fig. 7, are based on calculations carried out in [8] for certain types of soils and hard rocks. These curves have an estimative character and can, therefore, be used only for approximate calculations. The curves show that for soft soil (sand, loam, clay) there is a direct proportionality between the blast efficiency and specific consumption of explosives. This has also been proved through experiments [12]. Therefore, the following formula is applicable in general for soils up to a certain value of specific consumption of explosives q

$$\eta_0 = iq, \tag{1.22}$$

where η_0 is the blast efficiency of a slab charge of infinite linear dimensions; i is the proportionality factor which depends on the physicomechanical properties of the soil and explosives.

In approximate calculations for the ballistics of blasted earth, the factor i for sand, loam and clay, as is evident from Fig. 7, can be taken to be equal to 0.02; 0.04 and 0.06 kg/m³ respectively.

Substituting expression (1.22) in formula (1.20) the specific consumption of explosives for soils can be expressed by the following expression:

$$q = \sqrt{\frac{f(\alpha, \zeta) \rho g}{4 i e_{\exp} k}} D, \tag{1.23}$$

where the factor k accounts for edge effects (see Sec. 5).

4. RELATIONSHIP BETWEEN ANGLE OF INCLINATION AND COEFFICIENT OF RELATIVE ELEVATION

From the basic design formula (1.20), it follows that the specific consumption of explosives is proportional to the function of the angle of inclination and coefficient of relative elevation $f(\alpha, \zeta)$. Its analytical expression is given by the expression (1.17).

Table 1 contains the values of function $f(\alpha, \zeta)$ for a wide range of α and ζ. The coefficient of relative elevation ζ can be positive as well as negative. If the

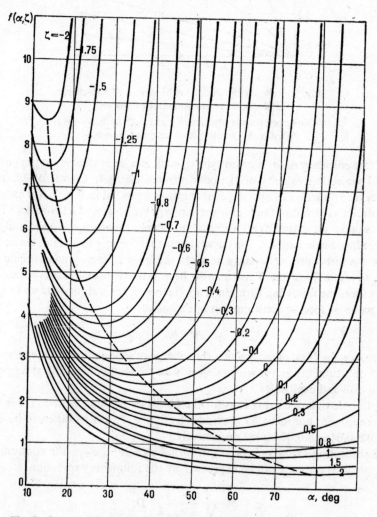

Fig. 8. Curves of function of angle of inclination and coefficient of relative elevation against the angle of inclination.

15

TABLE 1

Values of $f(\alpha, \zeta)$ at different values of ζ

α, deg	0	0.1	0.2	0.3	0.4	0.5	0.6	0.7	0.8	1.0	1.5	1.75	2.0
5	11.5	11.4	11.3	11.2	11.3	11.0	10.9	10.8	10.8	10.6	10.2	9.99	9.80
10	5.85	5.75	5.65	5.55	5.46	5.37	5.29	5.20	5.12	4.97	4.62	4.47	4.32
15	4.00	3.90	3.79	3.70	3.61	3.53	3.45	3.37	3.29	3.15	2.85	2.72	2.60
20	3.11	3.00	2.90	2.80	2.72	2.62	2.55	2.48	2.41	2.28	2.01	1.90	1.80
25	2.61	2.49	2.39	2.29	2.20	2.12	2.04	1.97	1.90	1.78	1.53	1.44	1.35
30	2.31	2.18	2.07	1.97	1.88	1.79	1.72	1.65	1.58	1.46	1.24	1.15	1.07
35	2.13	1.99	1.87	1.76	1.66	1.58	1.50	1.43	1.36	1.85	1.04	0.96	0.89
40	2.03	1.87	1.74	1.62	1.52	1.43	1.35	1.28	1.21	1.10	0.89	0.82	0.76
45	2.00	1.82	1.67	1.54	1.43	1.33	1.25	1.18	1.11	1.00	0.80	0.73	0.67
50	2.03	1.82	1.64	1.50	1.37	1.27	1.18	1.11	1.04	0.93	0.73	0.66	0.60
55	2.13	1.86	1.65	1.49	1.36	1.24	1.15	1.06	0.99	0.88	0.68	0.61	0.55
60	2.31	1.97	1.71	1.52	1.36	1.24	1.13	1.04	0.97	0.84	0.64	0.57	0.52
65	2.61	2.15	1.83	1.59	1.40	1.26	1.14	1.04	0.96	0.83	0.62	0.55	0.49
70	3.11	2.44	2.01	1.71	1.48	1.31	1.17	1.06	0.97	0.83	0.61	0.53	0.48
75	4.00	2.91	2.29	1.89	1.60	1.40	1.23	1.11	1.00	0.84	0.61	0.53	0.47
80	5.85	3.73	2.74	2.16	1.79	1.52	1.33	1.18	1.06	0.88	0.62	0.53	0.47
85	11.52	5.37	3.50	2.60	2.07	1.72	1.47	1.28	1.14	0.93	0.63	0.55	0.48
90		10.0	5.00	3.33	2.50	2.00	1.67	1.43	1.25	1.00	0.67	0.57	0.50

(Contd.)

Values of $f(\alpha, \zeta)$ at different values of ζ

α, deg	-2.0	-1.75	-1.5	-1.0	-0.8	-0.7	-0.6	-0.5	-0.4	-0.3	-0.2	-0.1	0
5	13.9	13.6	13.3	12.6	12.4	12.3	12.1	12.0	11.9	11.8	11.72	11.66	11.5
10	9.03	8.46	7.95	7.10	6.81	6.67	6.54	6.41	6.29	6.17	6.06	5.95	5.85
15	8.62	7.53	6.69	5.46	5.09	4.92	4.77	4.63	4.48	4.35	4.22	4.11	4.00
20	11.4	8.57	6.85	4.89	4.39	4.17	3.98	3.80	3.64	3.49	3.36	3.23	3.11
25	38.7	14.2	8.69	4.89	4.16	3.88	3.62	3.40	3.21	3.03	2.88	2.74	2.61
30	—	—	17.2	5.46	4.31	3.87	3.53	3.25	3.00	2.79	2.61	2.45	2.31
35	—	—	—	7.10	4.84	4.17	3.67	3.27	2.96	2.69	2.47	2.29	2.13
40	—	—	—	12.6	6.18	4.92	4.09	3.49	3.06	2.71	2.44	2.22	2.03
45	—	—	—	—	10.0	6.67	5.00	4.00	3.33	2.86	2.50	2.22	2.00
50	—	—	—	—	43.6	12.2	7.12	5.02	3.88	3.16	2.67	2.30	2.03
55	—	—	—	—	—	71.43	14.8	7.44	4.97	3.72	2.98	2.48	2.13
60	—	—	—	—	—	—	—	17.2	7.52	4.80	3.53	2.79	2.31
70	—	—	—	—	—	—	—	—	18.3	7.32	4.57	3.32	2.61
75	—	—	—	—	—	—	—	—	—	17.7	6.90	4.29	3.11
80	—	—	—	—	—	—	—	—	—	—	15.8	6.39	4.00
85	—	—	—	—	—	—	—	—	—	—	—	13.51	5.85
86	—	—	—	—	—	—	—	—	—	—	—	—	11.52

center of gravity of the blasted rock after settling is lower than the center of gravity of the rock before explosion, then both the relative elevation S and coefficient ζ are positive.

Conversely, when much rock is raised, for example, from an open-strip mine to higher levels, the values of S and ζ become negative.

Curves of function $f(\alpha, \zeta)$ have been plotted (Fig. 8) on the basis of the data contained in Table 1. Analysis of these curves shows that the function $f(\alpha, \zeta)$ for $\zeta < 0.3$ has a sharply defined minimum value corresponding to the optimum angle α_{opt}. This means that for optimum angles of inclination of a slab charge to the horizontal, the specific consumption of explosives, based on formula (1.20), will be minimum. Consequently, efforts should always be made to achieve the optimum angle α_{opt} with a view to minimizing the cost of blasting operations in the practical utilization of a slab-charge system.

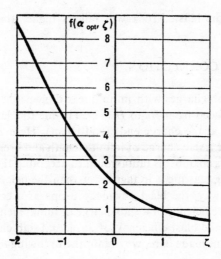

Fig. 9. Dependence of optimum angle of inclination of slab charge to the horizontal α_{opt} on coefficient of relative elevation ζ.

Figure 9 shows the curve of the optimum angle α_{opt} which depends on coefficient ζ. In Fig. 8 this curve is shown as a dotted line which connects the points of optimum angles α_{opt}. As is evident, the value of α_{opt} varies over a wide range: from 13° to 75° for corresponding variations of ζ between -2 and 2.

Figure 10 shows the curve of optimum values of function $f(\alpha, \zeta)$ for different values of the coefficient of relative elevation ζ. From the curve it is obvious that, for example, the ratio between the values of this function for $\zeta = -2$ and $\zeta = 0$ is equal to 4.2. From this, it follows that the specific consumption of explosives, and consequently, the cost of blasting operations will increase, if

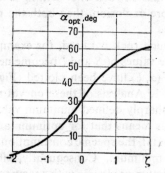

Fig. 10. Dependence of optimum value of function $f(\alpha_{opt}, \zeta)$
on coefficient of relative elevation ζ.

the surface where the blasted rock will fall, is higher than the site of the explosion.

5. 'GEOMETRY' CORRECTION

During firing of slab charges with finite linear dimensions, the height of the gas chamber is reduced from H_0 to H (Fig. 11) due to a lateral expansion of explosion products at the charge ends (end effect). Here H_0 is the height of the gas chamber for a slab charge of infinite length and breadth. Reduction in the height of the gas chamber is tantamount to accepting that the requirement of explosives during blasting is in fact lower than the calculated specific consumption. In this way, the blast efficiency drops as a consequence of end effects. If the blast efficiency of a slab charge of infinite length and breadth is denoted by η_0 and the blast efficiency of an actual slab charge by η, then in view of the statement made here, we obtain the following relationship:

$$k = \frac{\eta}{\eta_0} = \frac{H}{H_0} \, . \tag{1.24}$$

Coefficient k, characterizing the lowering in blast efficiency due to end effects, will be expressed, hereafter, as a correction for a limitation on linear dimensions of an actual slab charge or, briefly, as the 'geometry' correction.

Let us calculate the value of coefficient k by assuming that a slab charge explodes on an absolutely hard base which is a situation very close to reality. For example, in case of an explosion in limestone and granite, the downward charge action (rock subsidence under the charge) is 0.005 H_0. For loamy soils, this figure is (0.1 to 0.2) H_0. Hence, it is clear that the assumed hypothesis does not cause considerable errors.

In order to simplify calculations, let us assume that explosion products, while expanding to the sides, are restricted by round surfaces of a diameter equal to the gas chamber height H (see Fig. 11). In reality this diameter is less than H by 10–15%.

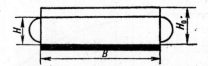

Fig. 11. Diagram for calculating the 'geometry' correction.

Let us calculate H under the condition that the volume of the gas chamber $U_{g.c}$ remains unchanged both under edge-effects as well as without them, and is equal to the volume of explosion products on their adiabatic expansion to atmospheric pressure.

If the length and breadth of a slab charge are denoted by L and B respectively, and the thickness of the rock layer to be blasted as W, then the mass of a slab charge for a specific consumption of explosives equal to q, will be $LBWq$ where LBW is the volume of the rock to be blasted. By multiplying the mass of the slab charge with the specific volume of explosion products u, we obtain the volume of the gas chamber: $U_{g.c} = ABWqu$.

On the other hand, from Fig. 11 it follows that this volume, after taking into account the gas expansion to lateral areas, will be equal to:

$$U_{g.c} = ABH + \frac{1}{4}\pi H^2 (L+B).$$

From the obtained equations we observe that

$$ABH + \frac{1}{4}\pi H^2 (A+B) = ABWqu. \tag{1.25}$$

If there were no edge-effects, then, as is evident from Fig. 11, $U_{g.c} = ABWqu = ABH_0$, whence, the height of the gas chamber H_0 for a slab charge with infinite dimensions is given by

$$H_0 = qWu. \tag{1.26}$$

By inserting a dimensionless parameter

$$\Omega = \frac{A+B}{2AB} Wuq,$$

from the expressions (1.24), (1.25) and (1.26), we obtain the formula for the coefficients of 'geometry' correction:

$$k = \frac{\sqrt{1+2\pi\Omega}-1}{\pi\Omega}. \tag{1.27}$$

It is convenient to express the parameter Ω, which characterizes the ratio between the dimensions of the slab charge and height of the gas chamber, through the relative length ξ and breadth χ of the charge.

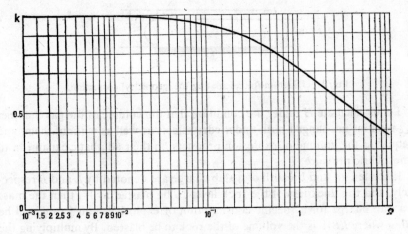

Fig. 12. Dependence of 'geometry' correction on parameter Ω.

Taking into account expressions (1.12) and (1.13), the parameter Ω can be written in a simpler form as follows:

$$\Omega = \frac{\xi + \chi}{2\xi\chi}. \tag{1.28}$$

In this way the efficiency of an actual charge according to expression (1.24) becomes:

$$\eta = k\eta_0. \tag{1.29}$$

For calculating the dimensionless parameter Ω, it is necessary to know the specific volume of the products of explosion u on their adiabatic expansion to atmospheric pressure. Calculations carried out on the basis of [16, 17] show that the density u for industrial explosives can be taken to be approximately 1 m^3/kg.

Relationship of coefficient k and parameter Ω is shown in Fig. 12.

6. DESIGN OF A SLAB CHARGE OF FINITE DIMENSIONS

On the basis of experimental and theoretical investigations [18–22] carried out during recent years, the various stages of the blasting process can be described in the following way. On detonating cylindrical or concentrated charges, which form a slab-charge system, a cylindrical or spherical gas chamber, respectively, is formed during the initial period and this subsequently expands intensely in all directions at a uniform rate. The muck deposited in the intermediate space is compressed and moves symmetrically with respect to the center of the charge in all directions. After a certain period, the gas chambers cease to expand downward (opposite to the free face) but continue to expand upward intensely. As a consequence of movement of the muck from the intermediate space, the individual gas chambers are linked, thereby forming a single gas chamber above the plane of charges. Here ends the first phase of the explosion. In the second phase, the gas chamber expands only in the upper space, thereby displacing the burden along the normal to the charge and in lateral directions. The burden receives a considerable part of energy of the explosion products at the moment when the pressure in the gas chamber approaches atmospheric pressure, and the process of accelerating the rock displacement ends. Here ends the second phase of explosion.

The rock detached by explosion continues to move as a compact mass for some time with the fragments compactly pressed against one another. With the movement in the air, the layer thickness decreases because the rock is scattered radially in all directions. But a moment comes when the thickness of the rock layer becomes equal to the size of individual fragments. As a result, the gas chamber disintegrates, and the fragments continue to fly independently in the air. In this way, the main feature of the third phase of explosion is the independent movement of the dome from the moment of its separation from the gas chamber till its disintegration into fragments. In the fourth phase of explosion, individual fragments start flying from the moment of disintegration of the gas chamber dome till they fall on the free face.

We will make the following assumptions while solving the problem of scattering of the rock.

1. At the moment of cessation of expansion of explosion products (end of the second phase), the rock starts moving along the normal to the surface of the gas chamber. This assumption is very close to reality, especially for cohesive soils. In the first phase of explosion of cohesive soils there is immense loss of energy because of irreversible plastic deformation of the compressed rock. Therefore, soft rocks attain an initial velocity of movement only when they become compact, chiefly in the second phase of explosion.

2. At the moment of maximum expansion the gas chamber attains such a shape that at its upper part, the sectional contours of its surface with the two principal planes are in the form of arcs of a circle, the centers of which lie on a line perpendicular to the plane of the charge and which passes through the center of the charge. The principal plane implies the plane perpendicular to the charge and passing parallel to one of the sides of the charge through its geometric center. Here, it is assumed that the slab charge has a rectangular cross section.

Hereafter the first principal plane will refer to the plane perpendicular to the horizontal and passing parallel to the principal direction of projection with respect to the center of the charge. The second principal plane passes

22

through the center of the charge and is perpendicular to the plane of the charge as well as to the principal plane. Figure 5, b shows the contours of the gas chamber and its dome in the section of the first principal plane. The second assumption is based on the fact that while plotting the blast curves on the basis of motion pictures, the dome contours of the slab charge explosion are practically the same as those of an arc of a circle.

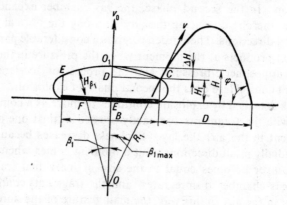

Fig. 13. Calculation of gas-chamber radius.

Let us calculate the radius of curvature of the upper part of the gas chamber in the cross sections by the two principal planes. In the first principal plane we shall determine the radius of curvature R_1 from a right angled triangle OCD (Fig. 13) bearing in mind that by virtue of the second assumption the contour of the gas chamber CO_1E is an arc of a circle. Therefore,

$$R_1^2 = (R - \Delta H)^2 + \frac{1}{4} B^2, \tag{1.30}$$

where

$$\Delta H = H - H_1.$$

In this case, the parameter H_1 represents the height of the gas chamber in contact with the semicylindrical cavity formed as a result of end effects. In order to simplify the calculation for the 'geometry' correction, it was assumed that the upper portion of the gas-chamber surface is flat and that for approximate calculations, coefficient k was fully acceptable because H, as shown in Table 2, differs from H_1 only by 10–15%.

Analysis of the movement of rock fragments shows that for a sufficiently wide range of specific consumption of explosives, the ratio $H : H_1$ may be taken constant:

$$\frac{H}{H_1} = c_H. \tag{1.31}$$

From the three expressions given above, it follows that

$$R_1 = \frac{B^2}{8H} \cdot \frac{c_H}{c_H - 1} \cdot$$

After taking into account the notations of (1.13) and (1.24), this expression becomes:

$$R_1 = c_R B \frac{\chi}{k}, \tag{1.32}$$

where

$$c_R = \frac{c_H}{8 (c_H - 1)} \cdot \tag{1.33}$$

The radius of curvature in the second principal plane can also be calculated in an identical manner:

$$R_2 = c_R B \frac{\xi}{k} \cdot \tag{1.34}$$

Let us determine the maximum angle β_{1max} at which the rock deposited in the peripheral region of a slab charge is blown away by the expanding explosion products. Since the contour CO_1E (see Fig. 13) is an arc of a circle, therefore $\angle DCO_1 = \frac{1}{2} \beta_{1max}$. From the right-angled triangle DO_1C, we have

$$\tan \frac{\beta_{1max}}{2} = 2 \frac{\Delta H}{B} \cdot$$

From Eqs. (1.30), (1.31) and (1.13) we obtain

$$\tan \frac{\beta_{1max}}{2} = \frac{2 (c_H - 1)}{c_H} \cdot \frac{k}{\chi} \cdot \tag{1.35}$$

Since angle β_{1max} is small for slab charges, therefore

$$\beta_{1max} = \frac{4 (c_H - 1)}{c_H} \cdot \frac{k}{\chi} \cdot \tag{1.36}$$

From Fig. 13, it is clear that during explosion of slab charges with finite dimensions, the rock is thrown in a radial direction. The rock fragments fly out along the radii emanating from the center O. As a result, the rock blasted by a horizontally-laid slab charge forms, on the free face, a muck pile with a surface area exceeding the area of the charge. Such a blast results in a *plane-parallel throw*. Let us find the conditions under which this type of movement of blasted rock can be realized. For this purpose it is essential that the place where the rock falls and which is along the length of the slab charge, should be a distance D commensurate with the length B of the charge (see Fig. 13). For a horizontal free face, according to the parabolic theory of exterior ballistics, we have

$$D = \frac{v^2 \sin 2\beta_{max}}{g}. \tag{1.37}$$

The initial velocity of projection v_0 can be expressed as:

$$\frac{v_0^2}{v^2} = \frac{H}{H_1} = c_H, \tag{1.38}$$

which follows from equations (1.21), (1.24) and (1.31). Therefore:

$$D = \frac{v_0^2 \sin 2\beta_{max}}{g c_H}. \tag{1.39}$$

Keeping in mind that angle β_{max} is small; let us express the obtained formula in the following form:

$$D = \frac{2 v_0 \, \beta_{max}}{g c_H}.$$

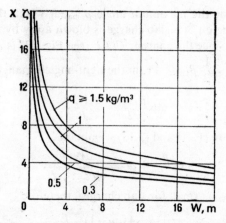

Fig. 14. Relationship of parameters χ and ζ and W for plane-parallel throw-out explosion of hard rock.

Substituting the dependences (1.21), (1.22) and (1.29) in this formula and taking into account the requirement that $D \leqslant B$ we obtain the following criterion for plane-parallel throw-out explosion:

$$\xi = \chi \geqslant 4 \sqrt{\frac{e_{exp} \, (c_H - 1) \, i k^2 q}{c_H^2 \, \rho g u W}}. \tag{1.40}$$

Figures 14 and 15 show the curves for quantities ζ and χ calculated by this formula for some of the average values of parameters included in it. For cohesive soils $i = 0.04$; for hard rock $i = 0.33$ kg/m³ ($q \leqslant 1.5$ kg/m³); if for hard rock the explosion takes place at $q > 1.5$ kg/m³, then during calculations we

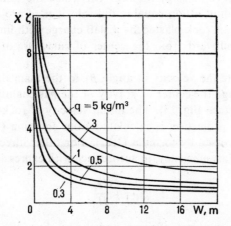

Fig. 15. Dependence of parameters χ and ζ and W for plane-parallel throw-out explosion of soft rock.

take $iq = 0.5$. From these curves it is easy to obtain conditions (1.12) and (1.13).

With a view to determining the quantity c_H, trial blasts of horizontally-laid slab charges were carried out in sand and loamy soils containing more than 50% of boulder rock and gravel. Cartridge-packed ammonite No. 6 was used as the explosive in the form of linear charges laid parallel to the predetermined horizontal surface. After explosion, the rock fell on the free face and formed a muck pile, the center of which coincided with the center of the charge. The mean distance D of the muck pile edge from the contours of the slab charge and the height of the rock in the vertical plane H_r were measured during the trials. If we take $H_r = v_0^2/2g$, then from expressions (1.36) and (1.40) it is easy to find the unknown parameter c_H. Table 2 contains experimental data for these explosions and the calculated value of c_H. Its average value is 1.15.

TABLE 2

Trial blast No.	$L \times B$, m	q, kg/m³	ξ	χ	Ω	k	H_r, m	D, m	c_H
1	6.7×6.1	2.44	2.77	2.52	0.39	0.75	41	17	1.15
2	6.1×6.0	1.64	3.72	3.65	0.27	0.81	20	4	1.10
3	6.1×6.0	2.11	2.88	2.84	0.35	0.78	26	12	1.12
4	6.0×5.8	2.11	1.9	2.0	0.48	0.73	29	27	1.23
5	6.3×6.0	3.00	2.1	2.0	0.50	0.72	48	27	1.14

All elementary volumes of the rock for which the centers of gravity lie along the same radius, attain equal initial velocities of projectile in the same direction because the rock blasted by a slab charge with limited dimensions, scatters radially outward from the center of curvature of the gas chamber (see Fig. 13).

Let us calculate the velocity at angle β_1 to the main direction of throw OO_1 passing through the center of gravity of the gas chamber and the center of the slab charge (see Fig. 13). The initial velocity of rock movement in the main direction of projection is calculated from equation (1.21) where blast efficiency η is expressed by formula (1.22). For other directions determined by angle β_1, the blast efficiency, in accordance with expression (1.24), will be:

$$\eta_{\beta_1} = \eta \, \frac{H_{\beta_1}}{H},$$

where η and H are the blast efficiency and the height of the gas chamber respectively, along the main direction of throw and H_{β_1} the height of the gas chamber along the direction under study. Introducing the notation:

$$k_{\beta_1} = \frac{H_{\beta_1}}{H},$$

we have

$$\eta_{\beta_1} = \eta k_{\beta_1}.$$

Considering that $\eta = k\eta_0$, we obtain

$$\eta = k k_{\beta_1} \eta_0. \tag{1.41}$$

Substituting η_{β_1}, in place of η in expression (1.21), we find the velocity of projection at angle β_1 to the main direction which lies in the first principal plane:

$$v = \sqrt{\frac{2e_{\exp}}{\rho} k k_{\beta_1} \eta_0 q}. \tag{1.42}$$

Let us calculate the coefficient k_{β_1}. From Fig. 13 it is obvious that $H_{\beta_1} = R_1 - OF$.

Since ΔOFE is a right-angled triangle, therefore,

$$OF = \frac{OE}{\cos \beta_1} = \frac{R_1 - H}{\cos \beta_1}.$$

Thus

$$H_{\beta_1} = R_1 - \frac{R_1 - H}{\cos \beta_1}.$$

Substituting R_1 in this expression in accordance with Eq. (1.32) and the relationships $k_{\beta_1} = H_{\beta_1}/H$, $\chi = B/kH_0$, we find

$$k_{\beta_1} = z_\chi \left[1 - \frac{1}{\cos \beta_1} \left(1 - \frac{1}{z_\chi} \right) \right],$$ (1.43)

where

$$z_\chi = c_R \left(\frac{\chi}{k} \right)^2.$$ (1.44)

While calculating the range of the projected rock fragment, it is convenient to use a dimensionless parameter

$$\delta_1 = 2 \frac{FE}{B},$$

which will be known as the *relative distance* between the rock being studied and the charge center in the first principal plane.
Since

$$\cos \beta_1 = \frac{1}{\sqrt{1 + \tan^2 \beta_1}}, \quad \tan \beta_1 = \frac{\delta_1 B}{2(R_1 - H)} \text{ and } \frac{B}{H} = \frac{\chi}{k},$$

by substituting expression (1.33) in these formulas we find

$$\cos \beta_1 = \frac{2 \left(c_R \dfrac{\chi}{k} - \dfrac{k}{\chi} \right)}{\sqrt{\delta_1^2 + 4 \left(c_R \dfrac{\chi}{k} - \dfrac{k}{\chi} \right)^2}}.$$ (1.45)

From expressions (1.43) to (1.45), we obtain the design formula for calculating coefficient k_{β_1} for the known parameters of a slab charge

$$k_{\beta_1} = c_R \left(\frac{\chi}{k} \right)^2 \left[1 - \frac{\sqrt{\delta_1^2 + 4 \left(c_R \dfrac{\chi}{k} - \dfrac{k}{\chi} \right)^2}}{2 \left(c_R \dfrac{\chi}{k} - \dfrac{k}{\chi} \right)} \left(1 - \frac{1}{c_R \left(\dfrac{\chi}{k} \right)^2} \right) \right].$$ (1.46)

The distance at which a blasted fragment, with a vector describing the initial direction at an angle β_1 to the principal direction of projectile (see Fig. 13) falls, can be obtained from the expression (1.21) by substituting coefficient of blast efficiency η_{β_1} in place of η, in accordance with correlation (1.41). In this case function $f(\alpha, \zeta)$ should be calculated for the following angle:

$$\alpha_1 = \alpha \mp \beta_1.$$ (1.47)

The clockwise direction is considered the positive direction for reading angle β_1.
In this way:

$$D_{\beta_1} = \frac{4\,kk_{\beta_1}\,\eta_0\,e_{exp}\,q}{f(\alpha_1,\zeta)\,\rho g}\,. \qquad (1.48)$$

The initial velocity of projection for the given angle β_1, in accordance with expression (1.21), will be

$$v = \sqrt{\frac{2}{\rho}\,e_{exp}\,kk_{\beta_1}\,\eta_0 q}, \qquad (1.49)$$

therefore, it follows that with an increase in β_1 (or its equivalent parameter δ_1), the value of v decreases as a consequence of a decrease in the coefficient k_{β_1}. If we denote the initial velocity of projectile for the main direction of throw by v_0, then at an angle β_1 to this direction, we have

$$v = v_0\,\sqrt{k_{\beta_1}}, \qquad (1.50)$$

where

$$v_0 = \sqrt{\frac{2e_{exp}\,k\eta_0}{\rho}}\,q. \qquad (1.51)$$

The blast efficiency η_0 used in this expression is evaluated from experiments, or its approximate value is used (Fig. 7).

The initial velocity in the main direction of throw for an explosion in rock, subject to relationship (1.22), is expressed through the following formula:

$$v_0 = q\,\sqrt{\frac{2}{\rho}\,e_{exp}\,ki}. \qquad (1.52)$$

The distance at which exploded rock falls for a given angle of projection can be determined from relatively simple geometric construction. For this purpose, it is necessary to trace the flight path of flyrock for an angle β_1 and for a computed value of v. The point of intersection of this trajectory with the free face determines the area of falling rock. Trajectory tracing is very

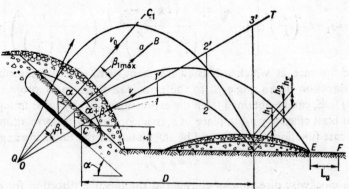

Fig. 16. Diagram of trajectories of rock fragments on firing
a system of slab charges.

simple if along the direction of explosion OT (Fig. 16), by setting the time t arbitrarily, we obtain the distance vt traversed by the fragment from the point C. As a result, we obtain the point $1'$ on the line OT. Thereafter, from this point vertically downward, we set aside a path $gt^2/2$ crossed as a result of uniform acceleration due to gravity. The end of this section will be the point 1 on the trajectory at a given moment t. In a similar way, we obtain a number of other points which form the trajectory. From expression (1.48), it follows that in the first principal plane, the range of the flight path is proportional to the ratio $k_{\beta_1}/f(\alpha_1, \zeta)$ which is the function of angle β_1. The value of $k_{\beta_1}/f(\alpha_1, \zeta)$ for a given β_1, can be calculated using formulas (1.17), (1.46) and (1.47).

After falling on the free face, rock fragments will either roll or slide and continue to move from point E to F (see Fig. 16), thereby expending the remaining kinetic energy in overcoming friction.

If we ignore air drag, then the path L_g taking retardation due to gravitational force into account can be calculated by the following formula:

$$L_g = \frac{v_{pr}^2}{2g \cos \varphi \, (k_{fr} - \tan \varphi)}, \tag{1.53}$$

where v_{pr} is the initial velocity of projection on the free face; φ the angle between the horizontal and the tangent to the free face; and k_{fr} the averaged coefficient of friction. The obtained formula allows us to take into account the movement of the rock pile in the direction of throw. This movement, in certain conditions, can be equal to the maximum range of rock throw and may even exceed it.

7. BUILD-UP OF PROFILE AND CONTOUR OF THE MUCK PILE

Let us find a method of constructing the profile of a muck pile in the first principal plane. Let us also consider the rock volume within the limits of a small solid angle (Fig. 17). Because of a radial spread, this volume covers the free face within the area of ΔF with an overburden thickness h. From the condition of equality of the blasted rock mass that falls on the free face, we get

$$\Delta F_{cone} \, W = k_{loos} \, h \, \Delta F,$$

where ΔF_{cone} is the area of the base of an elementary cone of thrown-out rock; W the thickness of the exploded layer of rock; and k_{loos} the coefficient of loosening.

Since

$$\frac{\Delta F_{cone}}{\Delta F} = \left(\frac{R_1}{L_g} \right)^n,$$

where L_g is the total length of the flight path from the origin to the point O,

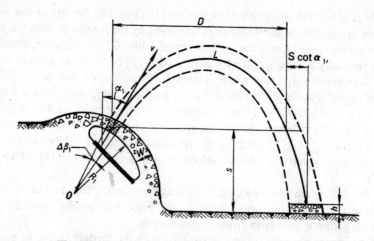

Fig. 17. Diagram for calculating height of muck pile.

n the exponent depending on the correlation between the trajectory length, and breadth B of the slab charge, therefore:

$$h = k_{loos} W \left(\frac{R_1}{L_g} \right)^n. \tag{1.54}$$

If $A \gg B$ then $n = 1$ (there is no radial spread along length L of the charge). When length L and breadth B of the slab charge are equal, $n = 2$. In case of peripheral charges (see below) when $A = B$, the exponent $n = 1$.

Length L_g of the trajectory can be determined with maximum precision by measuring it directly on the drawing with the help of an opisometer. The following formula can be used when $\alpha = 20$ to $70°$.

$$L \approx R + 1.1 D \left(1 + \cot^2 \alpha_1 \right)^{1/2} \mp \frac{S}{\cos \alpha_1}.$$

In this formula, the relative elevation S (see Fig. 17) is considered to be positive if the area where the exploded rock falls is lower than its escape point. It should be borne in mind that calculation using the aforementioned formula will have a large error if angle α_1 is greater than $70°$. It was stated earlier that at a given point of the pile in a principal plane, the muck falls from two different directions (see Fig. 16). Therefore, the aggregate thickness of the muck pile profile will be:

$$h_\Sigma = h_1 + h_2,$$

where h_1 and h_2 are calculated by formula (1.54) for the respective directions of projection. In this case we can make use of the approximation dependence:

$$\alpha' = 90 - \alpha,$$

where α' and α are the angles between the vertical plane and the lines of projection along which fragments fall at the same point.

Cross sections of the contour of a muck pile are determined by the area where the muck, flying in the plane of maximum range of the projectile, falls. This plane passes through the line OB, parallel to the longer side of the charge. Fig. 16 shows the plane of maximum range of a projectile Q (it is perpendicular to the plane of the drawing).

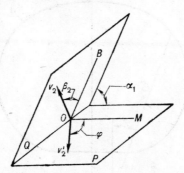

Fig. 18. Diagram for calculating the angle of rotation $\vec{v_2}$ describing the projection of velocity vector on the horizontal plane.

Plane Q is at an angle $\alpha_1 = 90 - (\alpha + \beta_{1max})$ to the horizontal plane P (Fig. 18), where β_{1max} is the angle between the line of projection and maximum range in the first principal plane (see Fig. 16). This direction, in the plane Q, is represented by the line OB. Any vector describing velocity of the projectile $\vec{v}$ in the plane Q can be projected on the horizontal plane and makes an angle φ with the main direction of throw OM, so that

$$\tan \varphi = \frac{\tan \beta_2}{\sin (\alpha + \beta_{1max})}, \qquad (1.55)$$

where β_2 is the angle between the direction of throw for maximum range and the velocity vector $\vec{v_2}$ in the plane Q.

For determining the range of spread of fragments and the initial velocity vector of projection in the plane Q, it is necessary to find angle α_2 between the vertical plane and velocity vector $\vec{v_2}$. This angle is determined from the following expression:

$$\sin \alpha_2 = \cos \alpha_1 \cos \beta_2. \qquad (1.56)$$

The value of angle β_2 for a given parameter δ_2 is calculated by formula (1.45) in which χ and δ_1, are substituted by ξ and δ_2 respectively.

The distance at which rock fragments fall, for a given δ_2, is calculated by a formula similar to formula (1.48).

$$D_{\beta_2} = \frac{4k\eta_0\, e_{exp}\, q}{\rho g} \cdot \frac{k_{\beta_2}}{f(\alpha_2, \zeta)} \cdot \tag{1.57}$$

Function $f(\alpha_2, \zeta)$ used here can be calculated from the curve shown in Fig. 8 at an angle α_2 which is calculated by formula (1.56). Here, the coefficient of relative elevation $\zeta = 0$ can be taken as a first approximation.

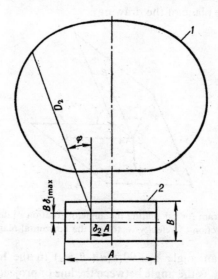

Fig. 19. Constructing the contours of a muck pile:
1—pile contour, 2—contour of slab charge.

Coefficient k_{β_2} is calculated by formula (1.46) in which χ and δ_1 are substituted by ξ and δ_2 respectively. For plotting the pile contour (Fig. 19), we set aside the quantity D_{β_2} (from the point determined by δ_2), at an angle φ to the main direction of throw. The end point of this segment corresponds to a point on the pile contour in its cross section. In the same way, we can find several points which form the pile contour when connected to each other. The maximum value of angle β_2 is determined by quantity $\delta_2 = 1$.

This method of constructing the pile contour and its thickness in a cross section of the principal plane is used when it is essential to, more or less accurately, determine the distribution of the blasted rock on the free face. In many practical instances (for example, in calculations connected with stripping of rock by blasting with a slab-charge explosion, it is not important to determine the distribution of fallen rock fragments with high accuracy. Often, it is necessary to have information only about approximate boundaries within which the main mass of rock will fall. If the blast area and the area where the rock mass falls are at the same level, then the following approximation method may be used.

The far end of the muck pile is determined by formula (1.48) in which we take $k_{\beta_1} = 1$ and $f(\alpha, \zeta) = 2$. This means that, according to our estimates, rock fragments will be projected to a maximum distance in the direction of throw which is at an angle $\alpha_1 = 45°$. In addition, it is presumed that $\zeta = 0$ (areas of throw and fall of the rock mass lie at the same level). Therefore, the maximum distance of the pile boundary from the center of the charge can be calculated by the following formula:

$$D_1 = \frac{2k\eta_0 \, e_{\exp}}{\rho g} \, q. \tag{1.58}$$

The nearer end of the pile can be calculated by the same formula (1.48), by taking $f(\alpha, \zeta) = 2$, and the coefficient k_{β_1} in accordance with formula (1.46). In this way, we obtain

$$D_2 = \frac{2k k_{\beta_1} \eta_0 \, e_{\exp}}{\rho g} \, q. \tag{1.59}$$

From formulas (1.58) and (1.59) we find the ratio

$$\frac{D_1}{D_2} = \frac{1}{k_{\beta_1}}. \tag{1.60}$$

Angle φ of the left and right ends of the pile (Fig. 20) is calculated by formula (1.55) by taking into account that angle β_2 corresponds to quantities $\delta_2 = 1$ and $\xi/k = L/kqWu$. Angle β_2 is calculated by formula (1.45). For the sake of simplification, in formula (1.55) we take $\beta_{1\,\max} = 0$. Therefore,

$$\tan \varphi = \frac{\tan \beta_2}{\sin \alpha_2}. \tag{1.61}$$

On the basis of this data, we draw the contours of the muck pile. The far end is elliptic, in which the minor axis is equal to $D_1 - D_2$ and the major axis is equal to the straight line AB drawn between lateral ends of the pile and passing through its center (see Fig. 20).

In the absence of peripheral charges (see below) the pile thickness at the center can be calculated by the following formula:

$$h = \frac{3W \, AB}{(D_1 - D_2) L_p},$$

where L_p is the pile width.

From Fig. 20, it is clear that

$$L_p = A + (D_1 - D_2) \tan \varphi.$$

If a slab charge is bounded on two sides, then $L_p = A$. Therefore

$$h = \frac{3WB}{D_1 - D_2}. \tag{1.62}$$

34

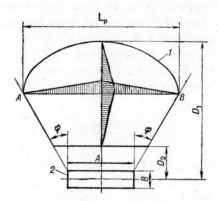

Fig. 20. Diagram of an approximate method of constructing
pile contours and profile:
1—pile contour, 2—contour of slab charge.

Experimental investigations show that the pile profile along two mutually perpendicular directions passing through its center, is approximately the same as shown in Fig. 20. Formulas (1.61) and (1.62) have been obtained on the basis of these considerations.

The method of approximation mentioned here can be used when the angles of inclination of the slab charge are within 25–60°. Beyond these limits this method will result in considerable errors. This method becomes progressively more accurate as the angle of inclination of the charge approaches 45°. For a plane-parallel throw-out explosion (see Fig. 5, a) quantities L_p and h are calculated by means of mine models, and for rough calculations we may take $L_p \approx B$ and $h \approx W$.

8. EDGE-EFFECT AND ITS ELIMINATION

It has already been mentioned that the effect of a slab-charge system of large cross-sectional dimensions on the medium is different from that of a single spherical or cylindrical charge. This difference lies in the fact that explosion products, in the case of slab charges, expand in a direction perpendicular to the plane of charges. As a result, the burden is thrown out, roughly, along the normal to this plane. However, the direction and accuracy of a throw-out explosion worsens with a decrease in the dimensions of the slab charge and simultaneous increase in the consumption of explosives. It is clear from Fig. 21 which shows the diagram of the outline of the rock dome in case of blasting by a slab charge having finite dimensions. It is evident that explosion products, while expanding, not only along the normal to the charge but also to the sides, form an expanding elliptic gas-dust cloud which initially bends the flat free face (Fig. 21, a). As a result, the muck spreads in a fan-shaped dome. Only a small part of blasted rock (unshaded rectangle *ABCD*) is moved along the normal to the charge, and the peripheral sections of the slab charge to the left of point *A* and to the right of point *D* throw the rock to the longer sides. This is known as the edge-effect of slab charges with finite dimensions.

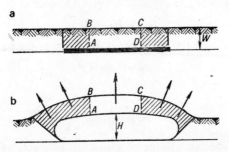

Fig. 21. Diagram showing outline of rock dome during blasting
of slab-charge systems:

a—before blasting; b—at the moment of maximum expansion
of gas-dust cloud.

With a view to achieving a throw of the entire rock mass lying along the normal to the charge (volume *ABCD* in Fig. 22) in a perpendicular direction (i.e. to achieve a plane-parallel throw-out explosion), it is necessary, as established through experiments [7], to surround the slab charge with charges of

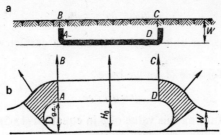

Fig. 22. Diagram showing spread of rock mass in case
of peripheral charges:

a—before blasting; b—at the moment of expansion
of gas-dust cloud.

strength p_0. The value of p_0 is such that the volume of the explosion products is equal to that volume of explosion products which expand to the longer sides of a usual (unsurrounded) slab charge. For this purpose, obviously, it is necessary that the diameters of the cylindrical gas-dust cloud $D_{g.c}$ formed by the explosion of peripheral charges, in the final stage of expansion, be equal to the height of the gas-dust cloud $H_0 = qWu$. The peripheral charges will form semi-cylindrical volumes equal to $1/8\pi D_{g.c}^2$. From these considerations, we obtain the equation

$$\frac{1}{8}\pi D_{g.c} = p_0 u.$$

Besides, on the basis of the explanation given above:

$$D_{g.c} = H_0 = qWu.$$

From here, we find that the mass per unit length of the peripheral charges is

$$p_0 = G_1 (qW)^2, \tag{1.63}$$

where

$$G_1 = \frac{1}{8}\pi u. \tag{1.64}$$

By taking the specific volume of explosion products as an average value $u = 1$ m³/kg, we find that $G_1 = 0.4$ m³/kg.

Formula (1.63) is applicable to an ideal slab charge (i.e. thin explosive plate). If an ideal slab charge is replaced with a linear charge system (see Fig. 3, a) then this formula is applicable for the longer sides DE and CF. On the shorter sides of this charge, there are linear charges CD and EF which partially fulfill the role of peripheral charges. In order to avoid edge-effects on these sides it is necessary to place more linear charges with a mass per unit length equal to:

$$p_{sub} = p_0 - \frac{1}{2}p, \tag{1.65}$$

where p is the mass per unit length of linear charges determined by formula (1.3).

Because, from formula (1.3), we have

$$p = qWb, \tag{1.66}$$

therefore, on substituting this value of p in equation (1.65), we find

$$p_{sub} = G_1 (qW)^2 - \frac{1}{2}qWb. \tag{1.67}$$

If a slab-charge system is made up of concentrated charges (see Fig. 3, b), then the average mass per unit length of concentrated charges deposited along sides CD and EF will be:

$$p = \frac{Q}{a}. \tag{1.68}$$

Along the longer sides DE and CF, this parameter is determined in an identical manner

$$p = \frac{Q}{b}. \tag{1.69}$$

By equating the obtained values of p to that of p_0 from expression (1.63), we can calculate the mass of peripheral charges Q_L and Q_B along the longer and shorter sides respectively of a slab-charge system (see Fig. 3, b).

$$Q_L = G_1 (qW)^2 b; \tag{1.70}$$

$$Q_B = G_1 (qW)^2 a. \tag{1.71}$$

With a view to eliminating the edge-effects in case of concentrated charges at the corners of a slab-charge system, their mass should be calculated using formulas (1.70) and (1.71).

From formula (1.63), it follows that the mass per unit length of peripheral charges is proportional to $(qW)^2$. Consequently, in case of a high specific consumption q of explosives and a low depth W, the mass of peripheral charges will increase. As a result, the actual specific consumption of explosives q_{actual} will be greater than the quantity q.

Let us find q_{actual} from the following equation

$$q_{actual} LBW = qLBW + 2 (L+B) G_1 (qW)^2,$$

where

$$\frac{q_{actual}}{q} = 1 + \frac{1}{2} \pi \Omega,$$

where parameter Ω is determined by expression (1.28).

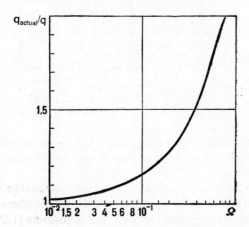

Fig. 23. Relationship of ratio q_{actual}/q and specific consumption of explosives q for an ideal slab charge.

Figure 23 shows the ratio curve q_{actual}/q. It is evident that the value q_{actual} increases with an increase in parameter Ω. However, it must be borne in

38

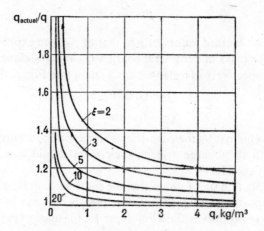

Fig. 24. Relationship of ratio q_{actual}/q and specific consumption
of explosives q for slab systems of linear charges.

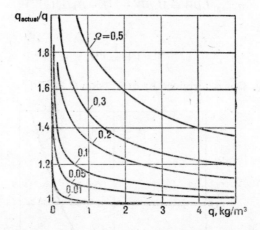

Fig. 25. Relationship of ratio q_{actual}/q and specific consumption
of explosives q for slab systems of concentrated charges.

mind that due to peripheral charges around the slab-charge system the blast
efficiency increases from η (for slab charges without peripheral charges) to η_0,
i.e. the blast efficiency increases, as is proved in formula (1.29), by $1/k$ times
(it should be noted that $k < 1$).

When a slab-charge system is prepared in the form of linear or concentrated charges, the distance between which is calculated by the expression
$a = b - W\sqrt{q/k_{exp}}$, the actual specific consumption of explosives can be calculated by the following formulas, respectively:

$$q_{\text{actual}} = q\left(1 + \frac{1}{u\xi\sqrt{qk_{\text{exp}}}}\right); \tag{1.72}$$

$$q_{\text{actual}} = q\left(1 + \frac{2\Omega}{u\sqrt{qk_{\text{exp}}}}\right). \tag{1.73}$$

It should be remembered that in expressions (1.63), (1.72) and (1.73), quantity q is the specific consumption of an ideal slab charge in the form of an explosive plate. It is calculated as the ratio between the mass of the charge and the rock volume deposited exactly along the normal to the charge. Figures 24 and 25 show the curves obtained from equations (1.72) and (1.73).

In the practical utilization of slab charge systems it is sometimes advisable to place peripheral charges only on opposite sides. In this case, depending on the manner in which peripheral charges are laid, the parameter Ω can be calculated by the following formulas:

$$\Omega = \frac{1}{2\chi};$$

$$\Omega = \frac{1}{2\xi}.$$

The first formula is for peripheral charging along the longer sides and the second formula is for peripheral charging along the shorter sides of the charge.

Chapter 2

Blast Efficiency

1. CALCULATION OF BLAST EFFICIENCY

The blast efficiency of a slab charge is the ratio between the kinetic energy of the blasted rock and the total blast energy.

It has been mentioned earlier that, depending on the physicomechanical properties of the rock to be blasted, the blast efficiency varies over a fairly wide range. It is, therefore, not possible to correctly design a slab charge without prior determination of a more or less accurate value of its blast efficiency. Most reliable data can be obtained by experimental computation of blast efficiency. However, it is not always possible to carry out such experiments. Moreover, they entail a relatively high requirement of manpower. Experiments for determining blast efficiency should be carried out only for making final calculations. At the initial stages of designing a charge, it is sufficient to know the approximate value of its blast efficiency. This can be found by computational methods.

Let us study an ideal slab charge (thin plate of infinite dimensions). In this case, the estimated blast efficiency for actual slab charges will serve as an upper limit. In practical calculations the edge-effects can be considered as 'geometry' correction factors (see Chapter 1, Sec. 5).

For calculation purposes, let us take the ideal elastoplastic model of the medium of which the dynamic compression diagram is shown in Fig. 26. Stress in the medium is shown along the ordinate and relative deformation due to axial compression—along the abscissa. In the section OB of the diagram, the medium undergoes elastic compression. This means that it returns to its initial state when the stress is removed from it. At yield stress Σ_y, a plastic deformation of the medium takes place. The medium remains in the same deformed state when the force is removed. At point C in the diagram, the medium is fully deformed and beyond this point it experiences an elastic compression (branch CD).

For simplifying the calculations, let us describe the process of deformation of a medium in the following manner. We shall assume that a slab charge is detonated instantly. The explosion products initially expand symmetrically

40

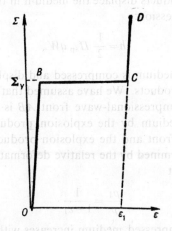

Fig. 26. Diagram showing stresses in an ideal
elastoplastic medium.

in the lower half-space along the x_2 axis as well as towards the free face along
the x_1 axis (Fig. 27) with an initial velocity equal to that of the mass velocity
v at the shock-wave front which is propagated in the medium under the effect
of explosion products. As is well known

$$v = \frac{\Sigma_0}{\rho D_w},$$ (2.1)

where Σ_0 is the initial pressure of explosion products; ρ the density of the
medium in the natural state; and D_w the velocity of propagation of the shock-
wave front in the medium.

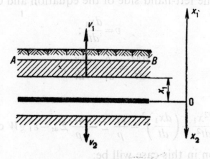

Fig. 27. Diagram for calculating blast efficiency.

In the lower half-space, the explosion products displace the medium to a
certain depth which depends on the springing index. Thereafter, the boundary
between the medium and explosion products becomes static. Depth h, to

which the explosion products displace the medium in the lower half-space, is determined by the expression:

$$h = \frac{1}{2}\, \Pi_{\mathrm{sp}}\, qW. \qquad (2.2)$$

At the same time the medium is compressed and displaced towards the free face by the explosion products. We have assumed that the medium is elasto-plastic, therefore, a compressional-wave front AB is formed as a result of compression of the medium by the explosion products (see Fig. 27). The medium between this front and the explosion products is compressed to a density ρ which is determined by the relative deformation ε_1 (see Fig. 26). It can be easily shown that

$$\frac{\rho_1}{\rho} = \frac{1}{1 - \varepsilon_1}.$$

The mass of the compressed medium increases with the movement of the compressional-wave front towards the free face. If explosion products cover a distance x_1 in the direction of the free face, then the mass M, covered by the compressional wave and attributed to a unit area, is

$$M = \frac{x_1}{\varepsilon_1}\, \rho.$$

The differential equation describing the movement of the medium caused by the compressional wave can be written as follows:

$$\frac{d}{dt}\,(vM) = \Sigma - \Sigma_{\mathrm{at}} - g\rho \cos \alpha, \qquad (2.3)$$

where v is the velocity of the medium and of explosion products at the interface, Σ the pressure of explosion products and Σ_{at} the atmospheric pressure. By differentiating the left-hand side of the equation and considering, that

$$v = \frac{dx_1}{dt};$$

$$\frac{dM}{dt} = \frac{\rho}{\varepsilon_1} \cdot \frac{dx_1}{dt},$$

we obtain

$$x_1 \frac{d^2 x_1}{dt^2} + \left(\frac{dx_1}{dt}\right)^2 = \frac{\varepsilon_1}{\rho}\, \Sigma - \frac{\varepsilon_1}{\rho}\, \Sigma_{\mathrm{at}} - \varepsilon_1\, gW \cos \alpha. \qquad (2.4)$$

The initial condition in this case will be:

$$t = 0; \quad x_1 = 0; \quad \frac{dx_1}{dt} = v_0 = \frac{\Sigma_0}{\rho D_{\mathrm{w}}}.$$

When the compressional wave reaches the free face, the mass of the compressed medium M becomes $W\rho$ and further compression ceases. As a result, the compressed layer is imparted a maximum velocity by the expanding explosion products.

In this way, the movement of the medium during passage of the compressional wave is defined by a differential equation (1.74). This has relevance for the section between $x_1=0$ to $x_1=W\varepsilon_1$. When $x_1>W\varepsilon_1$, i.e. when the rock layer of thickness W is fully compressed, the movement of the displaced medium in the upper half space is expressed by the following differential equation:

$$\rho W \frac{d^2x}{dt^2}=\Sigma-\rho gW-\Sigma_{at}-\frac{1}{2}c_x\,\rho_{air}\left(\frac{dx}{dt}\right)^2,\qquad(2.5)$$

where Σ_{at} is the atmospheric pressure, c_x the head drag coefficient, and ρ_{air} the density of air.

On the right-hand side of the equation, the second term takes gravity, and the last term the air drag, into account.

The following are the initial conditions for this equation: $t=0$; $x=x_1=\varepsilon W$; $\frac{dx}{dt}=\left(\frac{dx}{dt}\right)_{lim}$. In this case quantity $\left(\frac{dx}{dt}\right)_{lim}$ is two times the quantity obtained from equation (2.4) for $x_1=\varepsilon_1 W$. Here, doubling of the bulk velocity is considered when the compressional wave is repelled by the free face. It is obvious that the acceleration of blasted rock ceases when the right-hand side of equation (2.5) is reduced to zero. From this point onward, the medium is no longer accelerated and its independent movement commences with a maximum velocity ω_{max} under the force of gravity and air drag.

For constructing the trajectory of rock movement, it is better to take the center of gravity of the rock before blasting as the origin for purposes of computation. For this purpose, it is necessary to introduce some arbitrary initial velocity $v_{f\,max}$ which is finally reduced under the effect of gravity and air drag to v_{max} determined through equation (2.5). Let coordinate x, corresponding to quantity v_{max}, be equal to x^*. The quantity $v_{f\,max}$ can be determined from the following energy equation:

$$\frac{1}{2}\rho Wv_{f\,max}^2=\frac{1}{2}\rho Wv_{max}^2+\rho Wg\cos\alpha+\frac{1}{2}\rho_{air}\,c_x\left(\frac{v_{f\,max}+v_{max}}{2}\right)^2x^*,$$

which takes into account the losses in blast energy for overcoming the force due to gravity and air drag.

The blast efficiency will be equal to the ratio between the kinetic energy of the mass of the medium of thickness W lying above a unit area of the charge and the total energy of this charge.

The kinetic energy of the medium is

$$E_{kin}=\frac{1}{2}W\rho v_{f\,max}^2.$$

The total energy produced during explosion of a unit area of charge is expressed by the formula

$$E_{ch} = qWe_{exp}\, k,$$

where coefficient k takes into account the edge-effects (see Chapter 1, Sec. 5). By dividing the right- and left-hand sides of these equations, we get the expression for blast efficiency of a slab charge as follows:

$$\eta = \frac{\rho}{2qe_{exp}\, k}\, v_{f\,max}^2.$$

In order to simplify the solution, let us assume that the medium in the lower half space moves with a velocity which decreases linearly from its initial value to zero:

$$v = v_0\left(1 - \frac{x_2}{h}\right), \tag{2.6}$$

where

$$v = \frac{dx_2}{dt}.$$

The pressure of explosion products is expressed through a polytropic equation:

$$\Sigma u^\varkappa = \text{const},$$

where u is the specific volume of explosion products.

We will differentiate between the two periods. In the first period when the specific volume is between $u_0 - 2u_0$, the polytropic index $\varkappa$ varies between 3 and 1.25. However, in our approximate calculations we can use its mean value ($\varkappa = 2$). When $u > 2u_0$, the polytropic index is considered constant and equal to 1.25. The polytropic equation for the first period will be:

$$\Sigma u^\varkappa = \Sigma_{cr}\, u_{det}^\varkappa \left(\frac{\Sigma_{det}}{\Sigma_{cr}}\right)^{\frac{\varkappa}{2}} \left(\frac{\Sigma_{cr}}{\Sigma_{det}}\right),$$

where $\Sigma_{det} = \frac{1}{8}\rho_{exp}\, D^2$ is the pressure of charge detonation products; Σ_{cr} the critical pressure of explosion products which is determined on the basis of the specific volume u_{cr}; and u_{det} the specific volume of explosives.

From Fig. 27 it is evident that the specific volume is given by

$$u = \frac{x^* + x_2}{qW}.$$

Quantity x^* is calculated from successive solutions of Eqs. (2.4) and (2.5).

The downward distance covered by the explosion products up to a given moment t can be calculated from expression (2.6). On differentiating, we obtain the following equation:

$$\frac{d^2x_2}{dt^2} + \frac{v_0}{h} \cdot \frac{dx_2}{dt} = 0.$$

This equation is solved with initial conditions $t=0$; $x_2=0$; $dx_2/dt = v_0 = \Sigma_0/D_w \rho$ and possesses a physical significance when $x \leqslant h$.

From an analysis of the equations given above, it follows that blast efficiency will depend on the following parameters: q, W, e_{exp}, ρ, ε_1, ρ_{exp} and D_w. These parameters characterize the physicomechanical properties of the explosives and the medium to be blasted.

The method of finding the blast efficiency given above requires the application of numerical integration methods. The following approximation method can be used for finding a more accurate blast efficiency.

Let us write the following equation on the basis of the law of conservation of energy:

$$\eta = \eta_{lim} - \eta_{s.h} - \eta_{plast} ,$$

where η is the blast efficiency of a slab charge; η_{lim} the limiting value of blast efficiency (see Fig. 6); $\eta_{s.h}$ the part of energy utilized in moving the medium to the lower half space (i.e. in forming a sprung-hole cavity); and η_{plast} the part of energy utilized in plastic deformation when the medium moves to the upper half space.

Since η_{lim} is known for a given explosive, it is sufficient to know the approximate values of $\eta_{s.h}$ and η_{plast} for calculating η.

The energy expended in moving the medium to the lower half space can be calculated in accordance with the method given in [23, 24]. This part of the energy is approximately equal to half of the energy utilized by the charge in forming a sprung-hole cavity. From the expression for the work done in expansion of gases in an adiabatic process,

$$\eta_{s.h} = \frac{1}{2}\left[1 - \left(\frac{U_0}{U}\right)^{\varkappa-1}\right],$$

we can find the desired part of the energy by substituting in this expression $U = U_{s.h}$. In this expression U_0 represents the initial volume of explosion products. Coefficient $\frac{1}{2}$ shows that in the present case only the lower half-space is being considered. The work done by explosion products in bringing about a plastic deformation of the medium in the upper half-space accounts for energy equal to η_{plast}. Here $U_{s.h}$ is equal to the volume of the sprung hole; thus

$$U_{s.h} = qW\Pi_{sp}.$$

If we assume that U_0 is approximately equal to $2u_{det} qW$, then for $u > 2u_0$, the average adiabatic index can be taken as $\varkappa = 1.25$.

Since

$$u_{det} = \frac{1}{\rho_{exp}},$$

therefore

$$\eta_{s.h} = \frac{1}{2} \left[1 - \left(\frac{2q\,W}{U \rho_{exp}} \right)^{\varkappa - 1} \right].$$

By substituting $U = U_{s.h}$ we obtain

$$\eta_{s.h} = \frac{1}{2} \left[1 - \left(\frac{2q\,W}{U_{s.h}\,\rho_{exp}} \right)^{\varkappa - 1} \right].$$

In this manner

$$\eta_{s.h} = \frac{1}{2} \left[1 - \left(\frac{2}{\Pi_{sp}\,\rho_{exp}} \right)^{\varkappa - 1} \right].$$

Let us calculate the part of energy η_{plast}. According to the definition

$$\eta_{plast} = \frac{\Sigma_{plast}\,W \varepsilon_1}{q W\,e_{exp}},$$

where product $\Sigma_{plast}\,\varepsilon_1$ in the numerator represents the plastic deformation of a unit volume of an ideal elastoplastic medium and product $\Sigma_{plast}\,\varepsilon_1 W$ is the plastic deformation of a medium of thickness W at the moment when the compressional-wave front AB (see Fig. 27) reaches the free face. The denominator in the expression for η_{plast} is equal to the blast energy per unit area of the charge.

While calculating η_{plast} it may be found that $\Sigma_{plast}\,W \varepsilon_1 \geqslant q W e_{exp}$. This means that the blast energy is not sufficient for compressing the entire layer of the medium of thickness W. In this case the entire volume of the medium is not compressed. As a result an implosion takes place without a visible movement of the free face. The blast efficiency in this case is zero.

2. BLASTING WITHOUT STEMMING

Cinematographic and visual examinations of blasting with concentrated and deep-hole charges in hard rock reveal that, simultaneously with the throw of rock after escape of stemming, the explosion products release under pressure from the hole or adit in which the charge was placed. Naturally, it can be assumed that the gases escaping from the charge adit carry with them some part of the energy thereby reducing the blast efficiency. It is obvious that loss of blast energy will increase with an increase in the cross-sectional area of the charge adit.

Let us establish the correlation between the cross-sectional area of the charge adit and the charge mass at which the loss of energy through the adit can be ignored and the extent of such loss for given explosion parameters is large.

Since the slab charge is prepared, as a rule, in the form of a chamber-charge system for carrying out blasting on a large scale, it is better to solve the given problem for these very charges. For analyzing the results, it will be convenient to assume that stemming is altogether absent. In this case the obtained trial parameter will be calculated for the upper limit because in the presence of stemming, the effect of energy losses will be still lower.

Obviously, the problem can only be solved by setting up and solving a closed system of differential equations describing the temporal process of expansion of the charge chamber, movement of the blown-out medium and at the same time, the flow of explosion products through the charge adit. Let us describe this complex phenomenon in the following way. Consider that the process of expansion of the charge chamber and displacement of the medium to be blasted is completed in two successive phases. In the first phase, the spherically symmetrical expansion of the roof of the gas chamber (Fig. 28) takes place in the same manner as in the case of an implosion.

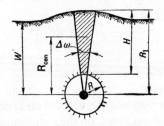

Fig. 28. First phase of expansion of gas chamber.

At the end of the first phase the gas chamber ceases to expand downward (opposite the free face). In the second phase the explosion products expand and the rock mass is displaced towards the free face (Fig. 29). Such a description of the process of explosion is nearer to actual conditions which is confirmed by a cinematographic study of the explosion process of a concentrated charge which was isolated from the camera by reinforced (iron-clad) glass [25].

Let us write a differential equation of movement of the displaced rock mass in the first phase of explosion. Let the spherical charge lie at a depth W below the free face and let us assume that its explosion products attain the volume of a sphere of radius R after a time t. Let us consider any small solid angle in the direction of the free face (Fig. 28). The mass of rock ΔM within the limits of this solid angle will be:

$$\Delta M = \frac{1}{3} \Delta \omega \, W^3 \rho.$$

48

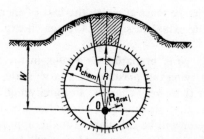

Fig. 29. Second phase of expansion of gas chamber.

The center of gravity of the volume under study is at a distance R_{cen} from the center of the sphere. A pressure equal to $\Delta\omega R^2 \Sigma$ is acting on this elementary volume of rock mass from the side of the gas chamber. Here, Σ is the pressure of explosion products at a given moment t. Let us assume that solid rock is an incompressible medium. The equation of movement after reduction to $\Delta\omega$ will be:

$$\frac{1}{3} W^3 \rho \frac{d^2 R_{\text{cen}}}{dt^2} = R^2 \Sigma - \frac{1}{3} W^3 g\rho \cos\alpha - R^2 \Sigma \mu_{\text{diss}} - R_1^2 \Sigma_{\text{at}}$$

$$- \frac{1}{2} c_x R_1^2 \rho_{\text{air}} \left(\frac{dR_1}{dt}\right)^2,$$

where μ_{diss} is the coefficient describing dissipation of explosive energy; R_1 the distance between the center of the gas chamber and the free face at a given moment; and α the angle between the vertical and the line of least resistance.

In this equation, the second term on the right-hand side takes into account the effect of gravity, the third term accounts for dissipation assumed to be proportional to the pressure in the gas chamber, the fourth represents atmospheric pressure, and the last term, the air drag.

The presumption that dissipation is proportional to the pressure in the gas chamber follows from the assumption made above in regard to the incompressibility of the medium and the law of Coulomb friction which states that the force of friction, and consequently, dissipation can be taken to be approximately proportional to the pressure (we ignore the medium for the breaking operation).

On dividing the left- and right-hand sides of the above equation by $1/3\, W\rho$, we obtain

$$\frac{d^2 R_{\text{cen}}}{dt^2} = \frac{3R^2}{W\rho} \Sigma - g\cos\alpha - \frac{3R^2 \mu_{\text{diss}}}{W^3 \rho} - \frac{3R_1^2}{W^3 \rho} \Sigma_{\text{at}}$$

$$- \frac{3c_x R_1^2 \rho_{\text{air}}}{W^3 \rho} \left(\frac{dR_1}{dt}\right)^2. \tag{2.7}$$

Let us find the relationship between R_{cen}, R_1 and R (see Fig. 28) from the condition that during the entire period of movement the rock mass in the elementary cone remains unchanged and is equal to $1/3\ \Omega\ W^3\rho$. On calculating we find the following expressions:

$$R_{cen} = \frac{3}{4W^3}[(W^3+R^3)^{4/3} - R^4];$$ (2.8)

$$R_1 = [W^3 + R^3]^{1/3}.$$ (2.9)

Differentiating the equation (2.8) twice and equation (2.9) once with respect to t, we find:

$$\frac{d^2 R_{cen}}{dt^2} = \frac{3}{W^3}\left\{[(W^3+R^3)^{1/3}\ R^2 - R^3]\ \frac{d^2 R}{dt^2}\right.$$

$$+[2\ (W^3+R^3)^{1/3}\ R + (W^3+R^3)^{-2/3}\ R^4 - 3R^2]\ \frac{dR}{dt}\right\};$$ (2.10)

$$\frac{dR_1}{dt} = (W^3 + R^3)^{2/3}\ R^2\ \frac{dR}{dt}.$$ (2.11)

For simplifying the solution of equations it is advisable to present them in a dimensionless form by selecting the following dimensionless parameters:

$$\left.\begin{array}{c} r = \dfrac{R}{R_0}; \\[2mm] r_{cen} = \dfrac{R_{cen}}{R_0}; \\[2mm] \tau = \dfrac{t}{t_0}; \\[2mm] \hat{W} = \dfrac{W}{R_0}; \\[2mm] r_1 = \dfrac{R_1}{R_0}; \\[2mm] \sigma = \dfrac{\Sigma}{\Sigma_0}; \\[2mm] \sigma_{at} = \dfrac{\Sigma_{at}}{\Sigma_0}, \end{array}\right\}$$ (2.12)

where R_0 is the radius of the charge chamber, and Σ_0 the initial pressure of explosion products.

For a unit time scale, we take time t_0 during which a point moving at a velocity equal to the mass velocity of particles v_0 beyond the shock-wave front at the initial moment of explosion, crosses the radius R_0 of the charge chamber. For solid rocks $v_0 = \Sigma_0/\rho D_w$, therefore

$$t_0 = \frac{R_0}{v_0} = \frac{R_0 \rho D_w}{\Sigma_0}.$$

From the equation of state for a unit mass of gas, we find

$$u_0 \Sigma_0 = e_{exp}(\varkappa - 1) \tag{2.13}$$

where $\varkappa$ is the isentropic index of explosion products. The radius of the charge chamber is calculated by the following known formula:

$$R_0 = 0.62 u_0^{1/3} \hat{Q} W, \tag{2.14}$$

where u_0 is the initial specific volume of explosion products; Q the mass of the charge, and

$$\hat{Q} = \frac{Q^{1/3}}{W}. \tag{2.15}$$

From the relations given above, we find

$$t_0 = \frac{0.62 \rho\, D_w u_0^{4/3} \hat{Q}}{e_{exp}(\varkappa - 1)} W. \tag{2.16}$$

Equation (2.7) in dimensionless parameters is transformed to:

$$\frac{d^2 r_{cen}}{d\tau^2} = 3m_1 \frac{r^2}{\hat{W}^3} \sigma - j_1 - 3n_1 \frac{r_1^3}{\hat{W}^3} \sigma_{at} - 3d_1 \frac{r_1^2}{\hat{W}^3} \left(\frac{dr_1}{d\tau}\right)^2, \tag{2.17}$$

where

$$m_1 = \frac{\rho D_w u_0 (1 - \mu_{diss})}{e_{exp}(\varkappa - 1)}; \tag{2.18}$$

$$j_1 = \frac{0.62 u_0^{7/3} \rho^2 D_w^2 \cos\alpha\, \hat{Q} W}{e_{exp}(\varkappa - 1)^2}; \tag{2.19}$$

$$n_1 = \frac{\rho D_w^2 \Sigma_{at} u_0^2}{e_{exp}^2 (\varkappa - 1)^2}; \tag{2.20}$$

$$d_1 = c_x \frac{\rho_{air}}{\rho}. \tag{2.21}$$

Obviously, the dimensionless quantities m_1, j_1, n_1 and d_1 are the criteria of similarity. From these criteria it is clear that only one of them (namely criterion j_1) depends on the scale of explosion because it includes depth W.

Consequently, in order to ensure that the small-scale explosion is similar to the large-scale explosion, it is necessary that other conditions being equal, the product gW included in criterion j_1 should be constant. This can be achieved with the help of centrifugal modeling [26].

By writing the expressions (2.10) and (2.11) in a dimensionless form and by substituting them in equation (2.7) and after simple transformations, we obtain the differential equation of rock movement in dimensionless parameters:

$$[(\hat{W}^3+R^3)^{5/3}\,r^2-r^3\,(\hat{W}^3+r^3)^{4/3}]\,\frac{d\hat{v}^2}{dr}+[4\,(\hat{W}^3+r^3)^{5/3}\,r$$

$$+2r^4\,(\hat{W}^3+r^3)^{2/3}-6r^2\,(\hat{W}^3+r^3)^{4/3}]\,\hat{v}^2=2m_1\,(W^3+r^3)^{4/3}\,r^2\sigma$$

$$-\frac{2}{3}\,j_1\,\hat{W}^3\,(\hat{W}^3+r^3)^{4/3}-2n_1\,(\hat{W}^3+r^3)^{6/3}-2d_1\,(\hat{W}^3+r^3)^{2/3}\,r^4\hat{v}^2,\quad (2.22)$$

where the dimensionless velocity is given by

$$\hat{v}=\frac{dr}{d\tau}.$$

Let us now write the differential equation for a specific volume of explosion products in an expanding gas chamber.

Specific volume u is equal to the volume of the gas chamber U divided by the mass Q of explosion products at a given moment by taking into account their flow through the charge adit:

$$u=\frac{U}{Q}.$$

Let us introduce the dimensionless parameters:

$$\hat{u}=\frac{u}{u_0};$$

$$\hat{s}=\frac{U}{U_0};$$

$$\hat{q}=\frac{Q}{Q_0}.$$

where u_0, U_0 and Q_0 are the initial specific volumes of explosion products, gas chamber and explosives respectively. Quantities $\hat{u}$, $\hat{s}$, $\hat{q}$ will be known as reduced quantities of the respective parameters.

In this way, the specific volume in dimensionless form will be

$$\hat{u} = \frac{\hat{s}}{\hat{q}}.$$

After differentiating with respect to r, we obtain

$$\frac{d\hat{u}}{dr} = \frac{1}{\hat{q}} \cdot \frac{d\hat{s}}{dr} - \frac{\hat{s}}{\hat{q}^2} \cdot \frac{d\hat{q}}{dr} . \tag{2.23}$$

The volume of gas chamber U in the first phase of explosion is

$$U = \frac{4}{3} \pi R^3,$$

which in dimensionless form can be written as

$$\hat{s} = r^3. \tag{2.24}$$

In the second phase, based on the assumption made above, the downward expansion of the gas chamber ceases and it expands only towards the free face. In this case the gas chamber is similar in shape to an ellipsoid of revolution. However, for the sake of simplicity of calculation, we shall consider the gas chamber shape in the second phase to be spherical with a radius R_{cham} the center of which with the progress of explosion, moves upward along the line of least resistance (see Fig. 29). As is evident, radius

$$R_{cham} = \frac{R + R_{first}}{2},$$

where R is the distance along the line of least resistance extending between the center of the charge and the boundary of the gas chamber; and R_{first} the radius of the gas chamber at the end of the first phase of explosion.

Thus, the dimensionless volume of the chamber in the second phase will be:

$$\hat{s} = \frac{1}{8} (r + r_{first})^3, \tag{2.25}$$

where r and r_{first} are the corresponding dimensionless radii.

From equation (2.25) it is clear that when $r = r_{first}$ (end of the first phase) $\hat{s} = r_{first}^3$ which coincides with the expression (2.24).

Differentiating equations (2.24) and (2.25) with respect to r we get:

$$\frac{d\hat{s}}{d\tau} = 3r^2 \text{ (first phase } r \leqslant r_{first}); \tag{2.26}$$

$$\frac{d\hat{s}}{dr} = \frac{3}{8}(r + r_{\text{first}})^2 \quad (\text{second phase } r > r_{\text{first}}). \tag{2.27}$$

Let us find the expression for the derivative $d\hat{q}/dr$. According to the theory of gas flow through a hole, we can write the following differential equation:

$$\frac{dQ}{dt} = -F_{\text{adit}} b_{\varkappa} \sqrt{\frac{\Sigma}{u}}, \tag{2.28}$$

where F_{adit} is the cross-sectional area of the charge adit.

$$b_{\varkappa} = \left(\frac{2}{\varkappa+1}\right)^{\frac{1}{\varkappa+1}} \sqrt{\frac{2\varkappa}{\varkappa+1}}.$$

Because the flow and expansion of explosion products are adiabatic, therefore, $\Sigma u^{\varkappa} = \Sigma_0 u_0^{\varkappa}$.

If $u > u^* = 3u_0$, then

$$\Sigma u^{\varkappa} = c_1,$$

where $c_1 = \Sigma_0 u_0^{\varkappa}$; $\Sigma_0 = (1/8) \, D_{\text{det}}/u_0$; $u_0 = 1/\rho_{\text{exp}}$; and D_{det} is the rate of detonation.

Substituting this dependence in equation (2.28) we find

$$\frac{dQ}{dt} = -F_{\text{adit}} b_{\varkappa} \sqrt{\frac{\Sigma_0}{u_0}} \left(\frac{\Sigma}{\Sigma_0}\right)^{\frac{\varkappa+1}{2\varkappa}}.$$

Considering that $Q = Q_0 \, \hat{q}$; $dt = t_0 \, d\tau$; $\Sigma = \Sigma_0 \, \sigma$; $\hat{v} = dr/d\tau$; $dt = dr/v$; $t_0 = R_0$ $\rho D_{\text{w}}/\Sigma_0$; $R_0 = 0.62 \, u_0^{1/3} \, \hat{Q} W$ and $\hat{Q} = Q_0^{1/3}/W$, the obtained equation can be written in the dimensionless form:

$$\frac{d\hat{q}}{dr} = -B_1 \frac{\sigma^{\frac{\varkappa+1}{2\varkappa}}}{\hat{v}}, \tag{2.29}$$

where

$$B_1 = 0.62 u_0^{1/3} \rho D_{\text{w}} b_{\varkappa} f_1 (\sqrt{e_{\exp}} (\varkappa - 2) \, \hat{Q}^2)^{-1}; \tag{2.30}$$

$$f_1 = \frac{F_{\text{adit}}}{W^2}. \tag{2.31}$$

The dimensionless parameter f_1 is the reduced cross-sectional area of the charge adit. From expression (2.29) it is clear that quantity B_1 can be viewed as a similarity criterion. If this criterion is constant, the flow of explosion products through a model adit and that under actual conditions will be similar.

Let us write a differential equation for blast efficiency taking into consideration the flow of explosion products through an adit. In the present case, blast efficiency implies the fraction of total energy η which is transferred to the medium by explosion products after their adiabatic expansion to atmospheric pressure. As we know, gases of mass Q, after adiabatic expansion to pressure Σ, can do work given by

$$A = Q \frac{\Sigma_0}{\varkappa - 1} \left[1 - \left(\frac{\Sigma}{\Sigma_0} \right)^{\frac{\varkappa-1}{\varkappa}} \right].$$

In accordance with expression (2.13) quantity $\Sigma_0 u_0/(\varkappa - 1) = e_{\exp}$. We will bear in mind that mass Q is the mass of gases remaining in the explosion chamber at a given moment during the process of their escape (flow).

Let us divide the left- and right-hand sides of this expression by the total blast energy $E_0 = Q_0 \, e_{\exp}$, the blast efficiency $\eta = A/E_0$ is then expressed by the following:

$$\eta = \hat{q} \left[1 - \sigma^{\frac{\varkappa-1}{\varkappa}} \right].$$

By differentiating this with respect to r, we obtain the equation in dimensionless form:

$$\frac{d\eta}{dr} \left(1 - \sigma^{\frac{\varkappa-1}{\varkappa}} \right) \frac{d\hat{q}}{dr} - \frac{\varkappa - 1}{\varkappa} \times \frac{\hat{q}}{\sigma^{\frac{1}{\varkappa}}} \frac{d\sigma}{dr}. \tag{2.32}$$

In order to obtain a closed system from the differential equations given above it is also necessary to consider the following two equations:

$$\varkappa \sigma u^{\varkappa - 1} \frac{du}{dr} + u^{\varkappa} \frac{d\sigma}{dr} = 0;$$

$$\frac{dr}{d\tau} = \hat{v}.$$

The first of these is the equation for adiabatic expansion in a differential form and the second that of velocity as a derivative of the path with respect to time.

Thus, the process of expansion of the gas chamber with a simultaneous flow of explosion products is described through the following system of differential equations in dimensionless forms:

$$\frac{d\eta}{dr} = \left(1 - \sigma^{\frac{\varkappa-1}{\varkappa}}\right) \frac{\hat{q}}{dr} - \frac{\varkappa-1}{\varkappa} \frac{\hat{q}}{\sigma^{1/\varkappa}} \cdot \frac{d\sigma}{dr}; \qquad (1)$$

$$\frac{du}{dr} = \frac{1}{\hat{q}} \cdot \frac{d\hat{s}}{dr} - \frac{\hat{s}}{\hat{q}^2} \cdot \frac{d\hat{q}}{dr}; \qquad (2)$$

$$\left.\begin{array}{ll}\dfrac{d\hat{s}}{dr} = 3r^2; & (r \leqslant r_{\text{first}}) \quad \text{(a)} \\[2mm] \dfrac{d\hat{s}}{dr} = \dfrac{3}{8}(r + r_{\text{first}})^2; & (r > r_{\text{first}}) \quad \text{(b)} \end{array}\right\} \qquad (3)$$

$$\frac{d\hat{q}}{dr} = -B_1 \frac{\sigma^{\frac{\varkappa+1}{2\varkappa}}}{\hat{v}}; \qquad (4)$$

$$[(\hat{W}^3 + r^3)^{5/3} r^2 - r^3 (\hat{W}^3 + r^3)^{4/3}] \frac{d\hat{v}^2}{dr} + [4(\hat{W}^3 + r^3)^{5/3}$$

$$+ 2r^4 (\hat{W}^3 + r^3)^{2/3} - 6r^2 (\hat{W}^3 + r^3)^{4/3}] \hat{v}^2 = 2m_1 (\hat{W}^3 + r^3)^{4/3} r^2 \sigma$$

$$- \frac{2}{3} j_1 \hat{W}^3 (\hat{W}^3 + r^3)^{4/3} - 2n_1 (\hat{W}^3 + r^3)^{3/2} - 2d_1 (\hat{W}^3 + r^3)^{2/3} r^4 \hat{v}^2; \qquad (5)$$

$$\varkappa \sigma u^{\varkappa-1} \frac{du}{dr} + u^\varkappa \frac{d\sigma}{dr} = 0; \qquad (6)$$

$$\frac{dr}{d\tau} = \hat{v}. \qquad (7)$$

$\Big\}$ (I)

The initial conditions of this system are: $r_0 = 1$, $\hat{u}_0 = 1$, $\eta_0 = 0$, $\sigma_0 = 1$, $\hat{q}_0 = 1$, $\hat{s}_0 = 1$, $\hat{v}_0 = 1$ and $\tau_0 = 0$.

In solving the system of equations (I) it is assumed that for small specific volumes the expansion of explosion products occurs in accordance with the Jones adiabatic relationship for which the index has a value $\varkappa = 3$. Hence the system of equations (I) must be solved in two stages. In the first stage, we consider $\varkappa = 3$, so that the integration of the system is closed and the specific volume of the explosion products increases to a certain critical value u_{cr} to give $u_{cr} \approx 2u_0$. In the second stage $\varkappa = 1.25$, and the respective values of the parameters at the end of the first stage are now taken as the initial conditions. Besides, when the dimensionless radius r reaches the value r_{cr}, equation (a) may be replaced by equation (b) in the system of equations (I).

The system of differential equations (I) was solved for explosion conditions

TABLE 3

No. of variant	r_{first}	$\hat{W}$	Criterion of similarity						Area of cross-section of charge adit F_{adit}, m²	Line of least resistance W, m	Reduced line of least resistance $W/\sqrt[3]{Q}$, m/kg$^{1/3}$
			d_1	n_1	j_1	m_1	B_1	$f_1 = \dfrac{F_{\text{adit}}}{W^2}$			
1	1.6	6.11		1.5×10^{-2}	2.3×10^{-2}	45	0.14	7.5×10^{-4}	5	82	2
2	2.0	11		6×10^{-3}	9×10^{-3}	45	0.18	6.2×10^{-4}	5	90	1.7
3	2.0	11		6×10^{-3}	5.5×10^{-3}	45	0.36	10^{-3}	2.5	50	1.5
4	1.6	11	4.6×10^{-4}	1.5×10^{-2}	1.2×10^{-2}	45	0.5	7.5×10^{-4}	5	82	1.11
5	1.6	11		1.5×10^{-2}	4.1×10^{-3}	45	3.4	2.3×10^{-3}	2.5	33	1.05
6	1.6	11		1.5×10^{-2}	1.2×10^{-2}	20	0.5	7.5×10^{-4}	5	82	0.74
7	1.6	11		1.5×10^{-2}	1.2×10^{-2}	80	0.5	7.5×10^{-4}	5	82	1.48

stated in Table 3. The majority of these conform to actual conditions which existed during explosions carried out for making the flood control dam in 1966 near Alma-Ata and the rock-fill dam of the Baipazinsk hydroelectric complex in 1968. Analysis of films of the explosion processs shows that explosion products flow intensively through a charge adit from the beginning of detonation till the end of explosion. It was not possible to calculate, in this case, the volume of gases escaping into the atmosphere because of intensive diffusion and turbulence of explosion products in the air. Due to the latter condition the calculated volume of gases is sometimes greater than the volume which could be obtained if all explosion products escaped.

While calculating the dimensionless criteria B_1, m_1, j_1, n_1 and d_1 contained in Table 3 the following values of parameters included in these were used: $b_x = 1.1$. $\rho = 2600$ kg/m^3; $D_w = 3500$ m/s (Baipazinsk explosion); $D_w = 4500$ m/s (Alma-Ata explosion), $e_{exp} = 4.27 \times 10^6$ J/kg; $\varkappa = 1.25$ (when $u_0 > 2.5 \times 10^{-3}$ m^3/kg); $\mu_{diss} = 0.3$; $g = 9.81$ m/s^2; $\Sigma_{at} = 10^5$ N/m^2; $\varkappa = 2.5$ (when $u_0 \leqslant 2.5 \times 10^{-3}$ m^3/kg); $c_x = 1$; and $\rho_{air} = 1.25$ kg/m^3.

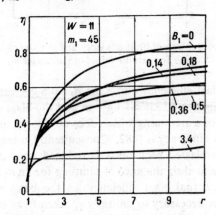

Fig. 30. Relationship of blast efficiency and reduced radius r of gas chamber for certain values of parameter B_1.

Graphs of the solutions have been shown in Figs. 30 and 31. If $r = 10$, the pressure of explosion products approaches that of atmospheric pressure and further expansion of the gas chamber ceases. Therefore, the maximum possible blast efficiency is determined by r, which roughly varies in the range 10–11. In case of large-scale explosions, e.g. if the line of least resistance is 50 m, the computed adit diameter, which may not be filled with stemming, and if blast energy losses were 5%, is equal to 0.35 m. However, in reality, the adit diameter is considerably larger. Such a difference arises because of mechanization of the charging process.

From the curves shown in Fig. 30, it is evident that blast efficiency η decreases with an increase in parameter B_1 which is proportional to the sectional area of the charge adit, because the energy losses with explosion products flowing from the charge adit increase with an increase in the ratio $F_{adit}/Q^{2/3}$. On the basis of the curves shown in Fig. 30, we can determine $F_{adit}/Q^{2/3}$ at which the harmful effect of the flow of explosion products through a charge adit can be ignored. Calculations show that energy losses will be about 5% when $F/Q^{2/3} \leqslant 5 \times 10^{-5}$ $m^2/kg^{2/3}$. For example, in this case, for $Q^{1/3}W = 1$ $kg^{1/3}/m$ and $W = 1$ m, the blast hole diameter d is 7 mm.

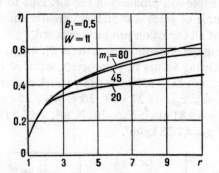

Fig. 31. Relationship of blast efficiency and reduced radius r of gas chamber for certain values of parameter m_1.

In Fig. 30 the blast efficiency curve for $B_1 = 0.5$ corresponds to an almost 2 m charge adit diameter $Q^{1/3}W \approx 1$ $kg^{1/3}$ m and the blast efficiency for $r = 11$ is 0.55. In the case of a stemming which does not fly out on explosion, i.e. for $B_1 = 0$, and the blast efficiency is 0.82. Consequently, a transfer of excavating explosion energy to blasted rock mass is reduced approximately 1.5 times. This reveals the practical significance of tamping for an excavating explosion. In order to obtain actual blast efficiency on the basis of the graphic data given in Fig. 30, it is necessary to multiply η, taken from curves for the given parameter B_1, by the coefficient of dissipation $1 - \mu_{diss}$. Calculations show that this coefficient is about 0.7 for hard rock. This value was used for solving the system of differential equations. The coefficient of dissipation is included as a multiplier in the dimensionless criterion m_1.

Other conditions being equal, blast efficiency depends on the dynamic hardness of the medium ρD_w^2 which is a function of the similarity criterion m_1. Blast efficiency decreases with a decrease in the value of m_1 (see Fig. 31).

The research work mentioned here had been carried out for concentrated (chamber) charges. In deep-hole charges, as explained in [13], the energy losses with explosion products flowing through the hole mouth, even in the presence of stemming covering 30–40% of the hole length, are far greater than the losses for chamber charges.

From an analysis of the solution of the differential equations (I) it became clear that a reduction in blast efficiency η_0, other conditions being equal, is determined by the ratio between the sectional area of the charge adit (or blast hole) F_{adit} and the surface area of the charge chamber F_{cham}.

Figure 32 indicates that the expression coefficient $k_{adit} = (\eta/\eta_0)_F$, where η is the blast efficiency without stemming, for the ratio F_{adit}/F_{cham}. From this expression, it follows that blast efficiency decreases by several times if $F_{adit}/F_{cham} = 0.05$. Energy losses with explosion products flowing out of the charge adit can be ignored if $F_{adit}/F_{cham} < 0.01$.

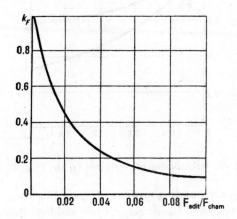

Fig. 32. Relationship of coefficient k_F and F_{adit}/F_{cham}.

It must be emphasized that the presence of a charge adit of sufficiently large volume comparable to the volume of the charge chamber, leads to the following undesirable effect. If the adit does not have stemming (or it is short in length), the explosion products, flying at a velocity of several km/s, fill the whole volume after which begins the process of a flow of these explosion products from the adit mouth. Since the rate of expansion of the gas chamber arches is less than the rate of dispersal of explosion products in the charge adit volume, therefore, during this process, the explosion products, present at this moment in the chamber, will transfer to the massif a small portion of their energy, especially if the charge chamber volume is greater than the volume of the charge itself. The explosion products, having filled the charge adit, begin to expand and transfer their own energy perpendicularly to the axis of the charge adit, i.e. in a direction coinciding with the line of least resistance. The final effect is such, as if the explosive charge was uniformly distributed in the chamber and adit. In this case a part of the charge in the adit does not transfer any part of its energy along the line of least resistance. Calculations show that a reduction in blast efficiency due to escape of gases into the charge adit depends on the ratio between the adit volume U_{adit} and chamber volume U_{cham} and the relative density of charging in the chamber Δ. If the volume of the charge chamber U_{cham} is greater than the volume of the charge U_{ch}, i.e. when the charging density is lower, the rate of expansion of the chamber arches reduces (compared to high density charging) and explosion products spread quickly in the adit volume, and do not thus fully transfer the

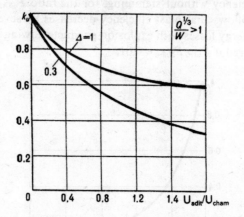

Fig. 33. Relationship of coefficient k_U and U_{adit}/U_{cham}.

blast energy to the surrounding medium. Figure 33 shows the relationship of coefficient $k_U = (\eta/\eta_0)_U$ and U_{adit}/U_{cham} for low and high density of charging Δ.

3. SELF-WEDGING STEMMING

During recent years, a great deal of original research [27–31] has been done for determining the effect of stemming on the blast efficiency. Nowadays, so-called blocking charges (Fig. 34) are laid in inert material at short intervals

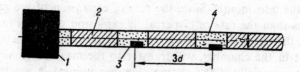

Fig. 34. Diagram showing self-wedging stemming:
1—main charge; 2—stemming section after explosion; 3—blocking charge;
and 4—explosion products of blocking charge.

with a view to prevent the blowout of stemming. These charges are fired simultaneously with the main charge [31]. The sections between blocking charges will be under high pressure. Due to lateral thrust in the stemming, great frictional forces τ_{fr} appear on the surface adjacent to the walls of the charge adit which are capable of overcoming the pressure of explosion products of the main charge. Calculations reveal that with a view to retaining the stemming in a static condition, the pressure of explosion products $\Sigma_{2\,max}$ for blocking charges (see Fig. 35), after compression of the stemming, may be 30–50 times less than the initial pressure $\Sigma_{1\,max}$ of the main charge. In this case, the length L of the stemming section between blocking charges should not be less than 3–4 times the diameter of the charge adit or hole.

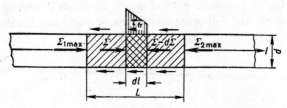

Fig. 35. Diagram for calculating self-wedging stemming.

During explosion with ordinary stemming, frictional forces develop in it due to the lateral thrust at the hole walls. However, their strength is considerably lower than in the case of stemming with blocking charges. The reason for this lies in the fact that the compressional wave produced by explosion of the main charge, pushes the material of ordinary tamping till it attains a high velocity. As a consequence, the coefficient of internal friction of stemming decreases to a small magnitude [32]. Therefore, the internal friction does not have as much effect on the scattering of ordinary stemming as on self-wedging stemming.

In this way, the advantages of blocking charges are: firstly, they form vapor locks in the stemming which prevent spreading of the compressional wave; secondly, they cause considerable static frictional forces by virtue of a lateral thrust. Stemming with blocking charges will be called self-wedging stemming. It should be noted that the role of stemming in excavating explosions is more important than in rock crushing by blasting inasmuch as the blasting process is completed when the explosion products expand fully (to atmospheric pressure) whereas, the crushing process comes to an end at the initial stage of expansion when explosion products experience high pressure. Therefore, in order to ensure better crushing, as stated in [28], the stemming should remain in the blast hole for a minimum period, approximately equal to the detonation time. In highly-fissured rocks, this time should be slightly greater. In case of an excavating explosion, the stemming retention period is equal to the period of full expansion of explosion products.

Let us analyze the process of flight of stemming from the charge adit in two extreme cases: during slow and rapid pressure buildup before tamping. We shall differentiate between these two cases by the following criterion. Suppose the pressure before tamping rises to a maximum $\Sigma_{1\,max}$ in time t_0. The stemming will be compressed simultaneously with the pressure buildup. If the time $t_w = L/D_w$ (where L is the stemming length and D_w the velocity of propagation of the wave), taken by the compressional wave to pass through the stemming is considerably less than time t_0, such a pressure buildup will be rapid and on the other hand, if $t_0 \gg t_w$, the pressure buildup will be considered slow (static).

Let us determine the conditions in which stemming remains static during a slow buildup of pressure. From Fig. 35, in which stemming is shown in a shaded rectangle, the following conditions of stemming immovability is evident:

$$\frac{\pi d^2}{4} \Sigma_{1\,max} = \frac{\pi d^2}{4} \Sigma_{2\,max} + \pi d \int_0^L \tau_{fr}(l)\,dl, \qquad (2.33)$$

where d is the stemming diameter; $\Sigma_{1\,max}$ and $\Sigma_{2\,max}$ are the pressure on the left and right ends of the stemming respectively; τ_{fr} is the tangential friction stress on the wall of the charge adit; and l the coordinate of length, measured from the left end of the stemming.

The dependence of τ_{fr} on l can be easily found from the differential equation of equilibrium of an elementary section of stemming of length dl (see Fig. 35).

$$\frac{\pi d^2}{4} d\Sigma = -\pi d\, \Sigma \gamma_{thrust} f_{fr}\, dl,$$

where Σ is the normal component of stress in the given section; γ_{thrust} the coefficient of lateral thrust; and f_{fr} the coefficient of static friction.

After integration, and incorporating the initial conditions $l=0$; $\Sigma = \Sigma_{1\,max}$ we obtain,

$$\Sigma = \Sigma_{1\,max}\, e^{-\frac{l}{l^*}},$$

where

$$l^* = \frac{d}{4\gamma_{thrust} f_{fr}}.$$

Because

$$\tau_{fr} = \gamma_{thrust} f_{fr}\, \Sigma,$$

by substituting the expression for Σ in this formula, we can determine the law of distribution of friction stress τ_{fr} along the charge adit length as follows:

$$\tau_{fr}(l) = \gamma_{thrust} f_{fr}\, \Sigma_{1\,max}\, e^{-\frac{l}{l^*}}.$$

Considering that friction of the adit walls develops because of an aggregate effect of pressures on the left and right ends of the stemming, we find

$$\tau_{fr}(l) = \gamma_{thrust} f_{fr} \left(\Sigma_{1\,max}\, e^{-\frac{l}{l^*}} + \Sigma_{2\,max}\, e^{-\frac{L-l}{l^*}} \right).$$

By substituting this relationship in expression (2.33) and integrating, we obtain the length L for which the stemming remains stationary:

$$L = l^* \ln \frac{1}{2} \left(\frac{\Sigma_{1 \, max}}{\Sigma_{2 \, max}} + 1 \right). \tag{2.34}$$

If, for static conditions, the approximate values of $\gamma_{thrust} = 0.5$ and $f_{fr} = 0.5$ then $l^* \approx d$. If pressure on the left end $\Sigma_{1 \, max}$ is 50 times higher than the pressure on the right end, i.e. $\Sigma_{1 \, max}/\Sigma_{2 \, max} = 50$, then $\ln \frac{1}{2}(\Sigma_{1 \, max}/\Sigma_{2 \, max} + 1) \approx 3$ and $L = 3d$. Thus, under the static effect of pressure, the stemming remains stationary if the ratio between its length and diameter is relatively small. Let us now determine length L at which the stemming will not fly out of the adit completely in case of a rapid buildup of pressure, i.e. in the case of a powerful blast.

In the case of a rapid pressure buildup the compressional wave will travel along the stemming at a velocity D_w. The pressure at the wave front will be considered approximately equal to $\Sigma_{1 \, max}$ at the left end of the stemming, whereas the stemming material attains a bulk velocity $v = \Sigma/\rho \, D_w$ in time $t = L/D_w$. The compressional wave will reverberate when it reaches the right end of the stemming. As a result, the rarefaction wave will travel in the opposite direction. The bulk velocity of the stemming material will increase twofold and will be equal to $2v$. At this moment, the pressure on the right end of the stemming will decrease from $\Sigma_{1 \, max}$ to atmospheric pressure Σ_{at}. After repeated reverberations of the compressional wave, the pressure along the stemming can be approximately taken as varying linearly. From this moment, the stemming acceleration will begin from the initial velocity $2v$ to a certain maximum velocity under the effect of pressure of the explosion products on the left end. Since the stemming mass will decrease as it flies off from the adit (part of the stemming which leaves the adit scatters laterally and does not hinder the movement of the remaining stemming in the adit), therefore, a progressive acceleration will take place. However, such a type of movement will take place only in case of a particular ratio between its length, diameter and initial velocity gained as a result of spreading of the compressional wave. Obviously, there is a critical length in traversing which the stemming (irrespective of the initial velocity) expends its kinetic energy in overcoming frictional forces without completely flying out of the adit. Let us calculate length l_1 of that part of the stemming which remains in the adit. It may be assumed that the stemming moves in the adit, and after exhausting its stored kinetic energy it comes to a halt under the effect of forces of friction. Its length (from the adit mouth to the rear section of stemming) will be equal to l_1. At the moment of halting, obviously, the following equation will be applicable:

$$\frac{\pi d^2}{4}(\Sigma_{1 \, max} - \Sigma_{at}) = \pi d \int_0^{l_2} f^*_{fr.d} \, \gamma_{thrust} \, \Sigma_{1 \, max} \left(1 - \frac{k_\Sigma}{\Sigma_{1 \, max}} - l \right) dl,$$

where Σ_{at} is the atmospheric pressure; $f_{\text{fr.d}}^{*}$ the coefficient of dynamic friction at low velocities (it is equal to the coefficient of friction at the moment the body is disturbed from its state of rest),

$$k_{\Sigma} = \frac{\Sigma_{1\,\text{max}} - \Sigma_{\text{at}}}{l_1}.$$

The left-hand side of this equation is the resultant force due to pressure and the right-hand side is the cumulative force of friction on the adit walls. After integration we obtain,

$$l_1 = \frac{d}{2\gamma_{\text{thrust}}\, f_{\text{fr.d}}^{*}}. \tag{2.35}$$

Let us now find length l_2 of the blasted part of stemming by solving the following differential equation of motion:

$$\frac{d}{dt}(mv) = \frac{\pi d^2}{4}\Sigma_{1\,\text{max}} - \pi d\, f_{\text{fr.d}}\, \gamma_{\text{thrust}} [(1-\varepsilon)\,L - x]\,\Sigma_{1\,\text{max}}, \tag{2.36}$$

where ε is the relative deformation during compression of the stemming due to shock wave; $f_{\text{fr.d}}$ the dynamic coefficient of friction at high velocities; and ρ_1 the density of the compressed stemming.

While writing equation (2.36), it was assumed that in case of slow movement the portion of stemming blown away from the adit will be disconnected from that part which is still in the adit because the force of friction does not act on the stemming which has been blown away (air drag is ignored). Therefore, during movement of the stemming, the inertial force will depend only on its mass m which, at a given moment, is in the adit. Hence, the value of m varies in time. In this equation the atmospheric pressure in comparison with the pressure of explosion products is ignored.

By differentiating the left-hand side we obtain

$$\frac{\pi d^2}{4}[(1-\varepsilon)\,L - x]\,\rho_1 \frac{d^2x}{dt^2} - \frac{\pi d^2}{4}\rho_1\left(\frac{dx}{dt}\right)^2$$

$$= \frac{\pi d^2}{4}\Sigma_{1\,\text{max}} - \pi d\, f_{\text{fr.d}}\, \gamma_{\text{thrust}} [(1-\varepsilon)\,L - \alpha]\,\Sigma_{1\,\text{max}}.$$

As we mentioned earlier, there can be three different cases of stemming movement covered by this equation. If the stemming length is considerably shorter than the critical length, the velocity keeps on increasing. If the stemming length is slightly less than the critical length, the velocity along the adit initially decreases but after reaching a minimum level it starts increasing as a result of a reduction in the inertial and frictional forces because of the blown-away portion of stemming. In case of a critical length of stemming, the velocity decreases continuously and the stemming movement ends when the velocity is reduced to zero. In this case the length of stemming remaining in the adit will be

$$l_1 = \frac{d}{2\gamma^*_{fr.d}\,\gamma_{thrust}} .$$

Equation (2.36) cannot be integrated in its final form. However, it can be integrated by taking $\left(\dfrac{dx}{dt}\right)^2$ with a mean value approximately equal to $0.5\,v_0^2$ where v is the mass velocity beyond the compressional wave front. For initial conditions $x=0$; $v=v_0=\Sigma_{1\,max}/D_w\rho$ the solution of the equation will be:

$$v^2 = v_0^2 + 2\left[\frac{\Sigma_{1\,max}}{\rho_1} + \frac{v_0^2}{2}\right]\ln\frac{L(1-\varepsilon)}{(1-\varepsilon)L-x} - \frac{4f_{fr.d}\,\gamma_{thrust}}{d\rho_1}\Sigma_{1\,max}\,x.$$

If $v=0$ (stemming came to a halt without being completely blown away) then $x=l_2$; $(1-\varepsilon)L-x=l_1$. Solving this equation for x when $v=0$ and bearing in mind that $v_0/D_w=\varepsilon$; $\rho/\rho_1=1-\varepsilon$ and $\Sigma_{1\,max}/\rho_1=(1-\varepsilon)\times v_0^2/\varepsilon$ we obtain

$$l_2 = l_1\frac{2-\varepsilon}{(1-\varepsilon)^2}\cdot\frac{f^*_{fr.d}}{f_{fr}}\ln\frac{L(1-\varepsilon)}{l_1} .$$

Since

$$(1-\varepsilon)L = l + l_1,$$

therefore,

$$L = \frac{l_1}{1-\varepsilon}\left[1 + \frac{2-\varepsilon}{2(1-\varepsilon)^2}\frac{f^*_{fr.d}}{f_{fr.d}}\ln\frac{L(1-\varepsilon)}{l_1}\right] .$$

Characteristically, the pressure $\Sigma_{1\,max}$ before tamping, is not expressed in the obtained solution. Physically speaking this can be explained as the variation in the face of friction in proportion to the variation in $\Sigma_{1\,max}$. Therefore, a drop in pressure with time, before tamping does not have any effect on the final result.

Stemming packed in sacks is laid in the adit usually in a loosely packed condition. Gaps are left between the sacks. Besides, a passage is also left in the adit for laying the explosion circuit. Therefore, the stemming density is usually 0.5–0.7 of its density in a natural state ($\rho\approx1500$ kg/m^3). Considering that the compressional wave increases the density of stemming to $\rho_1=2000$–2200 kg/m^3, we can take the relative deformation $\varepsilon=0.5$ in our circulations.

For evaluating the critical length L of a charge adit we may, on the basis of experiments, take the dynamic coefficient of friction $f^*_{f.rd}\approx0.5$ for low velocities of stemming. In the case of high velocities, the coefficient of friction $f^*_{fr.d}$, as explained in [32], decreases sharply. For calculation purposes, we may take $f^*_{fr.d}/f_{fr.d}\approx5$. With an approximate value of $\gamma_{thrust}=0.3$–0.7 in formulas (2.35) and (2.36) we obtain $l_1=2d$ and $L=(50$ to $100)\,d$.

In this way, stemming with a length several times greater than the charge-adit diameter can guarantee a complete blocking of the explosion chamber.

In practice, the charge adit length does not exceed (10 to 30) d. Consequently, in reality the stemming will always be ejected and open a path for the exit of explosion products, thereby reducing the blast efficiency. From the investigations made above, it follows that in order to prevent a scatter of the stemming it is necessary to, firstly, prevent the passage of the compressional wave through it and secondly, divide the stemming by several high-pressure vapor locks which ensure its self-wedging by virtue of the expression (2.34). It is not difficult to observe that the division of stemming by vapor locks prevents the passage of the compressional wave and an acceleration of the stemming by a shock effect. Vapor locks can be made with separate small explosives placed along the stemming at equal intervals.

Interestingly, such stemming has the property of self-regulation. In reality, if the first stemming section, with reference to the main charge, attains a high speed due to some reason, the first vapor lock is compressed and the pressure in it rises. This results in the development of a deceleration force which reduces the speed of this part of the stemming to practically zero. The last blocking charge will, in case of explosion, gradually displace the stemming section located adjacent to it. However, the velocity of this section, as has been established through calculations, is less than the velocity of stemming without blocking charges. As a result, the total time taken by the self-wedging stemming to be ejected is greatly extended and the period of expansion of explosion products of the main charge also increases.

Let us examine the design of charges intended for blocking the stemming. Let us assume that the stemming section, after being compressed with blocking charges, is equal to three times the diameter of the adit (hole). During explosion the blocking charge should compress the stemming material up to the maximum possible density by eliminating all gaps. The mass of a blocking charge can be calculated by the following formula:

$$Q_{\text{tamp } 1} = \frac{U_{\text{sec}} - U_{\text{stem}}}{u_{4000}},$$

where U_{sec} is the volume of the adit or hole section between blocking charges; and U_{stem} the volume of compressed stemming material in this section at a pressure of 4000 N/cm^2; u_{4000} is the specific volume of explosion products of a blocking charge at 4000 N/cm^2 pressure. The pressure of 4000 N/cm^2 is higher than the limiting strength of stemming material and is sufficient for consolidating and pressing it to the adit walls. Let us take the length of a stemming section after its compression by blocking charges to be $2d$, where d is the hole diameter. The length of this stemming section, before explosion of the blocking charges, is roughly equal to $3d$. In this case

$$U_{\text{sec}} = \frac{\pi d^2}{4} 3d;$$

$$U_{\text{stem}} = \frac{\pi d^2}{4}\, 2d.$$

In calculation, we may take $U_{4000} = 5 \times 10^{-3}$ m³/kg. Then

$$Q_{\text{tamp } 1} = 150d^3, \tag{2.37}$$

where Q is measured in kilograms and d in meters. The mass of subsequent blocking charges should be roughly 2–3 times less than that of the preceding charge. For example, for an adit of diameter $d = 2$ m, the mass of the blocking charge is equal to $Q_{\text{tamp } 1} = 150 \times 2^3 = 1200$ kg; $Q_{\text{tamp } 2} = 600$ kg and $Q_{\text{tamp } 3} = 200$ kg.

For deep-hole and blast-hole charges, the mass of the blocking charges, calculated by formula (2.37), can be so small that it may not detonate by the usual means of explosion. In this case the mass of the blocking charge is determined on the basis of its stable detonation.

For a reliable detonation in case of self-wedging stemming, it is necessary to have at least one or two stemming sections of $3d$ length (see Fig. 34). In this way, while using self-wedging stemming, the adit or hole can be filled with tamping material and blocking charges covering a length equal to 6 to 10 adit diameters. The remaining adit section may be left unfilled. In this case, it is necessary to place the stemming near the main charge.

Calculations show that the additional explosive mass required for blocking charges constitutes about one per cent of the explosive mass of the main charge. This negligible expenditure ultimately improves the blast efficiency for a propellent effect and, in fact, leads to a saving of 20–30 and sometimes even 50% in the mass of the main charge.

4. PROCEDURE FOR EXPERIMENTAL DETERMINATION OF BLAST EFFICIENCY

It is easy to determine the efficiency of a slab charge blasting if the initial velocity of throw for the given firing conditions is known. An initial velocity of throw implies a certain assumed velocity which will be applied to the center of gravity of the rock mass before blasting and will later impart an actual mechanical trajectory.

From formula (1.51) we obtain

$$\eta_0 = \frac{\rho v_0^2}{2e_{\text{exp}}\, qk}. \tag{2.38}$$

The most convenient method of determining the initial velocity v_0 is the cinematographic method. The camera installation should be at the following minimum distance from the center of the slab charge:

$$D = 4W\,(uq+1)\,\frac{f_{ob}}{L_h},$$

where f_{ob} is the focal length of the objective and L_h the height of the photographic frame.

In the above formula the product $W\,(uq+1)$ denotes the height of the upper point of the gas-chamber dome at the moment of maximum expansion of the gas chamber and multiplier 4 is used for ensuring a guaranteed filming of the explosion in case of a probable error in focusing the camera at the explosion site. As $u \approx 1$ kg/m³, we have

$$D = 4W\,(q+1)\,\frac{f_{ob}}{L_h}. \tag{2.39}$$

This distance of the camera from the explosion site ensures filming of the initial phase of blasting at the maximum possible scale. The film will cover the explosion process from the beginning till that moment when the gas chamber expansion is maximum and the rock mass attains a maximum velocity. The distance calculated by formula (2.39) is the minimum distance. In certain practical cases (for example, blasting of hard rock), it may happen that the camera installed at such a distance comes within the zone of scatter of stray fragments. In order to prevent probable damage to the camera, it is necessary to ensure its protection or to move the camera to a safe distance.

The filming speed should be adjusted in such a manner that at a given velocity of projection, at least five to eight slides can be made of the initial phase of blasting. The minimum speed of filming for the distance determined by formula (2.39) is calculated as follows:

$$v = \frac{5v}{W\,(q+1)}. \tag{2.40}$$

where velocity v is roughly calculated with the help of formula (1.51) and the graphs shown in Fig. 7.

We describe here the simplest and most reliable method of developing the film with a view to obtaining the initial velocity of projection. From the film we make slides of equal magnification on contrast photographic paper, beginning with the first. The first slide is that slide which filmed the signal for the beginning of explosion (usually, firing of the detonating cord section or a small charge of a few grams). If the speed of taking the shots is considerably greater than the minimal (for example, while filming on SKS-1) then the slides are chosen at equal intervals before printing. Each print is given a serial number of one of the film slides. By making prints on photographic paper, we can avoid the error of mistaking the expanding explosion products for the moving rock mass which leads to considerable overestimation of the initial velocity of projection. A standard relationship of velocity of projection and time is shown in Fig. 36. Velocity v_1 for $t=0$, corresponds to double the mass velocity of the rock when the compressional wave reaches the free face. Thereafter, the velocity starts increasing due to a subsequent expansion of explosion products (branch AB) and attains a maximum value v_{max} in time t_m. After this, the velocity decreases due to gravity and air drag (branch BC). Such a dependence of velocity on time is typical for rocks and has been proved experimentally [33]. However, the section of the graph AB can be found, especially for small-scale explosions, only at a very high filming speed. The acceleration time from v_1 to v_{max} is measured in this case, in hundredths of seconds. Therefore, if filming is done at the minimum speed calculated by formula (2.40), the accelerating section disappears and one can only succeed in recording velocity v_{max}. However, for practical purposes v_{max} is sufficient.

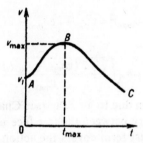

Fig. 36. Relationship of velocity of projection and time.

Velocity v_{max} should not be considered as the initial velocity of projection because in ballistic calculations, the latter can be conveniently related to the center of gravity of the rock before explosion. Moreover, the velocity v_{max} corresponds to a certain height h_{max} above the initial position of the center of the rock before explosion. A specific energy is required for overcoming this height. If we consider v_{max} to be the initial velocity of projection, relating it to the center of gravity of the rock before explosion we find that at height h_{max}, where velocity v_{max} was measured, the velocity is less than v_{max} because of gravity and the trajectory of the rock will differ from the actual one. Correction k_g in velocity v_{max} for gravity can be found from the following equation of energy:

$$(k_g v_{max})^2 = v_{max}^2 + 2gh_{max},$$

whence

$$k_g = \sqrt{1 + \frac{2gh_{max}}{v_{max}^2}}.$$

The left-hand side of the equation written above represents the kinetic energy and on the right-hand side the first term represents the kinetic energy of the rock at height h_{max} and the second term the potential energy of gravity. For example, if $k_{max} = 10$ m and $v_{max} = 10$ m/s by substituting them in the equation we get $k_g = 1.73$. Consequently, for relatively smaller values of v_{max} and greater values of h_{max} the correction for gravity is considerable.

Sometimes it is not possible to record the velocity v_{max} during blasting of hard rock because the stemming and explosion products are scattered as a result of which the gas dome is clouded. The slides show the rock which has already crossed the gas-dust cloud and is at an altitude which is considerably higher than h_{max}. In this case the gas dome had disintegrated into fragments which are subjected to the force of air drag. In order to determine the initial velocity of projection through correction $k_{g.d}$ it is necessary to take into account not only gravity but also the air drag. The corresponding equation of energy will be:

$$(k_{g.d} \, v_{max})^2 = v_{max}^2 + 2gh_{max} + 2j_{air} L,$$

whence

$$k_{g.d} = \sqrt{1 + \frac{2\,(gh_{max} + j_{air} L)}{v_{max}^2}}, \qquad (2.41)$$

where j_{air} is the retardation due to air drag (see Chapter 4); and L the length of the segment of trajectory between the free face and altitude h_{max}.

In this equation, the last term covers the action of air drag. It should be noted that L and h_{max} coincide in direction only for a slab charge laid horizontally. For inclined slab charges the height h_{max} varies along the vertical.

The obtained prints are processed as follows: In the first print we outline the original position of the free face and a subsequent calculation of the gas-dome height along the line of least resistance is made from this position. Thereafter, we make a graph for variation in altitude of the vertex of the dome with respect to time. The points on the graph are joined to form a smooth curve. Smoothing of the curve is necessary because the points on the graph are scattered due to fluctuation of the forward edge of the dome. From this graph, by evaluating the slope at each point of the curve, we can obtain values of the velocity at each point of time and thus plot a curve for $v = f(t)$. Time is calculated by the following formula:

$$t = \frac{N}{v};$$

where N is the serial number of the slide.

The dome height is calculated from the prints in the following way:

$$h = \frac{h_p L_h D}{L_{print} f_{ob}},$$

where h_p is the dome height in the print; L_h the height of the photographic frame; and L_{print} the print width.

When processing the film on explosion, it is necessary to take into account the angle φ between the lens axis and direction of projection (the so-called camera angle correction). If this angle is not equal to 90° then the true velocity of projection, after taking into account angle φ, will be:

$$v_0 = \frac{k_{g.d}\, v_{max}}{\sin \varphi}.$$

Angle φ is measured with the help of two hinged racks, one of which is in the direction of the line of least resistance and the other is leveled on the motion-picture camera.

Experimental calculation of blast efficiency should be carried out on the same rocks as are actually to be blasted. Preferably, the same explosives should be used in experimental and actual blastings. If self-wedging stemming is not used in experimental and actual blasting, it is desirable that the stemming in experimental blasting be geometrically identical to the stemming in actual blasting so that the error connected with the exit of explosion products through a shothole is prevented. The dimensions of a slab charge and thickness of the rock layer in experimental blasting should be geometrically similar to those in an actual blasting. Hence the following conditions should be fulfilled:

$$\xi = \frac{L}{qWu} = \text{idem};$$

$$\chi = \frac{B}{qWu} = \text{idem}.$$

Coefficient k used in formula (2.38) is found for experimental figures L, B, W and q from Fig. 12. Parameter Ω is calculated through formula (1.28).

If the initial velocity of projection does not exceed 20 to 30 m/s, the air drag on the rock movement is not heavy. In such a case, the blast efficiency can be determined for the maximum height of the gas dome H_{max}. From the equation of conservation of energy: $\rho g H_{max} = q e_{exp} k \eta_0$, we obtain the design formula as follows:

$$\eta_0 = \frac{\rho g H_{max}}{q e_{exp}{}^k}.$$

In practice, inclined and horizontal slab charges in an experimental field are laid by prior charging of the shotholes (inclined charge) or sprung-hole chambers made by springing (horizontal charge).

Since there is a direct proportionality between blast efficiency and specific consumption of explosives in the case of soil, therefore, in order to obtain an experimental blast efficiency curve, it is sufficient to make several blasts with different specific consumptions of explosives. In the case of rocks, the dependence of blast efficiency on the specific consumption of explosives is more complicated. Therefore, the number of experimental blasts for hard rocks is determined on the basis of the nature of the blast efficiency curve.

The blast efficiency of a slab-charge system can be calculated from the known initial velocity of projection of concentrated charges.

Experience in blasting operations reveals that for a blast index n_e the concentrated charges of mass Q at a depth W are laid over an area in such a way that the distance between them is

$$a = W\frac{n+1}{2} \approx Wn_e^{2/3},$$

the rock mass lying above these charges attains almost equal velocity v at all points. In this case blast efficiency is determined by the following formula:

$$\eta_0 = \frac{\rho v_0^2}{2q\,e_{\exp}}. \tag{2.42}$$

The initial velocity v_0 used in this formula can be calculated in the following manner. For concentrated charges the initial velocity v_{con} along the line of least resistance is expressed by the following expression:

$$v_{con} = A\left(\frac{Q^{1/3}}{W}\right)^m, \tag{2.43}$$

where A and the exponent m are calculated by the physicomechanical properties of the soil:

	A	m
Loess	8	3
Sand	9	2.4
Loam	16	2
Clay	22	1.8
Hard rock	40	1.5

The reduced values of the parameter A in formula (2.43) correspond to the following dimensional equalities: Q in kilograms and W in meters.

However, the rock velocity during blasting of a system of concentrated charges will be higher than the initial velocity of projection v_{con} in the case of blasting of a single concentrated charge because the charges nearer to it have a combined effect on the given volume of the rock. This situation can be taken into account by a correction factor ψ:

$$v_0 = A\left(\frac{Q^{1/3}}{W}\right)^m \psi. \tag{2.44}$$

Since the mass Q of a concentrated charge is

$$Q = k_e W^3 f(n), \tag{2.45}$$

where $f(n) \approx n^3$; k_e is the computed specific consumption of explosive charges for blasting [34], whereas the specific consumption of explosives for a system of concentrated charges is given by

$$q = \frac{Q}{a^2 W},$$ (2.46)

therefore, from equations (2.42) to (2.46), we obtain:

$$\eta_0 = \frac{k_e^{\frac{2m}{3}} A^2}{2e_{exp}} \left(\frac{q}{k_e} \right)^{m-1} \psi.$$ (2.47)

The value of correction factor ψ can be obtained by adding all components of the kinetic energy of motion of each charge lying closest to the given volume of the rock. For example, for an elementary volume of rock deposited on the free face symmetrically with respect to the four nearest charges, the total kinetic energy E_Σ relative to a unit mass of rock is calculated by the expression

$$E_\Sigma = \frac{1}{2} \rho v_\Sigma^2,$$

where

$$v_\Sigma^2 = 4v_\varphi^2 \cos \varphi^2.$$

Angle φ between the line of least resistance of the charge and direction of the elementary volume of rock under study is calculated by the following formula:

$$\tan \varphi = \frac{a}{\sqrt{2}W}.$$

The initial velocity of projection v_φ at angle φ to the line of least resistance is found from equation (2.43) by replacing W with $W/\cos \varphi$, the distance between the center of the charge and the center of the elementary volume of rock under study:

$$v_\varphi = A \left(\frac{Q^{1/3}}{W} \right)^m \cos^m \varphi.$$ (2.48)

The corresponding quantity v_Σ^2 for an elementary volume of burden on the free face exactly above the charge is determined while bearing in mind that the explosion of at least five charges will have an effect on this volume. These charges are: one charge placed directly under this volume of rock and the four nearest charges which are at a distance 'a' from the first charge.

Therefore:

$$v_\Sigma^2 = v_{con}^2 (1 + 4 \cos \varphi^{2\,(m+1)}),$$

where angle φ is calculated from the following expression:

$$\tan \varphi = \frac{a}{W} = n_e.$$

Corresponding calculations give the following expressions for factor ψ:

$$\psi = \frac{1}{\sqrt{2}} \sqrt{1 + \frac{2}{\left[\left(\frac{q}{k_e}\right)^{4/7} + 1\right]^{(m+1)}} + \frac{2}{\left[\frac{1}{2}\left(\frac{q}{k_e}\right)^{4/7} + 1\right]^{(m+1)}}}.$$

The dependence of blast efficiency on η_0 for a slab-charge system calculated according to formula (2.38) and given in Table 4 for different rocks is shown in Fig. 7. As is evident, the blast efficiency in sand, loam and clay (curves 5, 4 and 3 respectively) is less than in hard rock (curve 1). The reason for this is the massive dissipation of blast energy due to permanent plastic deformation of these soils.

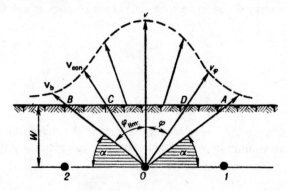

Fig. 37. Diagram for calculating blast efficiency of concentrated charges.

It was stated earlier that in case of an identical specific consumption of explosives, the blast efficiency of a single concentrated charge is less than the blast efficiency of a system of such charges. This is evident from Fig. 37 in which the horizontal shaded area shows a solid angle α. The blast energy of the concentrated charge in the compressional wave escapes into the surrounding medium without throwing out any explosion products within the limits of angle α. The latter incidence, in case of explosion of this charge, takes place only within the limits of the blast cone AOB. It is clearly seen that in a case of simultaneous explosion of a system of charges 1–0–2, the wasted portion of energy within the limits of solid angle α will be partially utilized in throwing the rock mass.

It is of practical interest to determine the blast efficiency of one concentrated charge because it will give us the lower limit of blast efficiency of a system of such charges.

The blast efficiency of one concentrated charge can be calculated by integrating the kinetic energy within the limits of the blast cone AOB (see Fig.

37). Here we make the hypothesis that the rock scatters radially at velocities in accordance with the following formula (2.48):

$$\eta = \frac{\displaystyle\int_0^{\varphi_m} \frac{1}{2} v_\varphi^2 \, dm}{Q e_{\exp}},$$

where

$$dm = \rho \, dU;$$

$$U = \frac{1}{3} \pi W^3 \tan^2 \varphi;$$

$$\cos \varphi_{\lim} = \frac{1}{\sqrt{\tan^2 \varphi_{\lim} + 1}};$$

$$\tan \varphi_{\lim} = \sqrt[3]{\frac{Q}{k_e k_w W^3}};$$

$$k_w = 1 + \frac{W}{50}.$$

On integration we obtain:

$$\eta = \frac{\pi^{\frac{2m}{3}}}{2 \times 3^{\frac{2m}{3}} (m-1) e_{\exp}} A^2 \rho k_w^{2m-3} q^{\frac{2m-3}{3}} \tan \varphi_{\lim}^{\frac{2}{3}(2m-3)} [1 - \cos \varphi_{\lim}^{2m-2}], \quad (2.49)$$

where quantities A and m depend on the type of rock. The specific consumption of explosives is determined by the formula:

$$q = \frac{Q}{U_{\lim}} = \frac{Q}{\dfrac{1}{3} \pi W^3 \tan^2 \varphi_{\lim}}.$$

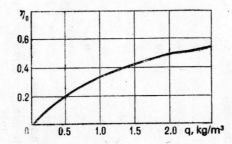

Fig. 38. Relationship of efficiency and consumption of explosives during blasting of concentrated charge in hard rock.

TABLE 4

Explosion No.	Date and place of explosion	Q, kg	W, m	v_m, m/s	h_m, m	j_w, m/s²	L, m	$k_{g.d}$	$F_{adit}/F_{cham.}$, m²/m²	k_F	$U_{adit}/U_{cham.}$, m³/m³	k_e	α	kW	$Q^{1/3}/W$, kg^{1/3}/m	v_r, m/s	v_{cal}, m/s
1	Baipaza, 1968	5	1.7	37	12.5	8	20	1.2	—	—	—	—	60	1.02	1	44	40
2	"	0.9	1.04	24	12	9	17	1.25	—	—	—	—	63	1.01	0.92	30	35
3	"	9	1.7	27	16	11	23	1.44	—	—	—	—	65	1.02	1.22	39	54
4	"	5	1.75	25.5	19	10	27	1.56	—	—	—	—	61	1.02	1.02	40	42
5	"	10	1.7	46	16	25	23	1.24	—	—	—	—	60	1.02	1.26	57	56
6	"	8	1.83	43	13	23	30	1.36	—	—	—	—	65	1.02	1.1	59	46.5
7	"	5	1.58	33	16	16	21	1.4	—	—	—	—	67	0.01	1.08	46	44.5
8	"	9	2.1	38	2	—	—	1.02	—	—	—	—	60	1.02	1	39	40
9	"	10	2.3	43	2	—	—	1.01	—	—	—	—	60	1.02	1.94	44	36.5
10	"	6.4×10⁵	82	22*	10	—	—	1.02	5/600	0.65	500/1000	0.39	65	1.69	1.25	46**	56
11	"	5.5×10⁴	32	24*	10	—	—	1.02	5/150	0.28	170/70	0.6	65	1.27	1.29	52**	58
12	Baipaza, 1965	9×10³	21.8	—	—	—	—	—	—	—	—	—	90	1	0.96	40***	38
13	Baiinchan, 1956	3.19×10⁵	37	157	50	—	—	1.02	—	—	—	—	0	1.74	2.22	157	132
14	"	2.29×10⁵	30	220	50	—	—	1.01	—	—	—	—	0	1.6	2.38	220	146
15	"	9.25×10⁵	47	150	60	—	—	1.02	—	—	—	—	0	1.94	2.56	150	166
16	"	9×10⁵	51	165	60	—	—	1.02	—	—	—	—	0	2.02	2.40	165	148
17	"	4.4×10⁵	37	145	50	—	—	1.02	—	—	—	—	0	1.74	2.47	145	156
18	"	4.1×10⁵	37	125	50	—	—	1.03	—	—	—	—	0	1.74	2.42	125	150

19	,,	5.97×10^5	38	200	50	—	1.01	—	—	—	0	1.76	2.68	200	176
20	,,	6.05×10^5	38	200	50	—	1.01	—	—	—	0	1.76	2.68	200	176
21 Altyn-Topkan, 1953		5.87×10^5	53	100	60	—	1.06	—	—	—	0	2.06	2.01	100	114
22	,,	144	3.3	75	5	—	1.01	—	—	—	0	1.07	1.63	76	83
23	,,	35	2.45	65	4	—	1.01	—	—	—	0	1.05	1.38	66	65
24 Alma-Ata, 1966		4.39×10^5	52.1	40*	40	0.63	1.25	4/400	300/550	0.7	4.5	1.74	1.75	75**	92
25 Voskresensk, 1963		18.5	2.4	37.5	4	—	1.03	—	—	—	0	1.05	1.12	37.5	47
26	,,	300	3	130	5	—	1.02	—	—	—	0	1.06	2.28	130	137

*Intense flow of gases through the charge adit was observed.

**Correction to flow of gases, volume of charge adit and filming angle (in explosion No. 24 $\varphi=30°$) must be considered.

***Calculated from rock throw range.

Figure 38 shows the dependence of efficiency of a concentrated charge on q, calculated by formula (2.49) for hard rock.

From formula (2.49), it follows that accuracy in the calculation of a quantity η largely depends on the accuracy of determining the constant A. The latter can be obtained only from experimental data for an initial velocity of projection.

Table 4 contains the results of cinematographic research on explosions of concentrated charges in hard rock. Quantity Q_{eff} in Table 4 is determined by the following formula:

$$Q_{\text{eff}} = \left(1 + \delta \frac{W \cos \alpha}{50}\right) Q, \qquad (2.50)$$

where α is the angle between the vertical and the line of least resistance.

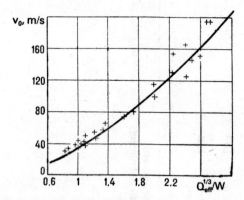

Fig. 39. Relationship of initial velocity of projection on $Q_{\text{eff}}^{1/3}/W$, during blasting of concentrated charge in hard rock.

The effective mass Q_{eff} of a concentrated charge has the following physical value. It is well known that in case of explosion of a concentrated charge at a depth of more than several scores of meters, the span of the throw funnel decreases in comparison with explosions at lesser depths (the charge mass O/W^3 being equal). The reason for this is, that in case of an explosion at greater depths a considerably high initial velocity of projection (vector $\vec{v}_C$) is required for moving the rock deposited directly on the lateral face of the formed funnel (see Fig. 37) in comparison with an explosion at relatively lesser depths (vector $\vec{v}_B$). It is evident that in case of an explosion at great depths the energy used for the blasting of rocks will be concentrated in a smaller solid angle ($\angle COD$) than in case of explosion at lesser depths ($\angle AOB$). As a result, the initial velocity of projection will increase with an increase in depth (when $Q^{1/3}/W = \text{const}$). This can be observed more vividly in case of explosion in hard rocks in which the so-called 'piston' effect of explosion products is very pronounced. The energy contained in the sprung-hole cavity (first phase of explosion) is almost completely utilized in blasting the rock situated within the limits of angle DOC. The volume of the sprung-hole cavity initially formed in case of explosion in loose soils is several times greater than in case of explosion in hard rock. As a result, the energy developed in a sprung-hole cavity at this moment is very limited and the 'piston' effect weakens. Therefore, factor δ in formula (2.50) for hard rock is closer to 1 and for loose soils δ varies from 0.2 to 0.3.

It has been proved through experiments (experiments No. 13 to 21, Table 4) that the initial throw velocity increases with an increase in the explosion depth in hard rock.

Due to the depth effect in these explosions, the initial velocity of projection is considerably higher than its value calculated using formula (2.43). However, if we take into consideration the depth effect in such a manner that in formula (2.43) the actual mass of charge Q is substituted by a certain effective quantity Q_{eff} according to formula (2.50), the experimental value of the initial velocity of projection for explosions at great depths will satisfy formula (2.43).

The measured initial rock velocity v_{max} is considerably lower than the value obtained from formula (2.43) when, during explosion, gas is ejected under pressure through the charge adit or cracks in the gas chamber dome. This is especially noticeable in case of explosions in hard rock at great depths (experiments No. 10, 11 & 24). If we introduce corrections $k_{g.d}$, k_F, k_v and k_W in the measured velocity, the difference between v_{con} and v_0 will be within the limits of natural deviation when measured with instruments.

Velocity v_0 should be calculated by the following formula:

$$v_0 = \frac{v_{max} \, k_{g.d}}{\sqrt{k_F k_v} \cos \varphi} \, .$$

Figure 39 shows the curve of the actual initial velocities of projection v_0 during blasting of concentrated charges in hard rock depending on the effective quantity $Q_{eff}^{1/3}/W$ $(kg^{1/3}/m)$. The effective mass of the charge was calculated by formula (2.50). The dotted line in the graph represents the curve derived from formula (2.43); when $A = 40$ and $m = 3/2$ (see Table 4). It is obvious that the experimental value v_0 corresponds to the value calculated by formula (2.43).

Chapter 3

Experiments in Blasting

1. PROCEDURE FOR CONDUCTING EXPERIMENTS.
SOME RESULTS OF EXPERIMENTS

Experiments for analyzing the blast patterns of slab-charge systems were carried out with a view to achieving the following objectives: working out a procedure for experimental determination of blast efficiency in respect of gravel-loam and gravel-sand mixtures for constructing embankments; experimental verification of the effects of peripheral charges for eliminating edge-effects; determining the criteria for a plane-parallel throw-out explosion; and experimental verification of the mining technique for calculating the throw and distribution of rock fragments in the muck pile.

Such were the problems that arose during construction of the rock-fill dam of the Baipazinsk hydroelectric grid.

The blasting method for moving the soil to the place where the screen, transition layer and apron were to be made, was dictated by the conditions that the time for constructing the said elements of the dam was limited to a few hours and the volume of burden to be removed was in hundreds of thousands of cubic meters. The slab-charge system proved to be the most rational of all the known methods. Soil meant for screen, transition layer and apron was earlier deposited on a slab charge which, on explosion, moved the former to the desired area. The chief aim of experimental explosion was to determine the specific consumption of explosives which ensures the removal of gravel-sand and gravel-loam mixtures to predetermined places, and to develop an optimal design of slab charge which could move rock with a satisfactory blast accuracy. Experimental explosions were conducted at the site of the Baipazinsk hydroelectric station in such rocks as were intended for making the screen, the transition layer and apron of the dam: for the Dyrgabsk open-pit mine, in a gravel-sandy mixture, and for a water drainage channel foundation pit, in a gravel-loam mixture.

From a geologic consideration, the foundation pit for the drainage channel was an alluvial-proluvial deposition in the form of coarse-grained material with loam-sand filler. The densities of sandy loam and loamy soils varied between 1.68–1.92 g/cm³ at an 11.7.% moisture content.

Deposits at the Dyrgabsk open-pit mine are found in the form of gravel and sand containing: 35.3% boulders, 42.3% gravel and 22.4% sand. The density of gravel and sand in loose condition was 2100 kg/m³.

Experimental explosions for determining blast efficiency were conducted on level ground. The ground was leveled with the help of a bulldozer and four pillars *1* (Fig. 40) of 15–20 cm in height erected above the free face. Four planks, of which *3* were in a perfectly horizontal plane, were attached to them from the outer side (the level was verified with a spirit level). The cross section of the planks formed a 6×6 m square. The space between the planks was filled with sand *4* which served as a cushion. Linear charges were laid from above in predug channels. The linear charges were prepared with ammonite cartridges No. 6, weighing 200 g (of 35 mm diameter and 230 mm in length) abutting each other and tied to a detonating cord with twine. In order to level the upper edge of the sand, four pillars *2*, each of which is two meters in height, were planted on the sides and four horizontal racks were secured to them at a distance of one meter above the upper end of the charge. The linear charges were covered from above with 40 mm-thick planks. From the top they were manually covered with a 20 cm-thick soil layer which protected the ammonite cartridge and detonating cord from the soil falling from the dump truck and the track pressures of the bulldozer. With the same objective in view, the main detonating cords coming out of the ground were also covered with a 30 cm-thick layer of soil.

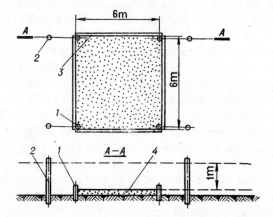

Fig. 40. Diagram of test area for exploding slab-charge system.

The gravel-loam mixture over the linear charges was prepared using dump trucks and was leveled with bulldozers. Thereafter, the soil was manually leveled in the area of the racks secured to the four lateral pillars. The experimental embankment prepared for blasting, is shown in Fig. 41.

Fig. 41. Embankment of gravel-loam mixture prepared for blasting.

A series of experiments with inclined slab charges in gravel-loam and gravel-sand and sandy soils were conducted in order to verify the procedure described here, for calculating the range of rock-scatter and the effectiveness of peripheral charges. The thickness of the soil layer to be blasted was kept within the limits of 0.2–2 m.

The blast process was recorded with a motion picture camera with a focal length $f_{ob} = 50$ mm. For duplication, the maximum height of the gas-chamber dome was also measured with the help of a theodolite and still cameras.

Experimental conditions and results are given in Table 5.

The initial velocity of projection in each experiment was determined by developing the film as explained in Sec. 4 of Chapter 2. Experimental data of blast efficiency calculated by formulas (2.38) and (2.41) for different soils are given in columns 21 and 22 of Table 5. Evidently the blast efficiencies calculated on the basis of the initial velocity and maximum height of the gas-chamber dome are close to each other. However, preference should be given to the blast efficiency calculated on the basis of the initial velocity because the air drag which cannot be accurately taken into account, influences the height of the gas-chamber dome.

Figure 42 shows the curves based on experimental data for blast efficiency. The efficiency for gravel-sand mixture was slightly higher than the efficiency for a gravel-loam mixture because the stone and gravel content were higher in these. The curves also show the relationship between efficiency and specific

TABLE 5

Expt. No.	Q_Σ, kg	Q_0, kg	W, m	q, kg/m³	a, deg	B, m	χ	L, m	ξ	Ω	k	b, m
1	2	3	4	5	6	7	8	9	10	11	12	13
Gravel-sand mixture, $\rho=2100$ kg/m³; $e_{exp}=4.27\times10^6$ J/kg												
1	99	0.0	1.0	2.42	0.0	6.10	2.53	6.70	2.77	0.38	0.75	0.40
2	80.4	6.0	1.0	2.07	14	6.00	2.70	6.00	2.70	0.37	0.75	0.40
3	60	0.0	1.0	1.64	0.0	6.00	3.66	6.10	2.72	0.27	0.81	0.55
4	334	19	1.8	2.92	36	5.00	0.87	12.0	2.10	0.82	0.65	1.20
Gravel-loam mixture, $\rho=1800$ kg/m³, $e_{exp}=4.27\times10^6$ J/kg												
5	75.6	0.0	1.0	2.11	0.0	5.90	2.80	6.10	2.9	0.36	0.67	0.40
6	104.8	30.8	1.0	2.11	0.0	5.80	1.94	6.00	2.0	0.5	0.73	0.40
7	91	19.0	1.0	2.21	23	4.60	2.00	5.80	2.06	0.5	0.73	0.40
8	134	19.0	1.0	3.00	0.0	6.30	1.80	6.00	1.72	0.57	0.70	0.27
9	6.1×10^4	7×10^3	8.0	2.00	35	32.0	2.28	120	8.6	0.27	0.81	3.00
Sand, $\rho=1500$ kg/m³, $e_{exp}=5.8\times10^6$ J/kg												
10	0.1	0.0	0.2	0.50	33	1.00	10.0	1.00	10.0	0.1	0.90	0.10
11	0.23	0.03	0.2	1.00	32	1.00	4.40	1.00	4.40	0.23	0.82	0.05
12	0.21	0.05	0.2	0.80	33	1.00	4.70	1.00	4.70	0.21	0.84	0.062
13	0.21	0.05	0.2	0.80	33	1.00	4.70	1.00	4.70	0.21	0.84	0.062
14	0.16	0.0	0.2	0.80	27	1.00	6.20	1.00	6.20	0.16	0.87	0.062
15	0.15	0.0	0.2	0.80	32	0.48	3.25	2.00	13.5	0.19	0.85	0.062
16	0.29	0.13	0.2	0.80	31	1.00	3.50	1.00	3.50	0.28	0.80	0.01
17	0.25	0.09	0.2	0.80	32	1.00	4.00	1.00	4.00	0.25	0.82	0.062
18	0.23	0.07	0.2	0.80	31	1.00	3.50	1.00	3.50	0.27	0.81	0.062
19	0.27	0.09	0.2	0.71	32	0.58	2.76	2.25	10.7	0.22	0.84	0.07
20	0.26	0.08	0.2	0.71	27	0.58	2.88	2.25	11.0	0.22	0.84	0.07
21	0.26	0.08	0.2	0.71	32	0.58	2.88	2.25	11.0	0.22	0.84	0.07
22	1.13	0.43	0.4	0.71	31	1.22	5.20	2.00	8.60	0.16	0.87	0.035
23	0.76	0.06	0.4	0.71	34	1.20	7.60	2.0	12.6	0.11	0.90	0.035

(Contd.)

Expt. No.	p, kg/m	P_{lat}, kg/m	P_{up}, kg/m	R, m	L_p, m	v, m/s	H, m	η_v	ηH	D_1, m	$D_{1\,cal}$, m	D_1/D_2	$1/k\beta_1$
1	14	15	16	17	18	19	20	21	22	23	24	25	26
Gravel-sand mixture, $\rho=2100$ kg/m³; $l_{exp}=4.27\times10^6$ J/kg													
1	0.97	—	—	20	—	26	41	12.8	13.5	—	—	—	—
2	0.97	—	1.33	—	25	—	—	—	—	40	37	2.0	2.2
3	0.85	—	—	7	—	20.5	20	9.5	9.5	65	99	1.85	1.43
4	6.3	5.3	—	—	36	30	—	17	—	—	—	—	—
Gravel-loam mixture, $\rho=1800$ kg/m³; $l_{exp}=4.27\times10^6$ J/kg													
5	0.85	—	—	15	—	23	26	8.7	8.5	—	—	—	—
6	0.85	—	2.0	30	—	30	29	11.8	7.6	—	—	—	—
7	0.88	2.65	—	—	8	28	—	11.2	—	70	78	1.85	2
8	0.81	2.3	—	30	—	—	48	—	13.6	—	73	—	—
9	49.0	147	—	—	120	—	—	0.08	—	60	—	—	—
Sand, $\rho=1500$ kg/m³; $l_{exp}=5.8\times10^6$ J/kg													
10	0.01	1.2×10^{-4}	1.2×10^{-4}	—	2.2	—	—	—	—	6.3	—	3.1	4.1
11	0.01	0.014	0.014	—	1.5	—	—	—	—	10.6	—	2.1	2.75
12	0.01	0.01	0.01	—	1.4	—	—	—	—	9.8	—	2.6	2.85
13	0.01	0.01	0.01	—	1.2	—	—	—	—	9.5	—	2.32	2.85
14	0.01	0.01	0.01	—	1.5	—	—	—	—	9.4	—	—	—
15	0.01	0.01	0.01	—	2.7	—	—	—	—	11.0	—	3.85	2.70
16	0.01	0.01	0.01	—	0.7	—	—	—	—	13.8	—	1.95	2.5
17	0.01	0.01	0.01	—	1.0	—	—	—	—	10.3	—	2.5	2.35
18	0.01	0.01	0.01	—	1.0	—	—	—	—	10.6	—	2.5	2.5
19	0.01	0.007	0.007	—	2.0	—	—	—	—	11.8	—	2.9	2.5
20	0.01	0.007	0.007	—	1.8	—	—	—	—	11.5	—	2.5	2.5
21	0.01	0.007	0.007	—	2.0	—	—	—	—	10.0	—	2.5	2.1
22	0.01	0.023	0.023	—	2.6	—	—	—	—	15.8	—	3.3	3.3
23	0.01	0.039	0.039	—	1.7	—	—	—	—	8.6	—	3.6	3.6

consumption of explosives. The obtained experimental data for blast efficiency helped in designing the slab charge intended for making the apron of the rockfill dam of the Baipazinsk hydroelectric grid.

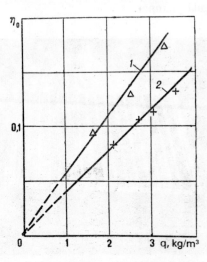

Fig. 42. Dependence of blast efficiency of a slab-charge system on specific consumption of explosives:

1—gravel-sand mixture; and 2—gravel-loam mixture.

The actual ratios of distances up to the far and near ends of the muck pile and the ratios calculated by formula (1.60) are given in columns 25 and 26 of Table 5. These parameters were measured for inclined slab charges only. The computed ratio D_1/D_2, equal to $1/k_{\beta_1}$ from formula (1.60), is fairly close to actual values. This proves that formula (1.60) can be used for calculating the ratio D_1/D_2 with reasonable accuracy.

Column 24 of the table indicates distances D_{cal} of the far end of the muck pile calculated by formula (1.58). These calculations were made on the basis of experimental blast efficiency (see Fig. 42) and actual values of parameters e_{exp}, ρ, etc. By comparing the calculated and actual values of D_1, we observe that they are reasonably close: a slight difference in experiment No. 4 is due to air drag, because distance D_1, calculated on the basis of an actual velocity of projection (30 m/s, see column 19), coincides with the value obtained by formula (1.16). Moreover, in this explosion the parameter χ exceeded the limits of condition (1.13), and the method for calculation explained above introduces considerable error. Processing of the experimental data reveals that the actual thickness h of the muck pile differs from the thickness calculated by formulas (1.54) and (1.62) by approximately $\pm 30\%$. The reason for such a difference lies in the fact that the rock fragments continue to move along the surface after touching the ground in view of the horizontal velocity component. This makes the profile of the muck pile nonuniform; humps and depressions are formed. Besides this, nonuniformity is also caused by peripheral charges which, while reducing the size of the muck pile, scatter rock fragments in an irregular manner.

Fig. 43. Slab-charge system made from segments of detonating cord.

Another aim of the experiments was to study the effect of peripheral charges intended for improving blast accuracy. A series of experiments were conducted in sand with a natural moisture content. A slab charge was made with detonating cord DSh-A, the threads of which were tied at equal intervals to a wooden frame (Fig. 43). Thereafter, the frame with the detonating cord was placed on an already prepared sloping sandy soil and more sand was spread over it to a given thickness. The sandy soil was later compacted till it achieved a natural density. The soil layer thickness over the charge was leveled with the help of two horizontal racks secured at the required height above the frame. The specific consumption of explosives in experiments with detonating cord was calculated by formula (1.3).

In actual experiments, the detonating cord is equipped with tetraethyl pyrophosphate calculated on an average basis at 10 g per meter. The specific energy $e_{exp} = 5.9 \times 10^6$ J/kg.

A typical view of ground throw-out in the case of a slab charge in sand is shown in Fig. 44. It shows the gas-dust cloud before it fell on the free face. One can see that the direction of soil movement is perpendicular to the plane of the charge which ensures a better directivity and accuracy of the blast. There is hardly any throw of sand on the right side.

The effect of peripheral charges was ascertained on the basis of the shape and dimensions of a muck pile by comparing the corresponding parameters of a pile made by blasting an ordinary slab charge without peripheral charges. By comparing the pile width L_p with the charge length L, one can see that L was equal to or less than A when the actual value p_{actual} of the peripheral charge approached the value calculated by formula (1.63). This implies that the peripheral charge was optimal along the end faces. It has been proved by analysis that formula (1.63) provides highly satisfactory results. However, peripheral charges placed above and below do not shorten the distance between the near and far ends of the muck pile. In fact, in certain cases the distance between D_1 and D_2 increases. The reason for this is that the blast symmetry in a vertical plane is disturbed because of gravity. It should be noted here that the given experiments were conducted at explosion depths of less than 0.5 m. It can be assumed that peripheral charges placed above and below will be more effective at greater depths.

Fig. 44. Typical ground throw-out picture on firing a system of slab charges.

The positive effect of peripheral charges along the end faces was checked for large-scale explosions in gravel-loam and gravel-sand soils with a rock thickness of 1 to 8 m. By analyzing experiments No. 4, 7 and 9, one can draw the conclusion that whenever a mass of unit length of peripheral charges conformed to the calculated value (see experiments 7 and 9), the pile width was closer to the longer side of the slab charge.

In the experiments, blast directivity was evaluated by the coefficient equal to the ratio between the mass of rock fallen on the horizontal surface and the entire mass of rock above the slab charge which was taken to be equal to LBW. In these experiments the free face in the zone of the contemplated pile was covered with a polyethylene film from which the entire soil was collected after blasting and its volume measured. Results of these experiments show that for angles $\alpha > 20°$, the directivity factor is equal to 90–100%. This means that practically the entire mass of rock above the slab charge is thrown in the required direction.

Analysis of the results of experiments proves that the slab-charge system designing procedure described above, satisfies conditions (1.12) and (1.13) and conforms to experience. This is also proved by the results of blasting of slab charges with which the apron of the rock-fill dam of the Baipazinsk hydroelectric grid was made in March, 1968. The blast parameters were as follows: dimensions of the area of the charge were: $L = 120$ m, $B = 32$ m, the thickness of gravel-loam mixture in accordance with the design was $W = 8$ m, and the density of rock to be exploded $\rho = 1800$ kg/m³. The angle of inclination of the horizontal was 35°, and the coefficient of relative elevation ζ was 0.33. The maximum distance of the pile boundary from the center of the slab charge D_1 varied from 55 to 60 m. Granulite AC with a specific energy $e_{exp} = 4.4 \times 10^6$ J/kg was used as explosive. Coefficient i, obtained from experimental data, was equal to 0.04 m³/kg (see Fig. 34).

Fig. 45. Slab system of elongated charges in the final phase of the earth fill.

Calculations for slab charges to be used for constructing the apron and preparations for blasting were carried out in the following manner. Geometry correction k was determined from the graph shown in Fig. 12. For this purpose, parameters ζ, χ and Ω were calculated from formulas (1.12), (1.13) and (1.28) by using $q = 1.8$ kg/m³ in the first approximation. From the calculations we get $\xi = 8.3$; $\chi = 2.22$; $\Omega = 0.28$ and $k = 0.81$. As $f(\alpha, \zeta) = 1.75$ therefore, by substituting the actual values of parameters in formula (1.23), we find $q \approx 2$ kg/m³. The distances between cylindrical charges were calculated by formula (1.3). As asbestos-cement pipes with an inner diameter of 0.28 m were used for preparing slab charges, and the loading density of explosives was 1000 kg/m³ therefore, for a relative charging density of 0.8, the mass per unit length of peripheral charge was $p = 49$ kg/m. Substituting these quantities in formula (1.3), where $W = 8$ m and $q = 2$ kg/m³, we get $b = 3$ m.

Quantity p of the peripheral charges was determined by formula (1.65). It is equal to 1000 kg/m. Since the capacity of standard pipe is 49 kg/m, therefore, two pipes were required for accommodating the peripheral charge. After placing the explosive-bearing pipes in a preplanned area the gravel-loam mixture was filled up to the calculated levels in such a way that the average thickness of the filled mixture was $W = 8$ m. This thickness was checked by a survey.

Filling of the gravel-loam mixture embankment was carried out by dump trucks and bulldozers. The beginning of filling of the second half of the embankment is shown in Fig. 45. The survey for explosion results, as well as analysis of motion picture film confirm that the procedure developed for designing slab charges provides satisfactory results.

2. A MODEL BLAST

From expressions (1.21) to (1.29), it follows that the initial velocities of projection in case of actual and model charges will be equal if a small-scale explosion is conducted in the same rock in which actual blasting is to be carried out. The explosives used in both cases are the same, and the following conditions are fulfilled:

$$\left.\begin{array}{l} \xi = \text{idem,} \\ \chi = \text{idem,} \\ uq = \text{idem.} \end{array}\right\} \tag{3.1}$$

Experimental determination of blast efficiency is only based on this factor.

With a known initial velocity and the relationships obtained above, we can calculate all parameters of an actual explosion which are required for outlining the boundaries and the profile of a muck pile. However, in practice it is very important to decide the shape of the muck pile to be made in an actual explosion on the basis of the data obtained from model explosions. Let us find the criteria under which the laws governing the distribution of the muck can be used in case of an actual explosion.

Obviously, the first condition should be a consistency in the ratios of all linear dimensions (necessary condition of geometric similarity):

$$\frac{A_f}{A_m} = \frac{B_f}{B_m} = \frac{W_f}{W_m} = S,$$ (3.2)

$$\alpha_m = \alpha_f,$$ (3.3)

where indexes 'm' and 'f' correspond to model and full-scale operations; and S represents the coefficient of geometric similarity.

In order to ensure the same direction of muck throw of identical volumes of rock in full-scale and model explosions, it is necessary to maintain a geometric similarity of the gas chambers. From the correlations given in Sec 5 of Chapter 1, it follows that gas chambers will have a geometric similarity if the conditions laid out in (3.1) are fulfilled. This implies that the explosives used in full-scale and model explosions should be the same.

The following condition should also be met, so that the range of movement of equal volumes of rock and thickness of the pile are equal in full-scale and model explosions:

$$\frac{D_f}{D_m} = \frac{\eta_{0f}\rho_m}{\eta_{0m}\rho_f} = S.$$ (3.4)

This condition follows from correlation (1.48), if we bear in mind that due to conditions (3.1) and (3.2), parameters, k, k_β, e_{exp}, $f(\alpha, \zeta)$ and q, do not vary when designing a model.

From condition (3.4), it follows that the coefficient of geometric similarity S cannot be selected arbitrarily. For example, if a model is to be designed for an explosion in hard rock ($\rho = 2700$ kg/m³, $\eta_0 = 0.5$) and, thereafter, a model explosion is carried out in loamy soil ($\rho = 1650$ kg/m³, $\eta_0 = 0.08$, and $q = 2$ kg/m³), then in accordance with condition (3.4), the coefficient of geometric similarity will be:

$$S = \frac{\eta_{0f}\rho_m}{\eta_{0m}\rho_f} = \frac{0.5 \times 1650}{0.08 \times 2700} = 3.8.$$

In case of model explosions in sand ($\rho = 1500$ kg/m³, $\eta_0 = 0.02$, $q = 2$ kg/m³), $S = 14$.

Since $\eta_0 = iq$ for soils, therefore, the coefficient of geometric similarity, other conditions being equal, decreases with an increase in specific consumption q.

Models of blasts help in checking the effectiveness of peripheral charges and correcting the quantity q.

It can, therefore, be concluded that model designing should precede experiments for determining blast efficiency in full-scale and model explosions.

While deciding upon the similarity criteria, it was presumed that air drag does not have any effect on movement of the muck (scatter). Obviously, under such conditions when this force cannot be ignored (see Chapter 4), simulation of explosion of slab charges is not possible.

Movement of Flyrock
Subject to Air Drag

1. EFFECT OF AIR DRAG

It is observed that air drag exerts a considerable influence on bodies of small linear dimensions as they move in the air at a high speed. This is clearly seen from Fig. 46. The figure shows trajectories of rock of 2700 kg/m³ density with an initial velocity of $v=76$ m/s and takeoff angle of 45° to the horizontal.

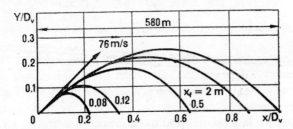

Fig. 46. Trajectory of fragments as a function of size.

Trajectories have been plotted in dimensionless coordinates. Along the x and y axes, the ratios between the respective coordinates and maximum range of movement in a vacuum along the horizontal are shown. For the given velocity and takeoff angle of 45°, this range is equal to 580 m. Each trajectory corresponds to a given mean dimension of a fragment of rock x_f. Trajectories have been plotted by solving the system of differential equations of motion on a computer. By comparing the trajectories shown in Fig. 46, it is clear that for a fragment of rock of diameter $x_f=0.08$ m the range along the horizontal in comparison with movement in a vacuum decreases by 4.5 times. For a fragment of 0.5 m it is decreased to almost half its value. Air drag has a noticeable effect even on fairly large bodies. For example, in the case of a linear dimension $x_f=2$ m, the maximum range of 580 m is reduced by more than 10%. It should be noted that initial velocity of 76 m/s used in calcula-

91

tions is achieved in the case of blasting of chamber charges with a blast efficiency of 1.8. Since the mean size of the rock fragments broken through blasting is about 10 cm when the line of least resistance is about 1 m, it is, therefore, evident that for a more or less accurate solution of the problem of fragment distribution in the muck pile it is very essential to take the effect of air drag into account.

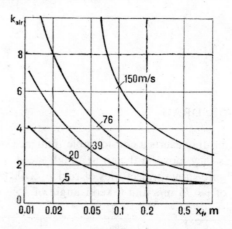

Fig. 47. Dependence of coefficient k_{air} on fragment size and velocity.

It was stated earlier that the degree of retarding effect of the air on a fragment in motion depends on the relation between its linear size, density and speed of movement. This relation is illustrated by graphs (Fig. 47) which show the dependence of the ratio k_{air} between the maximum range of scatter of fragments in a vacuum and the same range in the air on the dimensions x_f of the fragments and their velocity v. The density of fragments in this case is 2700 kg/m³. It is obvious that the rate of reduction in the range of scatter is sharply increased with an increase in velocity and a decrease in linear dimensions. For example, in the case of a fragment measuring $x_f =$ 10 cm, with an initial velocity of 76 m/s, the range in the air is 3 times less than the range in a vacuum. Even if the initial velocity of projection is 20 m/s a fragment of the same size will fall short of its maximum range by about 25%.

2. THE LAW OF RESISTANCE

All bodies moving in a gaseous or a fluid medium encounter a resistance to their movement. The resistive force depends on the viscosity of the medium as well as the pressure which develops on the surface of the body due to deflection and retardation of the flow of the medium. Resistance on account of viscosity is known as *friction drag*. The component of resistance describing the pressure due to retarded flow is called *pressure drag*. The predominant effect of each of these components depends on the correlation between velocity, linear dimensions of the moving body and viscosity of the medium. This correlation is characterized by the dimensionless Reynolds criterion:

$$\text{Re} = \frac{vx_f}{\nu}, \tag{4.1}$$

where ν is the kinematic viscosity of the medium.

In numerous experiments it has been proved that the viscosity of the medium begins to exert a predominant influence when the Reynolds number $\text{Re} < 1000$. The kinematic viscosity of air is $\nu = 0.14$ cm^2/s. By substituting these values of Re and ν in the expression for Re, we get the criterion value of the product vx_f into two parts. One of the two components of resistance becomes dominant in each of these parts:

$$vx_f = 140 \text{ cm}^2/\text{s}. \tag{4.2}$$

For $vx_f < 140$ cm^2/s, the viscosity of the medium becomes the determining factor and the drag is then directly proportional to the velocity. If $vx_f > 140$ cm^2/s, the effect of viscosity in comparison with the dynamic head is insignificant and therefore, the air drag is proportional to the square of the velocity.

We will calculate the expressions for air drag on rock fragments with average size exceeding 1 cm. The number of fragments of smaller size blown away is in fact very small and therefore, has no practical significance. For fragments measuring $x_f > 1$ cm, the maximum velocity v_{max}, at which air drag will vary according to the quadratic law, can be found if in expression (4.2), we substitute $x_f = 1$ cm:

$$v_{max} = 1.4 \text{ m/s}.$$

For rock fragments larger than 1 cm, the velocity will be still lower.

Thus, the quadratic law of resistance can be used for actual conditions of explosion. As we know the air drag in a quadratic law is expressed by the following dependence:

$$F_{air} = c_x S_f \frac{\rho_{air} v^2}{2}, \tag{4.3}$$

where c_x is the drag coefficient; S_f the area of the middle section of the fragment perpendicular to the velocity vector; and ρ_{air} the density of the air.

The deceleration due to air drag j_{air} can be obtained if the right-hand and the left-hand sides of this equation are divided by the mass M of the rock fragment:

$$j_{air} = b_d v^2, \tag{4.4}$$

where

$$b_d = \frac{c_x \rho_{air}}{2} \cdot \frac{S_f}{M}. \tag{4.5}$$

The drag factor b_d depends on the shape and mass of the fragment. From expressions (4.4) and (4.5), it is obvious that the lesser the quantity of j_{air} the greater will be the mass of the rock fragment for the same dimensions, i.e. the greater will be its density.

Let us consider the average size x_f of a rock fragment as the size of the edge of a cube of a volume equal to the volume of a parallelepiped which circumscribes the given fragment.

Studies [35] have revealed that the most probable shape of the fragment is that shape for which the ratio between its three dimensions a, b, and c is expressed by the proportion $a : b : c = 1.6 : 1 : 0.6$. As the true volume of the rock fragment is approximately 2.2 times less than the volume of the circumscribed parallelepiped, therefore:

$$x_f = 1.3 \sqrt[3]{\overline{U_f}},$$

where U_f is the true volume of the fragment.

The moving fragments rotate in an irregular manner. Therefore, the area of the middle section S_f perpendicular to the velocity vector will vary in time from a certain maximum value $S_{f\ max}$ to a minimum value $S_{f\ min}$. Quantity $S_f = 0.6 \times x_f^2$ can be taken as the mean value of S_f as proved in [35].

The mass of a rock fragment, on the basis of the law of fragmentation, is expressed by the following equation:

$$M = \frac{x_f^3}{2.2}.$$

Therefore

$$\frac{M}{S_f} = \frac{x_f \rho}{1.3}.$$

Substituting this ratio in formula (4.5) we find

$$b_d = 0.66 \frac{c_x}{x_f} \cdot \frac{\rho_{air}}{\rho}. \tag{4.6}$$

The drag coefficient for fragments of different shapes varies between 1.2 and 1.8 [36]. The mean value can be taken as: $c_x = 1.5$ and $\rho_{air} = 1.29$ kg/m³. Substituting these in formula (4.6) we get:

$$b_d = \frac{1.3}{x_f \rho}, \text{ m}^{-1}, \tag{4.7}$$

where x_f is measured in meters and ρ in kg/m³.

3. FORMULATION OF DIFFERENTIAL EQUATIONS OF MOTION OF A FRAGMENT

A detailed study of the movement of flyrock during blasting operations is replete with difficulties. These are explained below.

First of all, firing of charges in the soil and rocks crushes these into fragments of different sizes—from the smallest specks of dust to very large fragments which, in certain cases, are several meters in size. Besides, the shapes of the fragments obtained as a result of blasting differ greatly [35]. Since the extent of retarding force of the air depends on the shape and mass of the flyrock, it becomes very difficult to calculate air drag for an entire mass of rock blown away in view of the extremely wide range of rock breakage and uncertainty of rock shape.

Secondly, the initial velocity of projection of the fragment cannot be determined with adequate accuracy. Although we now have formulas for calculating the initial velocity of projection, they only hold good for the throw front. Velocities of individual rock fragments behind the throw front vary over a fairly wide range. This also applies to the direction of throw. For concentrated charges with an explosion effect index above 1.5, we still do not have any law for the projection of fragments in an exploded volume which are on one side of the line of least resistance.

Error in evaluating initial velocity leads to wide errors in calculating the range of rock scatter. For example, if the initial velocity of projection is known within a $\pm 10\%$ accuracy, it leads to an error of $\pm 20\%$ in the estimated value of the range of scatter in the case of very large fragments.

Thirdly, rock fragments moving in the air collide with each other. As a result the velocity changes abruptly in magnitude and direction. Besides, an induced effect on the nature of movement (a so-called interference) is observed when rock fragments move in the form of a solid mass. Large perturbations are caused by the bursting out of explosion products which are ejected at a velocity considerably greater than the velocity of individual fragments. The explosion products impart a very high velocity to the fragments emerging along with them. As a result, a cloud of fast-flying fragments is formed in front of the main mass of the exploded rock. This can be very clearly observed in enlarged photographs of a blasting operation, especially in hard rock.

Finally, it should be borne in mind that the air between the rock fragments moving at short distance from each other, is also set in motion which considerably changes the initial conditions of interaction between in-flight fragments and the medium as well as among themselves. In particular, it is clearly seen from high-speed photographs of blasts that a continuous stream of soil is usually divided into a number of cone-shaped jets, some of which move far ahead of the main body of the flyrock. A theoretical investigation of this process was first undertaken by Prof. G.I. Pokrovski [37].

It thus follows that the main problem of exterior ballistics (that of determining the law of movement of a body projected in the air) cannot be solved without a stagewise division in order to simplify the process of blasting.

In Chapter 1, the explosion and blast process was divided into four stages. With the breaking of the gas-dust dome into separate pieces, the fourth and last stage of explosion begins, namely the movement of individual fragments in the air without any interaction among them. In this chapter the fourth stage of explosion will be theoretically studied, i.e. the exterior ballistics of the fragments (in contrast to the interior ballistics which studies the initial stages of a blast during the period when a fragment attains a maximum velocity of projection).

We will make the following assumptions in order to simplify the solution of the basic problem of exterior ballistics.

1. The air-drag vector lies in a direction opposite to the velocity vector. This assumption is based on the fact that a rock fragment executes a rotary motion during the blasting process and this motion does not cease till it falls on the ground. Rotation of the fragment during the flight path leads to a slight symmetrical vibration of the drag vector about the velocity vector. Rotation of the fragment, in this way, levels the possible ripples which, as is well known, deflect the air-drag vector from the direction of the current.

2. Fragments moving in the air do not interact among themselves from the moment of breaking of the gas dome till they fall on the free face. This assumption is true for that stage of the blast when the rock fragments, as a result of radial movement, move away from each other to a sufficiently long distance, so that mutual interference ceases. This assumption does not hold good for relatively short distances from the center of the explosion when the rock blown away by blasting moves as a compact mass. However, it should be borne in mind that this stage of movement forms a very small fraction of the total length of the trajectory. Calculations show that for explosions of concentrated charges in excavation work, the length of trajectory at which interference of and interaction between fragments is considerably less than 10% of the total length of the trajectory. Therefore, the effect of these factors on the final results of the problem based on the law of movement can be ignored.

3. The atmosphere is considered to be static. The effect of wind on rock-fragment movement can be taken into account, if necessary, in the form of a correction.

4. The Magnus effect (emergence of a tangential force acting on a rock fragment which rotates while moving in the air) is not taken into account because in comparison with the effect of air drag, this effect exerts a noticeable influence only on large fragments moving at low velocities.

The equation of motion of a rock fragment in vector-form can be written as follows:

$$\frac{d\vec{v}}{dt} = -b_d \vec{v} v + \vec{g}, \tag{4.8}$$

where v is the rate of movement; g the acceleration due to gravity; and t the time.

The equation will be solved using an orthogonal system of coordinates.

The y axis is directed vertically upward and the x axis is horizontal. The origin of coordinates is the center of gravity of the rock under study.

By plotting the vectors of equation (4.8) on the x and y axes and denoting the velocity components along the axes as v_x and v_y, respectively, and joining two kinematic dependences to the equations of motion, we obtain the following system [36]:

$$\left.\begin{array}{l} \dfrac{dv_x}{dt} = -b_d v_x v; \\[2mm] \dfrac{dv_y}{dt} = -b_d v_y v - g; \\[2mm] \dfrac{dx}{dt} = v_x; \\[2mm] \dfrac{dy}{dt} = v_y, \end{array}\right\} \tag{4.9}$$

where

$$v = \sqrt{v_x^2 + v_y^2}.$$

The limits of integration are as follows: $t=0$; $v_x = v_0 \cos \theta_0$; $v_y = v_0 \sin \theta_0$; $x=0$; and $y=0$. Here, θ_0 is the angle of projection, being the angle between the velocity vector $\vec{v_0}$ and the horizontal.

The equations obtained are coupled, and their solution can be found only by numerical methods. However, for approximate calculations we can simplify the system of differential equations in such a manner, that they can be integrated in the final form.

We shall consider that, as in a vacuum, a body moving in the air also executes two independent movements: in the direction of the initial velocity vector, and in the direction of the force of gravity. Air drag acts on it in both cases. The magnitude of air drag corresponds to the velocity components in these directions.

A simplified oblique-angled system of coordinates zoy (Fig. 48) results in two independent differential equations as follows:

$$\left.\begin{array}{l} \dfrac{d^2z}{dt^2} = -b_d v_z^2; \\[2mm] \dfrac{d^2y}{dt^2} = g - b_d v_y^2, \end{array}\right\} \tag{4.10}$$

where z is the coordinate axis directed along the vector of initial velocity v_0; y the coordinate axis directed along the vector of gravity; and v_z and v_y are the velocities of independent motion along the coordinate axes.

The initial conditions of integration of this system are: $t=0$, $z=0$, $v=0$, $v_z = v_0$, $v_y = 0$. By solving equations (4.10), we obtain

$$z = \frac{1}{b_d} \ln [1 + b_d v_0 t];$$

$$y = \frac{1}{b_d} \ln \frac{e^{2t\sqrt{b_d g}} + 1}{2e^{t\sqrt{b_d g}}} \, . \qquad\qquad \left.\begin{matrix} \\ \\ \\ \\ \\ \end{matrix}\right\} \qquad (4.11)$$

By setting arbitrary values of time t, we can find the coordinates z and y by these formulas. We can plot the trajectory of a rock fragment along a few assigned points. The point of intersection of the trajectory with the ground determines the point of fall (see Fig. 48).

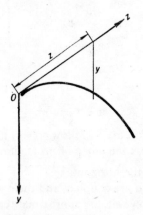

Fig. 48. Oblique-angled system of coordinates.

A comparison of solutions (4.11) with exact solutions of the system (4.9) shows that there is a $\pm$ (10 to 15%) error. Such an accuracy is acceptable for calculations.

4. BALLISTIC TABLES

The system of differential equations (4.9) and (4.11) has been solved on a computer for a wide range of input parameters.

Tables of solutions (4.11) are given in [18]. They cover the following ranges of parametric variations: $b_d = (10^{-4} - 0.3)$ m^{-1}; $v_0 = (2 - 150)$ m/s; $t = (0 - 90)$ s. These tables prove helpful in carrying out a quick calculation for determining the range of rock flight and drawing the profile of the muck pile on any configuration of the free face.

Ballistic tables prepared on the basis of a solution of the system (4.9) are given in the Appendix. They contain input parameters as follows: $b_d = (10^{-4} - 10^{-2})$ m^{-1}, $v = (14 - 248)$ m/s and $\theta = (-45)$ to $(+85°)$. The interval for parameter b_d is the same as given in the tables in [18], and has been selected in such a way that the error involved in rounding-off the actual parameteric value to the nearest value given in the table does not exceed 15%. The trajectory is plotted along a number of points in an orthogonal system of coordinates.

5. CRITERIA FOR ASSESSING THE RETARDING ACTION OF THE AIR

It has been stated earlier that the larger the rock fragment and lower the velocity, the lesser the retarding action of air. It is obvious that there exists some correlation between these parameters at which the effect of air drag can be ignored. In order to clarify this aspect it is essential to find a generalized criterion to cover the situation mentioned here. With this aim in view, let us express the system of differential equations (4.9) in dimensionless parameters:

$$
\left.
\begin{aligned}
\frac{d\hat{v}_x}{d\hat{t}} &= -\hat{v}_x\hat{v}; \\
\frac{d\hat{v}_y}{d\hat{t}} &= -\hat{v}_y\hat{v} = \frac{1}{\hat{j}}; \\
\frac{d\hat{x}}{d\hat{t}} &= \hat{v}_x; \\
\frac{d\hat{y}}{d\hat{t}} &= \hat{v}_y,
\end{aligned}
\right\}
\tag{4.12}
$$

where

$$
\hat{v} = \sqrt{\hat{v}_x^2 + \hat{v}_y^2}.
$$

The limits of integration will be as follows: $\hat{t}=0$; $\hat{x}=0$; $\hat{y}=0$; $\hat{v}=1$; $\hat{v}_x=\cos\theta_0$; and $\hat{v}_y=\sin\theta_0$. In the system (4.12), the dimensionless parameters, $\hat{x}$, $\hat{y}$, $\hat{v}_x$, $\hat{v}_y$ and $\hat{t}$ are expressed as follows:

$$
\hat{x}=xb_d; \quad \hat{y}=yb_d; \quad \hat{v}_x=\frac{v_x}{v_0};
$$

$$
\hat{v}_y=\frac{v_y}{v_0}; \quad \hat{t}=tb_dv_0; \quad \hat{j}=\frac{b_dv_0^2}{g}.
$$

Quantity $\hat{j}$ represents dimensionless acceleration and the product $b_dv_0^2$, in accordance with expression (4.4), is the deceleration due to air drag. In this way, the dimensionless number $\hat{j}$ indicates how much the deceleration due to air drag is greater in absolute magnitude as compared to the acceleration due to gravity. The value $\hat{j}$ is known as the surcharge.

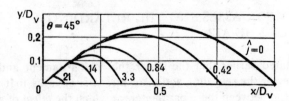

Fig. 49. Trajectory describing movement of rock fragments for several values of surcharge $\hat{j}$.

From the system (4.12) it follows that trajectories in dimensionless coordinates $\hat{x}$ and $\hat{y}$ will be similar when $\hat{j}=$const and $\theta_0=$const. The family of trajectories in dimensionless coordinates for certain values of $\hat{j}$ has been plotted in Fig. 49 in accordance with the data given in the ballistic tables in the Appendix. The ratio between corresponding coordinates x and y and the range of rock flight in a vacuum D_v is shown along the abscissa and the ordinates. As is well known [36]

$$D_v = \frac{v_0^2 \sin 2\theta_0}{g}. \tag{4.13}$$

As the quantity D_v during movement in the air possesses a limiting value, the maximum value of the dimensionless coordinate x/D_v cannot exceed 1.

From analysis it is evident that the dimensionless criterion

$$\hat{j} = \frac{b_d v_0^2}{g}, \tag{4.14}$$

used in the system (4.12) characterizes the degree of effect of air drag on the movement of a rock fragment in the air.

Calculations indicate that the trajectory of a fragment cannot be calculated with higher than 20% accuracy because neither the vector of initial velocity of projection $\vec{v_0}$ nor the drag factor b_d is correctly known. Therefore, air drag should be taken into account when the range of rock flight in the air is 20% less than the maximum range in a vacuum. Graphs of the ratio S_{air} between the range of rock flight D in the air and the range in a vacuum D_v with equal velocity v_0 for several throw angles θ_0 depending on surcharge $\hat{j}$ have been plotted in Fig. 50 on the basis of the solution of the system (4.9).

In this way, coefficient S_{air} characterizes the degree of retarding action of the air. It is easy to find that

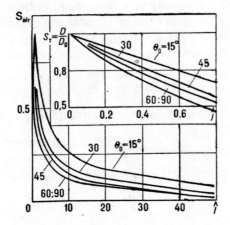

Fig. 50. Dependence of coefficient S_{air} on surcharge $\hat{j}$.

$$S_{air} = \frac{D}{D_v} = \frac{1}{k_{air}}. \tag{4.15}$$

The value of D/D_v for a wide range of surcharges $\hat{j}$, for angles $\theta_0 > 45°$, is the same as when $\theta_0 = 45°$.

From the graphs it is evident that for $\theta_0 = (15-60)°$ the rock-fragment trajectory is reduced by 10%, when $\hat{j} = (0.2-0.4)$. Hence it follows that while calculating the trajectory of movement of a rock fragment, the air drag should be taken into account provided criterion $\hat{j}$ satisfies the following inequality:

$$\hat{j} = \frac{b_d v_0^2}{g} > 0.3. \tag{4.16}$$

6. CORRECTION FOR AIR DRAG

Let us introduce a correction factor k_{air} which takes into account the retarding action of the air in calculating formulas (1.20) and (1.23) for a specific consumption q of explosives. It should be noted that in contrast to calculation of the trajectory, in which correction for air drag is made when inequality (4.16) is satisfied, the correction for a specific consumption of explosives is made in all cases.

From expression (1.19), we find that

$$q = \frac{\rho v_0^2}{2e_{exp}\, \eta},$$

where v_0 represents the velocity in a vacuum. In order to overcome the air drag and achieve a projection range equal to the range in a vacuum it is necessary to increase the initial velocity of projection v_0 to v_{con} to which will conform the following new value of specific consumption of explosives:

$$q_{con} = \frac{\rho v_{con}^2}{2 g e_{exp} \eta} .$$

Let us introduce the following correction factor k_{air} for air drag:

$$k_{air} = \frac{q_{con}}{q} . \tag{4.17}$$

By substituting in Eq. (4.17) the expressions for q_{con} and q, we get

$$k_{air} = \frac{v_{con}^2}{v_0^2} .$$

If the body moves at a velocity v_{con} in a vacuum, its range, from the correlation (1.16), will be

$$D_v = \frac{2 v_{con}^2}{g f(\alpha, \zeta)} ,$$

where

$$\alpha = 90 - \theta_0 .$$

Because of air drag, the range D_v is reduced to D which is calculated by a similar relationship in case of a specific consumption q without taking drag into account:

$$D = \frac{2 v_0^2}{g f(\alpha, \zeta)} .$$

From these two correlations we find that

$$\frac{v_{con}^2}{v_0^2} = \frac{D_v}{D} .$$

Since $k_{air} = v_{con}^2 / v_0^2$, therefore,

$$k_{air} = \frac{D_v}{D} .$$

Earlier, we have found that $S_{air} = D/D_v$ (see Fig. 50), therefore,

$$k_{air} = \frac{1}{S_{air}} .$$

From analysis of the system (4.12), it follows that S_{air} depends on two parameters $\hat{j}$ and θ_0. In general, for an arbitrary outline of the free face, the graph for function S_{air} should be drawn on the basis of ballistic tables (see

Appendix). For this purpose, it is essential to draw the end sections of the trajectory and calculate the range D for a number of values of the surcharge $\hat{j}=b_d v_0^2/g$ on the point of intersection with the free face. Thereafter, the value of D_v should be calculated by formula (4.13), and the value of $S_{air}=D/D_v$ found for a given value of $\hat{j}$. Function $k_{air}=1/S_{air}=f(\hat{j},\,\theta_0)$ is found for several points on the graph of S_{air} and $\hat{j}$ for a given value of θ_0. Figure 51 shows graphs of factor k_{air} for throw angles equal to 15, 30, 45, 60 and 90°. Here it is presumed that the point of escape and the point of fall of the rock are in the same plane. These graphs, however, can be used as fairly accurate approximations and for a very general case when the free face has an arbitrary outline. By taking into account the correction factor the formulas (1.20), (1.23) and (1.48) become

$$q=f(\alpha,\zeta)\,\frac{\rho g}{4\eta e_{exp}}\,k_{air}\,D; \tag{4.18}$$

$$q=\sqrt{\frac{f(\alpha,\zeta)\,\rho g}{4ie_{exp}\,k}}\,\sqrt{k_{air}\,D}; \tag{4.19}$$

$$q=\frac{f(\alpha_1,\zeta)\,\rho g}{4kk_{\beta_1}\,\eta_0 e_{exp}}\,k_{air}\,D\beta_1; \tag{4.20}$$

$$q=\sqrt{\frac{f(\alpha_1,\zeta)\,\rho g}{4ie_{exp}\,kk_{\beta_1}}}\,\sqrt{k_{air}\,D\beta_1}\,. \tag{4.21}$$

From the graphs in Fig 51 it is evident that in case of greater values of surcharge $\hat{j}$ the correction factor k_{air} also increases. As a result, the specific consumption of explosives for overcoming the air drag increases sharply. For example, if $\hat{j}=10$ then for hard rock the specific consumption of explosives

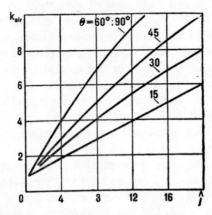

Fig. 51. Dependence of correction for air drag k_{air} on surcharge $\hat{j}$.

increases by 1.7 times when $\theta = 60°$. For soils, the specific consumption of explosives q is proportional to the square root of the correction factor k_{air} and its influence is, therefore, slightly weaker than in the case of hard rock.

This indicates that air drag, in case of a high surcharge $\hat{j}$, can be overcome only at a very high cost. Hence, there arises a persistent need to analyze and find such means which can reduce the adverse effects of air drag.

7. WAYS OF REDUCING AIR DRAG

In Fig. 17 the dotted lines indicate the limiting trajectories within which lies a certain portion of the flyrock initially located within the area covered by cross shading. Deceleration of this portion of the flyrock during the flight path due to the air drag at a given moment is determined by the following expressions:

$$j = c_x \frac{\rho_{air} v^2}{2} \cdot \frac{S}{M} ; \qquad (4.22)$$

$$M = W\rho, \qquad (4.23)$$

where M is mass of the rock under study; and S the cross-sectional area perpendicular to the velocity vector (middle cross section).

Quantity M/S is called the *specific load* and is calculated by the quantity of rock mass per square meter of the middle cross section. It is obvious that the specific load is inversely proportional to the air drag. The mean value of specific load at a given point of the trajectory can be calculated by the following formula:

$$\frac{M}{S} = \rho h, \qquad (4.24)$$

where h is the thickness of the moving volume of a rock fragment in the direction of the tangent to a point on the trajectory.

From Fig. 17 it is clear that if a rock mass moving in the air is further compacted, the specific load will correspondingly increase and consequently, the air drag will have less effect. Concentration (compactness) of movement of the rock blown away by blasting can be achieved with a relatively greater radius of curvature R_1 of the gas chamber (see Fig. 17).

A plane-parallel throw ($R_1 \to \infty$), which can be achieved by fulfilling criterion (1.40), is an ideal method of overcoming (or sharply reducing) the adverse effects of air drag. This is one of the major advantages of the slab-charge system.

If under given conditions a plane-parallel throw is not possible (length L and breadth B of the slab system are relatively small and the specific consumption of explosives is high), peripheral charges can be used for achieving a compact throw (see Sec. 8, Chapter 1). From expressions (4.22) and (4.23) it is evident that the specific load for a rock moving in the air is proportional to the thickness W of the rock layer to be blasted. Therefore, one of the methods of reducing the adverse effects of air drag is to increase the scale of explosion.

Let us determine the value of W for a given range D at which air drag can be ignored for a plane-parallel throw of rock.

Bearing in mind that for a plane-parallel throw, in case the trial condition (1.40) is fulfilled, we have

$$h_m = \frac{1}{5}W; \tag{4.25}$$

$$b_d = \frac{1.3}{h_m \rho}. \tag{4.26}$$

We find from the condition (4.16) and expressions (4.13), (4.25) and (4.26), that

$$\frac{W}{D} \geqslant \frac{22}{\rho \sin 2\theta_0}. \tag{4.27}$$

Air drag can be ignored while satisfying this inequality for a plane-parallel throw and a horizontal free face (in this case the inaccuracy in directing the blasted rock, due to air drag, is about 10% of the calculated range). In a particular case when $\theta_0 = 45°$ and $\rho = 2500$ kg/m³, the thickness of the rock layer to be blasted for which the effect of air drag is negligible;

$$W \geqslant \frac{D}{100}. \tag{4.28}$$

In addition to plane-parallel throw-out explosion method and application of peripheral charges, another very effective method of achieving a concentrated throw is directional blasting which is carried out by a so-called subcritical short-delay blasting technique (see Chapter 7). In directional blasting, the flyrock is concentrated in the direction of throw by means of a cumulative effect during fulmination of an entire slab-charge system from the periphery toward the center. The specific load during rock movement in the air can be increased by a multirow short-delay firing on a system of deep-hole charges. In the latter case, the exploded rock is thrown as a compact mass which increases with an increase in the number of rows of charges for a single short-delay firing. The specific load M/S, in the given case, increases in direct proportion to the number of exploded rows of charges.

8. EFFECT OF GAS FLOW ON FLYROCK

It is of practical interest to calculate the effect of wind or gases emanating from the charge adit on the rock fragments blown away by explosion. Let us suppose that a fragment measuring x_f moves at a velocity v_f through a

stream of gases issuing forth from a charge adit at an initial velocity v_0 and
with an angle of spread α (Fig. 52). Let us assume that velocity v of the gas
stream, of that section in which the rock moves, and density ρ_g are constant
for the duration of movement. Let us calculate the additional velocity u gained by the fragment under the effect of gases escaping along the x axis. Let us
also denote the mass of the fragment by m, the area of its middle cross section
by S_f, the drag coefficient by c_x, and the velocity of the gas stream by v. The
differential equation of describing the movement of this rock fragment will be
expressed in the following form:

$$m \frac{du}{dt} = \frac{1}{2} c_x S_f \rho_g (v-u)^2, \tag{4.29}$$

the initial conditions being $t=0$ and $u=0$.

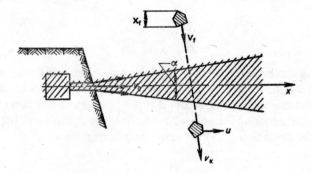

Fig. 52. Diagram for calculating effect of gas flow on movement of a rock fragment.

As is evident, this equation belongs to the class of equations with dividing
variables.

By denoting

$$b_g = \frac{1}{2} c_x \rho_g \frac{S_f}{m}, \tag{4.30}$$

we obtain the equation

$$\frac{du}{(v-u)^2} = b_g \, dt.$$

For this solution we introduce a new variable $y=v-u$. Equation (4.29)
then takes the following form:

$$\frac{dy}{y^2} = -b_g \, dt.$$

The solution of this equation is

$$\frac{1}{y} - c = b_g \, t,$$

where

$$c = \frac{1}{v} .$$

Let us introduce relative velocity $\bar{u} = u/v$.

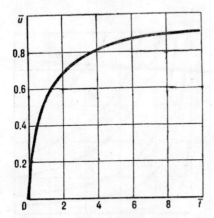

Fig. 53. Dependence of relative velocity $\bar{u}$ on dimensionless time $\bar{t}$.

The final form of the solution of the differential equation (4.29) will be

$$\bar{u} = 1 - \frac{1}{1+\bar{t}} , \qquad (4.31)$$

where dimensionless time

$$\bar{t} = \frac{t}{t_{\text{eff}}} ;$$

$$t_{\text{eff}} = \frac{2}{c_x \rho v} \cdot \frac{m}{S_f} .$$

Because

$$\frac{m}{S_f} = \frac{x_f \rho}{1.3} ,$$

therefore

$$\bar{t} = \frac{1.6 x_f \rho}{c_x \rho v} . \qquad (4.32)$$

From solution (4.31) we can easily determine the physical meaning of effective time t_{eff}. If $t = t_{\text{eff}}$ then $\bar{u} = \frac{1}{2}$ and $u = \frac{1}{2} v$. Consequently t_{eff} is that time during which the rock fragment in the gas stream gains half of its velocity. For the given case, when the rock fragment is under the effect of wind or escaping gases, we can take the following average quantities: $\rho = 2500 \text{ kg/m}^3$;

$c_x = 1.5$; $\rho_g = 1.3$ kg/m³. In this case

$$t_{\text{eff}} = 2000 \frac{x_f}{v}. \tag{4.33}$$

Dependence $\bar{u}$ on $\bar{t}$ and t_{eff} on x_f and v have been shown in Figs. 53 and 54 respectively.

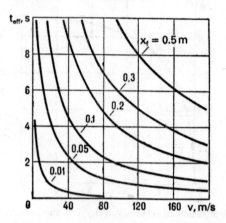

Fig. 54. Dependence of effective time t_{eff} on velocity and size of rock fragment.

From expression (4.31) and the related graphs, it is clear that a fragment measuring 0.1 m moving with an initial speed of 50 m/s and at an angle of projection to the horizontal $\theta_0 = 45°$, at a wind velocity $v = 10$ m/s, will be in the air for 7 s. For this example, the effective time $t_{\text{eff}} = 20$ s and $\bar{u} = 0.25$. Consequently, this part of the rock gains a horizontal vector component of 2.5 m|s and an additional displacement of 17.5 m along the horizontal due to wind effect. In this way, wind has practically no effect on a fragment of 0.1 m in comparison with its ballistic range of about 150 m (taking air drag into account).

Let us calculate the effect of explosion products escaping through a charge adit on a rock fragment measuring $x_f = 0.3$ m moving at a velocity $v_f = 50$ m/s at a distance $x = 100$ m from the adit mouth (see Fig. 52). To consider a specific example let us assume that the gases escape from the entrance (5 m²) of the adit at a velocity $v = 200$ m/s and the angular spread of the gas stream, according to experimental data, is $\alpha = 6°$. A rock fragment, blown through the gas stream, gains a velocity u as shown in Fig 55. Clearly, the value of this velocity drops sharply as the distance from the adit mouth increases, but does not exceed 0.2 m/s at a distance of $x > 100$ m. From this account it can be seen that the gas flow of explosion products cannot significantly affect the projection of fragments measuring $x_f > 0.3$ m. However, fragments measuring only a fraction of a centimeter can be carried by the gas flow to a considerable distance. Estimations for small fragments appearing in the gas stream during the blasting of deep-hole charges can be carried out on the basis of these dependences.

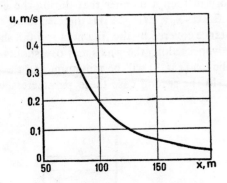

Fig. 55. Dependence of additional velocity u
on distance from adit mouth.

9. ACCURACY IN BALLISTIC CALCULATIONS

From Fig. 39 it is evident that the actual initial velocity of projection, during firing of charges, may deviate from the calculated value within limits of $\pm 10\%$. Let us calculate the error in the calculated range of rock movement as a result of inaccuracy in obtaining the initial velocity of projection. Let us study this problem for a particular instance of projection of a body having an initial velocity v_0 at an angle $\theta_0 = 45°$ without taking air drag into account. For the sake of simplification let us assume that the point at which the rock fragment is blown away and that point at which it falls on the free face lie in the same horizontal plane. The range of movement in such a case will be

$$D = \frac{v_0^2 \sin 2\theta_0}{g}.$$

After differentiating we find

$$\frac{\Delta D}{D} = 2\left(\frac{\Delta v}{v_0} + \cot 2\theta_0 \, \Delta\theta_0\right).$$

Hence, we find that for $\theta_0 = 45°$, $\Delta D/D = 2\Delta v/v_0$.

In this manner, error in the velocity of projection $\Delta v/v_0 = \mp 0.1$ leads to an error in range of $\Delta D/D = \mp 0.2$.

However, the movement of a rock fragment in the air predetermines the presence of wide errors in the calculated range of rock displacement. The reason for this can be found in the uncertain shape and mass of the fragment as well as its irregular rotation. Let us study these in more detail.

Variation of the actual from the calculated trajectory will be determined, as is evident from equation (4.8), by the variation of actual values of the drag factor b_d and initial velocity of projection v_0 from the calculated values. In principle it is possible that a fragment with linear dimensions a, b and c may move in a trajectory in such a manner that during the entire period of its movement one of its axes x, y, z (Fig. 56) is directed along the velocity vector. The maximum distance covered by the fragment will fall short of the calculated point when it advances along the y axis and overshoots the x axis. In the first case the specific load (M/S) will be minimum and the drag coefficient c_x will be maximum. In the second case these parameters will have opposite external values.

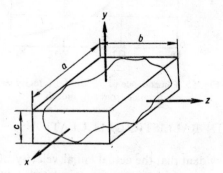

Fig. 56. Possible orientation of a rock fragment
during its movement in the air.

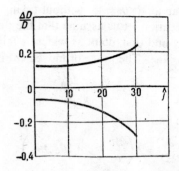

Fig. 57. Dependence of maximum error in
calculated range on surcharge $\hat{j}$.

Figure 57 shows a graph of possible maximum relative errors in $\Delta D/D$ while calculating the range of displacement of a fragment depending on the relative surcharge $\hat{j}$. From this graph it is clear that the probable errors during calculation of the trajectory will increase with an increase in parameter $\hat{j}$. In this case the smaller the size of the fragment and the greater the velocity, the lower will be the accuracy in calculation of its range of fall on the free face. In Fig. 57 the difference in the range of rock displacement corresponds to a negative value of $\Delta D/D$.

Surcharge $\hat{j}$ for the given explosion condition is calculated as follows:

$$\hat{j} = \frac{b_d v_0^2}{g} = \frac{1.3}{\rho g h_m} v_0^2, \tag{4.34}$$

where h_m is the mean thickness of the rock layer as it moves in the air.

This parameter can be calculated by the following formula:

$$h_m = \frac{W}{2}\left(1 + \frac{S_{ch}}{S_p}\right); h_m \geqslant x_m, \tag{4.35}$$

where S_{ch} is the cross-sectional area of the charge; S_p the area of the muck pile and x_m the mean size of rock fragment.

If $h_m \leqslant x_m$ we can use the quantity x_m instead of h_m. In this case the gas dome is divided into separate segments, the size of which depends on air drag. Methods of evaluating the mean dimension x_m of a fragment will be given in Chapter 5.

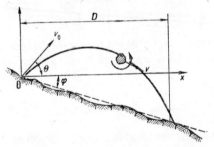

Fig. 58. Diagram for calculating slant distance
according to a general formula.

It is very easy to calculate the range D of displacement of the fragment of size x_f with the help of ballistic tables. However, the following general formula can be used for this purpose:

$$D = S_{air}(f, \theta_0)\frac{2v_0^2}{g}(\tan \varphi + \tan \theta)\cos^2 \theta. \tag{4.36}$$

Here, function $S_{air}(j, \theta_0)$ is determined from the graphs (Fig. 50) and the mean angle of inclination of the free face to the horizontal φ is determined from the drawing of a section of the area in the plane of the trajectory followed by the fragment (Fig. 58). It should be noted that the graphs of function

S_{air} shown in Fig. 50 are, strictly speaking, true for the cases where the points of escape and fall of fragments are at the same level. In other cases they provide a fairly good approximation of the true value of S_{air}.

10. EXPERIMENTAL DETERMINATION OF THE COEFFICIENT OF AIR DRAG

If the line of least resistance coincides with the normal to the horizontal, the movement of a fragment will be described by the second equation of the system (4.9):

$$\frac{dv_y}{dt} = b_{\text{d}}v_y v - g.$$

Because, in this case $v_y = v$, we obtain

$$\frac{dv}{dt} = -b_{\text{d}}v^2 - g.$$

The initial condition will be $t = 0$ and $v = v_0$.

The solution of this differential equation with the given initial conditions is

$$\frac{\sqrt{\dfrac{b_{\text{d}}}{g}}\,(v_0 - v)}{1 + \dfrac{b_{\text{d}}}{g}\,v_0 v} = \tan\,(t\sqrt{b_{\text{d}}g}).$$

At the culmination points, i.e. at the maximum height attained by the fragment after explosion, we have $v = 0$, and $t = t_{\text{f}}$. Substituting these quantities we obtain

$$v_0 \sqrt{\frac{b_{\text{d}}}{g}} = \tan\,(t_{\text{f}}\,\sqrt{b_{\text{d}}g}). \tag{4.37}$$

The values of v_0 and t_{f} can be calculated from the film of the blasting process. Substituting these in equation (4.37), and solving it by a part-graphic and part-analytic method, we can calculate the experimental value of the drag factor b_{d}.

Chapter 5

Exterior Ballistics and Breakage of Rock through Blasting

1. METHODS OF SOLVING THE BASIC PROBLEM OF EXTERIOR BALLISTICS

The main problem of exterior ballistics in mining, i.e. determination of the law of motion of a body projected into the air and the point of its fall on the free face, as applicable to the rock mass blown away by blasting, has (as already indicated in Chapter 4) a number of specific peculiarities and associated difficulties.

Practical experience and investigations carried out so far show that by firing charges in hard rock and connecting soils, the medium is crushed to fragments of widely differing sizes. Since the retarding action of the air depends on the size and mass of the moving fragment, therefore, in case of equal vectors of initial velocity, the rock fragments projected in radial directions after blasting, will scatter on the free face in such a way that the smaller fragments fall near the point of explosion and the larger ones fall at different distances. The larger the size of the fragment, the greater the range of its flight from the point of blast.

The point of fall of each fragment on the free face can be calculated by the method given in Chapter 4 if the granule-size composition of the rock mass is known. The thickness of the muck pile at a given point is calculated by adding all those elementary volumes of rock mass which fall at a given point.

This method of calculation of rock distribution in a muck pile is very accurate. However, it involves a large number of calculations. This method is justified in those cases where a very high accuracy is desired in the calculation of the profile of a muck pile and the initial velocities of projection and granule-size composition of the rock crushed by blasting are known for the entire volume of rock to be blasted.

A simple method of calculating the profile of a pile is the part-graphic and part-analytic method. According to this method the entire volume of rock to be exploded is divided into i elementary volumes and the entire spectrum of

113

114

granule-size composition is divided into j (roughly five to ten) fractions. By assuming that each fraction moves independently of the others (without interference), a trajectory is plotted for an average fragment of each fraction in the ith elementary volume. The size of this fragment x_j is calculated as the arithmetic mean with respect to the limiting sizes of the jth fraction:

$$x_j = \frac{x_j + x_{j+1}}{2}.$$

By substituting the calculated value of x_j in place of x_f in formula (4.7), we calculate the drag factor b_d. Thereafter, from actual values of the drag factor b_d, the initial velocity of projection v and the angle of projection θ, one can look for corresponding parameters in the ballistic tables (see Appendix) closest in values to these values. In case a higher accuracy is desired, values can be found by interpolation. Thus the trajectory of an average-size fragment of each fraction can be plotted (Fig. 59).

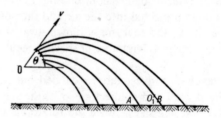

Fig. 59. Trajectories for rock fragments of different
sizes projected in the air.

The thickness of the muck pile h_{ij} made by the jth fraction falling at point O_1 of the free face can be determined by dividing volume U_{ij} of the jth fraction from the ith elementary volume covering an area S_j, within the limits of which falls the given volume of muck. The area S_j may be determined in the following manner. The distance between the two adjacent trajectories is divided into two and points A and B are determined (see Fig. 59). In a similar manner points C and D are found in the plane perpendicular to the plane of the diagram. The area S_j will then be equal to the area of rectangle $ABCD$:

$$S_j = AB \times CD.$$

The volume of rock of the jth fraction is

$$U_{ij} = U_i [F(x_j) - F(x_{j-1})],$$

where $F(x)$ is the law of cumulative distribution of the granule-size composition of the rock crushed on blasting; x_j and x_{j-1} the sizes of fragments at the boundaries of the jth fraction; and U_i the ith elementary volume of the blasted rock. In this way,

$$h_{ij} = \frac{U_{ij}}{S_j} .$$

The thickness of the pile made by all the ith elementary volumes at point O'_1 on the free face will be expressed as follows:

$$h = \sum_{i=1}^{i} \frac{U_{ij}}{S_j} .$$

If the profile of the muck pile is determined approximately, then the following simplified method can be used.

The weighted mean of the size of fragment x_m is mined for a given granule-size composition. Air drag is calculated for this size in accordance with formula (4.7). The trajectory of the center of gravity of an elementary rock volume U_i is plotted with the help of ballistic tables for this factor b_d and the actual values v and θ. In this case, the pile thickness at a given point on the free face is calculated by the method explained in Chapter 1, Sec. 7.

Let us determine the criteria for finding the conditions of explosion under which the retarding effect of the air can be ignored. We shall proceed from the assumption that the accuracy of ballistic calculations does not exceed $\pm 20\%$ of the true flight range of the fragment. Therefore, if air drag reduces the range of flight by about 20% of its range in a vacuum, it can be ignored within the limits of natural errors.

From Fig. 50, it is evident that a reduction in the range of trajectory is equal to 20% of the range D_v in a vacuum, when $\hat{j} = b_d v_0^2/g \leqslant 0.4$, which corresponds to $S_{air} = 0.8$. In this way, if surcharge $\hat{j}$ for a given size x_m of fragment and the actual initial velocity v_0 are less than 0.4, the air drag can be ignored:

$$\hat{j} = \frac{b_d v_0^2}{g} \leqslant 0.4.$$

By substituting b_d in accordance with formula (4.7) in the above expression

$$\hat{j} = \frac{1.3 v_0^2}{x_f \rho g} \leqslant 0.4.$$

From this expression we obtain the following criterion:

$$\frac{v_0^2}{x_f \rho} \leqslant 3. \tag{5.1}$$

From this inequality, we can calculate the maximum size of a fine fraction x_{fine} for which the effect of air drag becomes substantial. From the given volume of rock mass, we can separate the relatively small volume of a fine fraction (the so-called 'fines') which do not have any significant effect while calculating the thickness of the muck pile. Within the limits of given natural errors constituting $\pm 20\%$ of the true values of linear dimensions of the pile, we can separate the volume of a fine fraction which can be ignored while calculating the pile parameters. This volume can be taken to be equal to about 10% of the entire volume of rock blasted.

Let us express the granule-size composition of a rock mass crushed by blasting, according to the cumulative law of distribution $F\dfrac{x}{x_{max}}$, where x is the size at any given instant of time, and x_{max} the maximum size of a fragment in the given rock mass. Function $F\dfrac{x}{x_{max}}$ represents the volume of rock mass measuring less than the original size x in the entire volume of the rock to be blasted. Therefore, quantity x_{fine} found in the condition (5.1) is calculated by the following inequality:

$$F\left(\frac{x_{fine}}{x_{max}}\right) < 0.1. \tag{5.2}$$

In this way, conditions (5.1) and (5.2) determine the criteria for the fulfilment of calculations in regard to the pile parameters without taking air drag into account.

In certain practical cases it is necessary to know only the thickness of the pile for calculating its parameters. In order to consider the effect of air drag on pile width, it is necessary to determine pile width as the range of scattering of the largest fragments in motion, i.e. fragments of size x_{max}. The drag factor b_d is calculated with the known value of x_{max} and the trajectory of scattering of the largest fragments is plotted with the help of ballistic tables (see Appendix). The distance from the recess to the point of intersection of this trajectory with the free face determines the pile width.

2. GRANULE-SIZE COMPOSITION OF ROCK FRAGMENTED THROUGH BLASTING

From the preceding section it follows, that in addition to the initial velocity of projection v_0 and the angle of escape θ_0, it is necessary to know the following major characteristics of the granule-size composition of a rock mass moving in the air for carrying out ballistic calculations: the law of cumulative distribution $F(x/x_{max})$, the maximum size x_{max} and the mean size x_m of the fragments. The following integral law of distribution for a large monolithic mass has been derived from theoretical considerations [38] which are supported by laboratory and commercial trials:

$$F\left(\frac{x}{x_{max}}\right) = \left(\frac{x}{x_{max}}\right)^c.$$

The value of exponent c for deep-hole charges is 1/3. Let us introduce the following parameter:

$$\hat{x}=\frac{x}{x_{\max}},$$

then

$$F(\hat{x})=\hat{x}^c,$$

where $\hat{x}$ is the relative size of a rock fragment.

The mean size of a fragment $\hat{x}_m = x_m/x_{\max}$ is calculated as an expectation value with respect to density distributions:

$$f(\hat{x})=\frac{d}{d\hat{x}}F(\hat{x}),$$

i.e.

$$\hat{x}_m=\int_0^1 f(\hat{x})\,\hat{x}\,d\hat{x}.$$

The distribution function of density $f(\hat{x})$ (differential law) characterizes the size distribution of a rock mass crushed during blasting. There are three types of differential laws of distribution (Fig. 60): 1—with predominantly small, 2—with uniform, and 3—with predominantly large fragments.

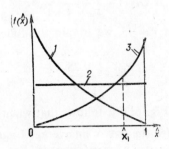

Fig. 60. Three types of differential laws of distribution.

In the graph, the relative size of a fragment $\hat{x}$ is shown along the abscissa and fragments measuring between $\hat{x}$ and $\hat{x}+d\hat{x}$ along the ordinates. From the graph it follows that the area of exit of the large fragments equals the area under the respective curves between sections $\hat{x}=\hat{x}_1$ and $\hat{x}=1$, where $\hat{x}_1$ is the relative size of a large fragment. Evidently, the exit of a large fragment is heavily dependent on the nature of the law of distribution $f(\hat{x})$.

Because

$$f(\hat{x}) = c\hat{x}^{c-1},$$

therefore,

$$\hat{x}_m = \int_0^1 c\hat{x}^{c-1}\,\hat{x}\,d\hat{x} = \frac{c}{c+1}.$$

Therefore, considering that $c = 1/3$ for a monolithic mass,

$$x_m = \frac{c}{c+1}\,x_{max} = \frac{1}{4}\,x_{max}. \tag{5.3}$$

The characteristic of a rock mass in the percentage of exits of large-size fragments most widely used in open-pit mining is:

$$\hat{U}_1 = 1 - F(\hat{x}_1) = 1 - \hat{x}_1^c, \tag{5.4}$$

where $\hat{U}_1$ is the ratio between the volume of large-size fragments and the total volume of the rock mass.

Because, in accordance with (5.3)

$$x_{max} = \frac{c+1}{c}\,x_m,$$

therefore, substituting this expression in equation (5.4), we obtain the following dependence

$$\hat{U}_1 = 1 - \frac{x_1}{x_m} \cdot \frac{c}{c+1}, \tag{5.5}$$

whence, considering that $c = 1/3$, we obtain the following well-established equation [38]:

$$\frac{x_m}{x_1} = \frac{0.25}{(1 - \hat{U}_1)^3}. \tag{5.6}$$

From this equation (Fig. 61), it follows that when the exit of large-size fragments $\hat{U}_1$ is very limited the size of an average fragment for a given rock mass, is a quarter of the size of a large fragment. For average and large-size fissured rocks, as explained in [38], the relationships between U_1, x_1 and x_m are given by the following formulas:

$$U_1 = U_e^+ \left[1 - \left(\frac{x_1}{x_m} \cdot \frac{c}{c+1} \right)^c \right], \tag{5.7}$$

$$\frac{x_m}{x_1} = \frac{0.27}{\left(\dfrac{U_e^+ - U_1}{U_e^+} \right)^2}, \tag{5.8}$$

where U_e^+ is the content of large fissured blocks in the rock mass before

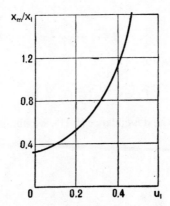

Fig. 61. Dependence of mean-size fragments on the percentage
of exits of large-size fragments.

blasting. For practical calculations and theoretical analysis, it is very important to know the dependences of the exit of large-size fragments on the specific consumption of explosives. The equations obtained in [38], have the following forms:

for a monolithic mass

$$\hat{U}_{exit} = 1 - k_b q; \tag{5.9}$$

for a fissured massif

$$U_1 = U_e^+ (1 - k_b q), \tag{5.10}$$

where k_b is the size coefficient which depends on quantity x_1. The value of k_b is the inverse value of specific consumption q_0.

From expressions (5.9) and (5.10) it is evident that $U_1 = 0$ when $q = q_0$. Consequently, the computed specific consumption q_0 of explosives is of such magnitude that the exit of large-size rocks is either absent or $x_1 = x_{max}$. Quantity q_0 depends on the properties of the rock, size of large fragments x_1 and characteristics of the explosive. Quantity q_0 is calculated by the following empirical expression [39]:

$$\frac{q_{01}}{q_0} = \left(\frac{x}{x_1}\right)^{2/5}, \tag{5.11}$$

where q_{01} is the computed specific consumption of explosives when the standard size of the fragment is x_1.

Considering that the integral law of distribution $F(x)$ and the yield of large-size fragments $\hat{U}_1$ are related as follows: $F(x) = 1 - \hat{U}_1$, we obtain the following equation from expressions (5.9) and (5.10):

120

for a monolithic mass

$$F(\hat{x}) = k_b q;\qquad(5.12)$$

for a fissured massif [39, 40]

$$F(\hat{x}) = 1 - \hat{U}_e^+ (1 - k_b q).\qquad(5.13)$$

By taking into account the dependence (5.11), we obtain the following dependence for monolithic rocks:

$$F(\hat{x}) = C\hat{x}^{2/5},\qquad(5.14)$$

where,

$$C = \frac{1}{q'_{01}\,\hat{x}_1^{2/5}}.$$

3. MECHANISM OF ROCK BREAKAGE THROUGH BLASTING

The mechanism of rock breakage through blasting is highly complicated. It is, therefore, not surprising that till now scientists working on this problem have not been able to come to a uniformly acceptable viewpoint on the mechanism of breaking even in the simplest type of blasting in a monolithic mass.

However, certain conclusions of a general nature in regard to the mechanism of fragmentation under some ideal blasting conditions can be drawn on the basis of a large volume of experimental data on the crushing of rock in the laboratory as well as under real working conditions. Theoretical analysis carried out in this field is also of great significance.

The process of fragmentation by blasting, its progress in time and space close to the center of explosion, and at a distance from the explosion center, depend on the physicomechanical properties of the medium and the shape of the explosive. To a lesser degree, the mechanism of breaking depends on the energy and other similar physicomechanical properties of the explosives. Therefore, for a given explosive and a well-defined shape of charge, the technique of crushing rock by blasting will be determined solely on the basis of the physicomechanical properties of the medium in which the explosion takes place.

Rocks, greatly differing from each other from the viewpoint of propagation of the explosion process and transfer of energy to the medium, can be divided into three classes: monolithic rocks; highly fissured rocks with open fissures filled with air; and fissured rocks in which the fissures are filled with loose soil or water.

Monolithic rock implies rock which is not fissured. Besides, it is presumed that in the case of monolithic rocks, the physicomechanical properties viz. density, acoustical stiffness, Young's modulus, etc., which determine the passage of the compressional wave, are constant at least within that volume where the rock is crushed.

The explosive energy is transferred to the surrounding monolithic medium through a *compressional wave* which is propagated at supersonic speed. The compressional wave in the nearby zone of a few radii of the charge (depending on the specific energy of explosives and acoustical stiffness of the medium) has a shock effect with a steep forward front; in the rear zone, it loses its shock effect and induces a stressed state in the medium with a gradual increase in all parameters beyond its forward front. If blasting takes place near the free face, the compressional wave is deflected and a rarefaction wave spreads in an opposite direction which causes the so-called 'break away' phenomenon. The compressional wave induces a complex stressed state in the medium. Cracks appear in that elementary volume in which the stress exceeds the breaking limit. After the passing of the compressional wave, a number of cracks appear in a certain part which determines the general breaking of the rock. If blasting is carried out on a limited scale then these cracks open and the monolithic block is blown away in the form of fragments of varying sizes and shapes.

From this consideration it follows that in the case of blasting of a charge in monolithic rock, the latter is crushed because of the compressional wave.

A fundamentally different technique is involved in case of blasting of a charge in fissured rock with large fissures filled with air. In such types of rock the compressional wave cannot be propagated because there are numerous fissures in its path. However, instead of a compressional wave a *condensation wave* is propagated in highly fissured rock. One can vividly understand this concept by observing the impact made by a locomotive on a train of which the wagons have a clearance l_{fis} between them. On being pushed by the locomotive the wagons will, one following another, come in close contact with the next wagon and at every such moment there will be an impact between two adjoining wagons. The point of contact between wagons (also point of impact) moves along the rails at a certain mean speed D_w.

An identical process will take place when a charge is blasted in a rock comprising individual fragments. Denoting the average dimensions of a rock fragment by x, the distance between two adjoining fragments (fissure width) by l_{fis}, the bulk velocity of the medium at a given distance from the center of the charge by v and the velocity of sound in the rock fragment by c_{rock} we can write the following equation

$$\frac{l_{fis}}{v} + \frac{x}{c_{rock}} = \frac{l_{fis} + x}{D_w},$$

from which we find the following expression for the velocity of propagation of the condensation wave D_w :

$$D_w = \frac{l_{fis} + x}{vx + l_{fis}\, c_{rock}}\, c_{rock}\, v. \qquad (5.15)$$

After the passage of the condensation wave, the rock fragments, which constitute the exploded mass, come into close contact with each other and the fissure width is sharply reduced. The rock volume covered by the condensation wave is transformed into a conglomerate consisting of individual fragments compactly touching each other. At the moment of passage of the condensation wave this conglomerate gives a rigid push to the fragments located ahead of the wave. As a result of impact, the fragments located ahead of the condensation wave as well as those in the conglomerate, on the surface of which are located the points of impact, will be crushed.

In this way a highly fissured rock massif is crushed due to a rigid push caused at the moment of passage of the condensation wave.

In the rock massif in which fissures are filled with a solid medium, the compressional wave produced due to firing of explosive charges can be propagated to a relatively greater distance. However, the compression wave in fissured rock attenuates more rapidly than in monolithic rock because of increased dissipation. Nevertheless, the compressional wave, passing through individual fragments, breaks them in the same manner as in the case of a monolithic medium.

If the fissures in the rock massif are partly filled with loose soil then in the proximity of the charge, the compressional wave will crush the rock whereas, at a greater distance the rock will be crushed due to passage of the condensation wave. In case of a slightly delayed blasting, the condensation wave, propagated to the depth of the massif, presses the filling material in the fissures, and thereby facilitates propagation of the compressional wave which is produced by the firing of the next row of charges. Obviously, the optimum time-delay will be that interval during which the condensation wave succeeds in covering the distance between two adjacent rows of deep-hole charges.

From the analysis given above, it is clear that there can be two basically different mechanisms of rock crushing depending on the extent of fissures in the massif, width of the fissures and physicomechanical properties of the substance filling the fissures. The first is determined by the compressional wave passing through the massif and the second by an inelastic impact caused by the propagation of a condensation wave. A detailed account of these mechanisms will be given below.

4. MECHANISM OF FRAGMENTATION OF UNBROKEN ROCK

A compressional wave is propagated in the surrounding medium after detonation of a charge. At a distance of a few radii from the charge, the wave pressure is equal to the pressure of the detonation products, which reaches several thousand atmospheres. At such pressure, a plastic flow of the medium takes place as a result of which the medium moves like a fluid on account of the expanding explosion products. With the propagation of the wave the medium is set in motion in radial directions. If detonation takes place near the free face, the medium is moved by the explosion products to the upper half-space after the deflection of the compressional wave.

In a monolithic medium, the blast energy is transferred by the wave and expanding explosion products in the form of elastic and kinetic components. It has been proved through calculations and experimental measurements [25] that the elastic and kinetic components are equal. The question arises, as to which of these components is responsible for crushing the medium. From experimental and theoretical analysis it follows that the rock situated at a given distance from the center of the charge does not start breaking when the compressional wave is propagated through it. Experiments conducted in optically transparent mediums and observations made during working conditions [41–43] show that a given rock starts breaking after a lapse of some time following the passage of the wave. This observation is proved by the fact that the rate of spread of fissures for general fragmentation in a given rock mass is less than the velocity of propagation of the compressional wave. The rate of spread of fissures, depending on the physicomechanical properties of the medium, varies between 30 and 90% of the velocity of sound.

These factors give rise to the opinion that crushing of rock does not depend on the elastic component of the compressional wave and it should not be computed for that state of stress in rock which is caused by this wave.

The following two factors also favour this view on the role of the elastic component of the compressional wave in the process of breaking of the medium.

The medium situated nearer to the zone of blasting behaves as a fluid because of high pressure (about several thousand atmospheres). Therefore, all three principal stresses σ_1, σ_2, σ_3 are equal. According to the theory of elastic potential energy, which conforms to experiments, rock is broken by the potential energy of deformation U_{def} which is proportional to $[(\sigma_1-\sigma_2)^2 + (\sigma_2-\sigma_3)^2+(\sigma_3-\sigma_1)^2]$. In view of equation $\sigma_1=\sigma_2=\sigma_3$, we obtain $U_{def}=0$. From this it follows that the medium in the zone nearer to the point of blast under the effect of the elastic component of the wave as well as pressure of the explosion products, will not be broken.

In the adjacent zone the reduced stress σ_{red} in the wave, according to theory of elastic potential energy, is equal to $(1-2\mu)\ \sigma/\ (1-\mu)$ where σ is the pressure due to the compressional wave. Calculations show that quantity σ_{red} in the next zone is less than the yield point of the rock (at least for rocks of which the coefficient of rigidity $f>5$). As a consequence, rock in the adjacent zone will also not be broken under the effect of the elastic component of the wave.

One could have assumed that on interacting with the incident wave the rarefaction wave coming from the free face causes a 'break-away' phenomenon which is the breaking factor for mediums with high acoustical stiffness. However, experiments and theoretical analyses [42, 44–47] reveal that for these mediums, the breaking is caused by the charge, and the volume of breaking, caused by the 'break-away' phenomenon, constitutes a negligible part of the entire volume of broken rock.

Research carried out on blasting in monolithic mediums proves that the mean size of the largest fragment of the broken mass increases with an increase in the scale of blasting (i.e. with an increase in the direction of the line of least resistance). However, we will not observe such a trend if we assume that the elastic component of the compressional wave is the major factor in crushing rock. In fact, according to the theory of similarity, the reduced stress σ_{red} will be equal at equal relative distances measured in terms of charge radii from the center of the charge. Consequently, for these conditions, the elastic component will be equal in a unit volume of rock, which should break the rock into fragments of equal size, but this conclusion contradicts the results of practical experience.

Calculations show that at the moment when the entire volume of rock is involved in a breaking action, the stress wave, as a consequence of a sharp lag of the front of general breaking of the rock, moves away to a distance which exceeds by several times the size of the rock mass being broken. If, for this moment, we calculate the elastic component of wave energy by integration over the entire volume of the rock being broken, it will be found that it is several times less than the kinetic component for the same volume. From this, it clearly follows that the elastic component is not responsible for breaking the medium. With a view to illustrating the process of breaking of the rock due to the kinetic component of wave energy, let us study a symmetrical explosion when spherical or cylindrical charges are laid in the center of the spherical, or along the axis of cylindrical shaped volumes of rock respectively. A semistatic symmetrical movement of the medium towards the free face is observed after blasting of the charge and deflection of the stress wave from the free face and the expanding gas chamber. The elementary layer of the medium, having a thickness equal to dr and which is at a distance r from the charge, will be in a complex stressed state which is characterized by compressing radial and tensile tangential stresses σ_r and σ_t respectively (Fig. 62). At the same time, this layer will move away radially from the center of the charge with a velocity v_r and expand in a perpendicular direction with a velocity v_t (Fig. 63). Since the velocity of stress increases in a direction toward the center of the charge, this elementary layer of rock will be subjected to a nonuniform stress along its thickness: particles at the lower end of the layer will be displaced faster than particles at the upper end. As a result of this the elementary layer of the rock will be simultaneously subjected to a deformation due

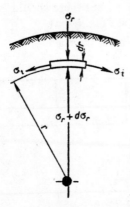

Fig. 62. Diagram of complex stressed state of an elementary
volume of rock on firing of charge.

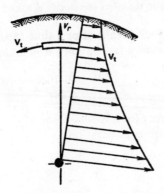

Fig. 63. Velocity field formed in a medium due to firing of a charge.

to shock stress as well as shock shear.

Shock shear implies that a certain imaginary plane rotates at an angle γ_{thrust} from position a–a to b–b (Fig. 64, a) because of a stress velocity gradient. The term 'shock' shear has been used here because the layer of the medium deposited nearer to the charge seemingly shifts with respect to the layer deposited above it due to a nonuniform velocity of stress. This shear simultaneously causes a skew (angular displacement in a fixed plane).

Let us see which of these deformations (strain or shock shear) is more damaging. For this purpose, it is obviously sufficient to determine as to which deformation breaks the rock quicker.

In one case, let us subject a rock having a layer thickness Δr and length l, to a shock stress with velocity v_t (see Fig. 64, b), and in another, to a shock shear with average velocity v_t, but there should be a gradient $\Delta v_t / \Delta r$ which gives rise to a shock shear (see Fig. 64, a).

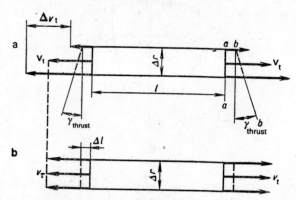

Fig. 64. Two types of deformation of elementary volume of a medium: a—shock shear due to the effect of velocity gradient; b—uniaxial stress.

In the first case, the rock breaks after a time lapse Δt_b, when the relative elongation ε in the elementary layer attains a breaking value ε_b. Because

$$\varepsilon_b = \frac{\Delta l}{l} = \frac{v_t \Delta t_b}{2\pi r}, \tag{5.16}$$

where l is the circumference for a radius r, therefore, by taking the following kinematic dependence into consideration

$$v_t = 2\pi v_r, \tag{5.17}$$

we obtain

$$\varepsilon_b = \frac{v_r \Delta t_b}{r}.$$

We know that

$$\varepsilon_b = \frac{\sigma}{E},$$

where σ is the limiting strength of the rock for shock stress; and E the modulus of elasticity. Therefore,

$$\Delta t_b = \frac{\sigma}{E} \cdot \frac{r}{v_r}. \tag{5.18}$$

In the second case, rock is broken when the relative shear γ (deviation of plane from position a–a to b–b) attains the breaking value γ_{thrust}. From Fig. 64, a, it is clear that

$$\gamma_{\text{thrust}} = \frac{\Delta v_{\text{t}} \, \Delta t_{\text{sh}}}{\Delta r},$$

where Δt_{sh} is the time during which an elementary layer of rock will be broken under the effect of shock shear deformation.

Because, for shear

$$\gamma_{\text{thrust}} = \frac{\tau_{\text{b}}}{G_{\text{sh}}},$$

where τ_{b} is the tangential stress at which the rock breaks and G_{sh} the shear modulus.

Therefore, by using the known dependences:

$$G_{\text{sh}} = \frac{E}{2\,(1+\mu)};$$ (5.19)

$$\tau_{\text{b}} = \sigma,$$ (5.20)

where μ is the Poisson ratio, we obtain:

$$\Delta t_{\text{sh}} \frac{2\,(1+\mu)\,\sigma}{E \frac{dv_{\text{t}}}{dr}}.$$ (5.21)

Differentiating Eq. (5.17) with respect to r, we obtain

$$\frac{dv_{\text{t}}}{dr} = 2\pi \frac{dv_r}{dr}.$$

Substituting it in Eq. (5.21), we have

$$\Delta t_{\text{sh}} = \frac{(1+\mu)\,\sigma}{\pi E \frac{dv_r}{dr}}.$$ (5.22)

The ratio $\Delta t_{\text{b}} : \Delta t_{\text{sh}}$ is obtained from expressions (5.18) and (5.22):

$$\frac{\Delta t_{\text{b}}}{\Delta t_{\text{sh}}} = \frac{\pi}{1+\mu} \cdot \frac{r}{v_r} \cdot \frac{dv_r}{dr}.$$ (5.23)

The law of variation of v_r with distance r can be approximated with the equation

$$v_r = v_0 \left(\frac{r_0}{r}\right)^n,$$ (5.24)

where v_0 is the initial velocity of the rock at the boundary adjoining the gas hole; n the exponent which is dependent on the type of explosion symmetry and physicomechanical properties of the rock and r_0 the radius of the charge.

By differentiating we can obtain an absolute value

$$\left|\frac{dv_r}{dr}\right| = n\,\frac{v_0}{r^{n+1}}\,r_0^n. \tag{5.25}$$

Substituting formulas (5.24) and (5.25) in expression (5.23), we have

$$\frac{\Delta t_b}{\Delta t_{sh}} = \frac{n\pi}{1+\mu}.$$

Reference values of n can be taken to be equal to 2 and 1 respectively for spherical and cylindrical symmetry. The value of μ in comparison with unity can also be ignored for calculations of expectation values. We then find that the ratio $\Delta t_b : \Delta t_{sh}$ for spherical and cylindrical explosions is equal to 2π and π respectively. From this it is evident that rock is broken more readily due to the strain caused by shock shear than from the deformation due to tensile stress. It can be proved by a simple calculation that the ratio between the potential energy due to stress and potential energy due to shear at the moment of breaking of rock, because of shock shear strain, is equal to $(1+\mu)/2\,(n\pi)^2$. For cylindrical $(n=1)$ and spherical $(n=2)$ symmetries these ratios are equal to 1/16 and 1/64 respectively. Consequently, shear energy can be ignored in calculations for the crushing of rock.

From these observations it follows that the process of breaking monolithic medium by blasting of concentrated and deep-hole charges is based on shock shear caused by a radial velocity gradient which appears in the form of skewing for the rock being broken. The velocity gradient is absent for ideal slab charges in cases of established quasi-stationary motion. Consequently, shock shear strain will also be absent. Therefore, ideal slab charges (explosive plates) in comparison with concentrated and deep-hole charges break rock less effectively. This factor partly explains the reason for deterioration in the quality of massif fragmentation with a reduction in the distance between the holes (when the specific consumption of explosives remains constant). In the given instance, the row of deep-hole charges is transformed into an ideal slab charge with the distance reduced to zero. In the case of blasting of ideal slab charges rock is crushed only under the effect of certain compressive stresses (in the nearer zone) and the 'breaking away' effects (in the farther zone) which are developed when the compressional wave is deflected from the free face. The breaking effect of ideal slab charges is not analyzed in the present book.

In the process of rock fragmentation by blasting of concentrated and deep-hole charges the compressional wave causes instant movement of the medium which gives rise to shock shear due to a radial velocity gradient. Shock shear is the major factor in the crushing of rocks.

The theory of fragmentation presented in [68] is also based on this process. However, an error was committed in formulating the basic energy equation. The reason for this error will be explained in the following paragraph.

Only breaking of friable rock was considered when carrying out the calculations given above. It was also assumed that Hooke's law is applicable up to the limiting strength of the rock. In reality, such simple laws do not hold good. It is well known that some rocks have an elastoplastic (viscous) nature of breaking which, in contrast with the brittle type of breaking, has a fluidity area in the 'stress-deformation' diagram. It will be explained in the following article that the conditions given above are immaterial if the calculations for rock fragmentation are based on the law of conservation of energy.

5. LAWS OF FRAGMENTATION OF UNBROKEN ROCK

Let us calculate the work done u_1, required in the fragmentation of a unit volume of rock. The dimensions of the fragments are subject to a localized law of distribution $f_{loc}(\hat{x}_{f.loc})$. This law is the quantitative characteristic of the situation that the fragments in the given volume of rock at a given distance r from the center of the charge vary in dimensions from a negligible to a maximum size $x_{r\ max}$. This kind of diversity in rock fragments is subject to a certain localized differential law of distribution $f_{loc}(\hat{x}_{f.loc})$ (depending, in general, on the distance r from the charge). This law indicates how much of the total crushed volume of rock is contained in a unit volume of fragments. If, for example, it is necessary to determine the volume of flyrock in the interval between x_f and $x_f + dx_f$, then $dU = U f_{loc}(\hat{x}_{f.loc})\, d\hat{x}_{f.loc}$ where U is the total volume of crushed rock at a distance r from the charge.

If S_{Σ} represents the surface area of all fragments in a unit volume of broken rock; S_{crack} the surface area of microfissures in the same volume; and e the specific energy required in forming a unit area of the re-formed surface, then the work done u_1 is calculated by the following product:

$$u_1 = (S_{\Sigma} - S_{crack})\, e. \tag{5.26}$$

Area S_f of a rock fragment of an average size x_f is expressed by the following dependence:

$$S_f = 6 k_{f.f}\, x_f^2,$$

where $k_{f.f}$ is the form factor which takes into account the deviation of the actual shape of the fragment from the ideal cubic shape.

The number of rock fragments of sizes between x_f and $x_f + dx_f$ in a unit volume of rock will be

$$dN = \frac{k_{loc}^3}{x_f^3}\, f(\hat{x}_{f.loc})\, d\hat{x}_{f.loc}, \tag{5.27}$$

where $\hat{x}_{f.loc} = x_f / x_{r\ max}$ is the relative size of a rock fragment, and k_{loc} the ratio between the maximum and actual average sizes of the fragments [35].

The surface area of these fragments will be

$$dS_\Sigma = S_f\, dN = 6k_{f.f}\, \frac{k_{loc}^3}{x_f}\, f(\hat{x}_{f.loc})\, d\hat{x}_{f.loc}.$$

In future calculations we shall bear in mind that, in accordance with [35]

$$k_{f.f} = \frac{2}{k_{loc}^2}.$$

Quantity k_{loc} may be taken to be equal to 1.3.

The total surface area of all pieces in a unit volume will be:

$$S_\Sigma = \frac{6k_{loc}}{x_{r\,max}} \int_0^1 \frac{f_{loc}(\hat{x}_{f.loc})\, d\hat{x}_{f.loc}}{\hat{x}_{f.loc}}, \qquad (5.28)$$

Let us calculate the area of microfissures in the same way:

$$S_{crack} = \frac{6k_{loc}}{x_{crack\,max}} \int_0^1 \frac{f(\hat{x}_{crack})\, d\hat{x}_{crack}}{\hat{x}_{crack}}, \qquad (5.29)$$

where $f(\hat{x}_{crack})$ is the differential law of distribution of natural fragmentation in the massif formed by the system of microfissures, $x_{crack\,max}$ the size of the largest natural fragment; and, $\hat{x}_{crack} = x_{crack}/x_{crack\,max}$ the relative size of a natural fragment.

From expressions (5.26) to (5.29) we find:

$$u_1 = 6k_{loc}\, e \left[\frac{1}{x_{r\,max}} \int_0^1 \frac{f_{loc}(\hat{x}_{f.loc})\, d\hat{x}_{f.loc}}{\hat{x}_{t.loc}} - \frac{1}{x_{crack\,max}} \int_0^1 \frac{f(\hat{x}_{crack})\, d\hat{x}_{crack}}{\hat{x}_{crack}} \right].$$

In a particular case $S_{crack} = 0$ (monolithic rock without microfissures), so that

$$u_1 = \frac{6k_{loc}\, e}{x_{r\,max}} \int_0^1 \frac{f_{loc}(\hat{x}_{f.loc})\, d\hat{x}_{f.loc}}{\hat{x}_{f.loc}}. \qquad (5.30)$$

Let us calculate the kinetic energy E_{f1} which is accumulated due to shock shear in a unit volume of rock and represents the work done in a break operation u_1.

For this purpose let us consider a rock fragment in the form of a cube in which the lower surface is static and the upper surface moves at a velocity v (Fig. 65). As a result, a tensile stress (σ_1) and compressive stress (σ_2) develop along the diagonals of the cube. In view of the smallness of size x_f let us assume that all layers of the rock between the upper and lower surfaces move at a velocity which varies linearly. The kinetic energy dE_f of an elementary layer of rock of thickness dz at a distance z from the cube base will be:

$$dE_f = \frac{1}{2}\,\rho x_f^2 \left(\frac{v}{x_f}\,z\right)^2 dz.$$

In this case the ratio v/x_f represents the velocity gradient of stress dv_t/dr, which can be determined after differentiating expression (5.17):

$$\frac{dv_t}{dr} = 2\pi\,\frac{dv_r}{dr}\,. \qquad (5.31)$$

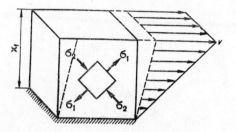

Fig. 65. Diagram showing shock shear.

By substituting this value of dv_t/dr in the expression for dE_f in place of gradient v/x_f we obtain

$$dE_f = 2\pi^2 \rho x_f^2 \left(\frac{dv_r}{dr}\,z\right)^2 dz.$$

By integrating with respect to z in the limits between O to x_f we find the kinetic energy of shock shear in a rock equal in volume to that of a fragment of size x_f,

$$E_f = \frac{2\pi^2 \rho}{3} \left(\frac{dv_r}{dr}\right)^2 x_f^5. \qquad (5.32)$$

Since the number of fragments dN of sizes between x_f and $x_f + dx_f$ in a unit volume of rock is determined by expression (5.27), the kinetic energy of a unit volume E_{f1} in case of shock shear will be

$$E_{f1} = \int\limits_0^{x_{r\,max}} \frac{2\pi^2 \rho}{3} \left(\frac{dv_r}{dr}\right)^2 x_f^5\,\frac{k_{loc}^3 f_{loc}(\hat{x}_{f.loc})\,d\hat{x}_{f.loc}}{x_f^3}$$

$$= \frac{2\pi^2 \rho k_{loc}^3}{3}\,x_{r\,max}^2 \left(\frac{dv_r}{dr}\right)^2 \int\limits_0^1 \hat{x}_{f.loc}^2\, f_{loc}(\hat{x}_{f.loc})\,d\hat{x}_{f.loc}. \qquad (5.33)$$

This energy represents the work done for breaking the rock. Equating the right-hand sides of expressions (5.26) and (5.33), we get the basic energy equation of crushing

$$E_f = u_1. \qquad (5.34)$$

From this equation, we find the relation between the maximum size of a fragment $x_{r\,max}$ and the gradient of radial velocity dv_r/dr in the case of rocks which do not have microfissures:

$$x_{r\,max} = \frac{B_1}{\left(\dfrac{dv_r}{dr}\right)^{2/3}}, \qquad (5.35)$$

where

$$B_1 = \left[\frac{9ek_{loc}^2}{\pi^2 \rho}\, d_1\right]^{1/3}; \qquad (5.36)$$

$$d_1 = \frac{\displaystyle\int_0^1 \frac{f_{loc}\,(\hat{x}_{f.loc})}{\hat{x}_{f.loc}}\, d\hat{x}_{f.loc}}{\displaystyle\int_2^0 \hat{x}_{f.loc}^2\, f_{loc}\,(\hat{x}_{f.loc})\, d\hat{x}_{f.loc}}. \qquad (5.37)$$

Some scientists [68] equate kinetic energy E_{f1} accumulated in a unit volume of rock on account of shock shear with $0.5\,\rho v_{cr}^2$, where v_{cr} is the critical velocity for breaking the rock. However, this is not correct because the critical velocity v_{cr} is inversely proportional to $x_f^{1/2}$ (see Sec. 11).

Quantity e, entering in expression (5.26), can be expressed by the following equation:

$$e = \beta_b\, \frac{\sigma^2}{2E}, \qquad (5.38)$$

where coefficient β_b, with a linear dimension, takes into account the explosion symmetry, the dynamics of application of the load and also the nature of brittle or elastoplastic breaking of the given rock.

Since, in the case of blasting of a spherical charge, a rock expands in two mutually perpendicular directions, therefore, other conditions being equal, coefficient β_b for a spherical charge is only half of that for a cylindrical charge. Initial velocity v_0 used in expression (5.24) is calculated by the following known formula:

$$v_0 = \frac{\Sigma_0}{\rho D_w}, $$

where Σ_0 is the initial pressure of explosion products; and D_w the velocity of the shock wave front.

According to [25] D_w is approximately equal to

$$D_w = k_{w1}\, c_{rock},$$

where c_{rock} is the velocity of sound in the rock; and k_{w1} the proportionality factor greater than unity, therefore:

$$v_0 = \frac{\Sigma_0}{\rho k_{w1} c_{rock}}.$$

Let us express initial pressure Σ_0 in terms of specific energy e_{exp} and density ρ_{exp}:

$$\Sigma_0 = (\varkappa - 1)\, e_{exp}\, \rho_{exp}.$$

In this way

$$v_0 = \frac{(\varkappa - 1)\, e_{exp}\, \rho_{exp}}{\rho k_{w1} c_{rock}}. \tag{5.39}$$

By substituting this expression in formula (5.25) we obtain:

$$\left| \frac{dv_r}{dr} \right| = \frac{n(\varkappa - 1)\, e_{exp}\, \rho_{exp}}{\rho k_{w1} c_{rock}\, r^{n+1}}\, r_0^n. \tag{5.40}$$

Let us express radius r_0 of the charge in terms of the relative loading density $\varDelta$ and density of explosive ρ_{exp}. Calculations give the value of r_0 as follows:

for a cylindrical charge

$$r_0 = \sqrt{\frac{p}{\pi\, \varDelta \rho_{exp}}};$$

for a spherical charge

$$r_0 = \sqrt[3]{\frac{3 Q_{sp}}{4\pi\, \varDelta \rho_{exp}}},$$

where Q_{sp} is the mass of the spherical charge.

For the sake of convenience in future calculations, let us express the dependences obtained above in the following form:
for a cylindrical charge:

$$r_0 = \sqrt{\frac{W^2}{\varDelta \pi \rho_{exp}}} \cdot \frac{\sqrt{p}}{W}; \tag{5.41}$$

for a spherical charge:

$$r_0 = \sqrt[3]{\frac{3 W^3}{4\pi\, \varDelta \rho_{exp}}} \cdot \frac{\sqrt[3]{Q_{sp}}}{W}. \tag{5.42}$$

Taking into account that $c_{rock} = (E/\rho)^{1/2}$, we find from expressions (5.24), (5.35), (5.40) and (5.41) the size dependence of the largest fragment $x_{r\,max}$ at a given distance r from the charges:

for a cylindrical charge ($n = 1$)

$$x_{r\,max} = \psi_{cy}\,\omega_{cy}\,\frac{r^{4/3}}{\left(\dfrac{p^{1/2}}{W}\right)^{2/3} W^{2/3}};\qquad (5.43)$$

for a spherical charge ($n = 2$)

$$x_{r\,max} = \psi_{sp}\,\omega_{sp}\,\frac{r^2}{\left(\dfrac{Q_{sp}^{1/3}}{W}\right)^{4/3} W^{4/3}},\qquad (5.44)$$

where

$$\left.\begin{aligned}
\psi_{cy} &= \left[\frac{4.5\,k_{w1}^2\,\beta_b\,k_{loc}^2}{\pi\,(\varkappa - 1)^2}\,d_1\right]^{1/3};\\[6pt]
\psi_{sp} &= \left[\frac{1.6\,k_{w1}^2\,\beta_b\,k_{loc}^2}{\pi^{2/3}\,(\varkappa - 1)^2}\,d_1\right]^{1/3};\\[6pt]
\omega_{cy} &= \left[\frac{\sigma}{e_{exp}}\right]^{2/3}\cdot\frac{\varDelta^{1/3}}{\rho_{exp}^{1/3}};\\[6pt]
\omega_{sp} &= \left[\frac{\sigma}{e_{exp}}\right]^{2/3}\cdot\frac{\varDelta^{4/9}}{\rho_{exp}^{2/9}}.
\end{aligned}\right\}\qquad (5.45)$$

The largest fragments $x_{f\,max}$ in the entire massif are obtained at distance $r = W$. Substituting $r = W$ in formulas (5.43) and (5.44) we obtain:

$$x_{cy\,max} = \psi_{cy}\,\omega_{cy}\,\frac{W^{2/3}}{\left(\dfrac{p^{1/2}}{W}\right)^{2/3}};\qquad (5.46)$$

$$x_{sp\,max} = \psi_{sp}\,\omega_{sp}\,\frac{W^{2/3}}{\left(\dfrac{Q_{sp}^{1/3}}{W}\right)^{4/3}}.\qquad (5.47)$$

Since the mass per unit length of a cylindrical charge of diameter d is:

$$p = \frac{1}{4}\,\pi d^2 \rho_{exp}\,\varDelta,\qquad (5.48)$$

therefore,

$$x_{cy} = \psi_{cy}\,\omega_{cy}\left[\frac{4}{\pi\rho_{exp}\,\varDelta}\right]^{1/3}\frac{W^{4/3}}{d^{2/3}}.\qquad (5.49)$$

Results of an experimental verification of the obtained design formulas are given in Table 6. The value of x_{max} is calculated by formulas (5.46) and (5.47) and actual sizes of the largest rock fragments are determined.

It must be borne in mind that this verification is of an approximate nature because the material used during experiments did not fully conform to the

135

TABLE 6

Experimental parameters

Medium	W, m	p, kg/m	q, kg/m³	Q_{sp}, kg	$\rho \times 10^{-3}$, kg/m³	$\sigma \times 10^{-4}$, daN/cm²	$e_{exp} \times 10^{-6}$, J/kg	$x_{max\,b}$, m	$x_{max\,ff}$, m	Reference
Coal	0.08	—	—	1.6×10^{-3}	1.5	15	5	0.20	0.19	[60]
Concrete	0.13	—	—	1.5×10^{-3}	1	50	4.3	0.09	0.10	[44]
Basalt	0.13	—	—	2×10^{-2}	1	180	4.3	0.06	0.05	[99]
Sand-cement mixture	0.13	—	—	2×10^{-3}	1.5	27	5.2	0.05	0.05	[99]
Concrete	0.15	—	—	2.5×10^{-3}	1.5	14	5.6	0.06	0.05	[100]
”	0.18	—	—	5×10^{-3}	1	38	5.6	0.05	0.09	[57]
”	0.26	—	—	5×10^{-2}	1	50	4.3	0.09	0.10	[60]
Dolomite	0.40	—	—	0.6	1	340	4.3	0.25	0.28	[99]
Concrete	0.65	—	—	0.2	1	40	3.7	0.29	0.45	[101]
Quartzite	1.00	1.67	—	—	1	170	4.3	0.43	0.30	[58]
”	7.50	0.46	—	—	1	100	4.3	2.1	2.00	[102]
”	8.00	—	0.37	—	1	100	4.3	1.8	1.50	[95]
Granite	8.00	0.75	—	—	1	100	4.3	1.4	1.50	[103]
Limestone	60.00	—	—	5.9×10^{-5}	1	130	4.3	5.0	4.00	—

conditions of spherical and cylindrical symmetry for which we obtained formulas (5.46) and (5.47). Besides, in a majority of cases the limiting strength of the rock (for blasting) was not known. The approximate value of the limiting strength for compression was taken to be 0.1.

A charge was considered to be cylindrical (elongated) if the length l, diameter d and the value W satisfy the condition $l/W > 1$ and $l/d > 30$.

By comparing the figures given in columns 9 and 10 of Table 6, it is clear that the calculation formulas give highly satisfactory results when the explosion depth varies between 0.08 m and 60 m, i.e. almost 1000 times. It should be noted that formulas (5.46) and (5.47) give the value of x_{max} at the end of a general breaking of rock. The fragments are further crushed by the kinetic energy accumulated during blasting when the flyrock falls on the free face, especially if it is hard rock. The values of coefficients ψ_{cy} and ψ_{sp} in formulas (5.46) and (5.47) were calculated on the assumption that the localized law of distribution $f_{loc}(\hat{x}_{f.loc})$ conforms to the Rozin-Rammler type of distribution

$$f_{loc}(\hat{x}_{f.loc}) = A_1 \hat{x}_{f.loc}^{n_1-1} \times e^{-b1}(\hat{x}_{f.loc} - a_1) \, n_1 \text{ when } n_1 = 2, \, b_1 = 5, \, a_1 = 0.5 \text{ and } A_1 =$$

2.8 are the most probable values. Coefficient d_1 used in expression (5.45) is equal to 4.4 for these conditions; the remaining coefficients are: $\varkappa = 2.5$; $k_{w1} = 1.25$; $\beta_b = 1$ (cylinder); $\beta_b = \frac{1}{2}$ (sphere); and $k_{f,f} = 1.4$. Calculations made by formulas (5.45) give the value of $\psi_{cy} = 1.8$ and $\psi_{sp} = 1.15$. More precise value of these coefficients for a given rock can be obtained only by means of experiments.

6. FRAGMENTATION OF UNBROKEN ROCK ON FIRING OF CONVERGING DEEP-HOLE CHARGES

Theoretical and experimental analyses [48–55] carried out during recent years reveal that the average size of a rock fragment in the massif obtained as a result of blasting is noticeably reduced by using a pair of converging deep-hole charges in monolithic and slightly fissured rocks.

The essence of the converging charges technique is that if one deep-hole charge of diameter d_1 is replaced by two charges of smaller diameter, but with an equal total mass of explosives in them and these charges are placed parallel to each other at a distance of four-to-six diameters, the rock fragments obtained by their explosion are smaller in size than those by using a single deep-hole charge (Fig. 66, a and f).

The reason for obtaining smaller fragments in converging deep-hole charge blasting lies in the fact that, in accordance with formula (5.49), the size of the largest rock fragment when $W = \text{const}$, is reduced with an increase in the diameter of the cylindrical charge. However, only the peripheral part (about 10–15%) of the entire deep-hole charge is used in the development of the initial velocity field which is responsible for shock shear and breaking; the

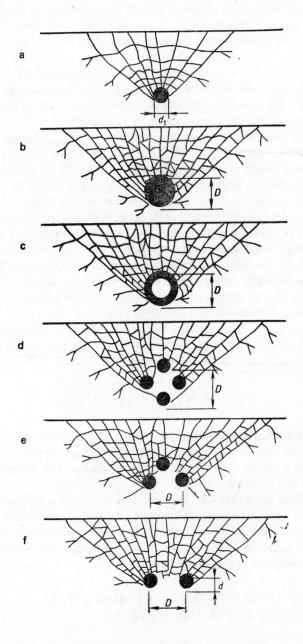

Fig. 66. Diagram showing fragmentation of rock on firing single and converging charges of equal masses.

remaining part of the charge energy (85–90%) is expended in the flight of the rock and is partially retained in the explosion products in the form of heat. Therefore, if the middle part of the deep-hole charge is removed and replaced by a monolithic inert medium then the remaining cylindrical charges (see Fig. 66, c) will break the same volume of rock as is crushed by a continuous deep-hole charge of the same diameter (Fig. 66, b). The effect will be identical if instead of a hollow cylindrical charge we use deep-hole charges of an equal total mass covering approximately the same area (Fig. 66, d). The diameter of this circle should be equal to the diameter of the hollow cylindrical charge. Other variants of the charges are: charges converging in a half-circle (Fig. 66, e) and charges converging in pairs (Fig. 66, f). In all these variants the effective distance between the deep-hole charges should not exceed six times the charge diameter because at greater distances the dissipation of energy is considerable (mainly for breaking the medium) and the charges cease to interact. The physical significance of this factor is that the shock wave emanating from any charge, and being propagated in the medium in which a compressional wave has been produced by the prior firing of another charge, does not lose its energy during further breaking of rock fragments. The reason for this is that at very high pressures rock behaves like a fluid.

When using the technique of converging deep-hole charges the size of the largest fragment can be calculated by formula (5.49) by replacing the hole diameter d with a circle of diameter D at which converging deep-hole charges are laid:

$$x_{max} = \psi_{cy}\, \omega_{cy} \left(\frac{4}{\pi \rho_{exp}\, \Delta} \right)^{1/3} \times \frac{W^{4/3}}{D^{2/3}}. \qquad (5.50)$$

It has been experimentally proved that the distance between two converging charges or between two neighboring charges (in the case of a set of holes) should be equal to 3.5 times the charge diameter d. Fragmentation is optimum at this distance.

Calculations show that the size of crushed fragments is reduced by approximately two times using the technique of converging deep-hole charges in comparison with the ordinary method. A similar reduction in size of fragments can be achieved by the usual method of blasting only by increasing the specific consumption of explosives by 6–8 times.

7. LAWS OF DISTRIBUTION FOR SYMMETRICAL BLASTING IN UNBROKEN ROCK

Let us calculate the integral law of distribution of fragments when a cylindrical charge is placed along the axis of the cylindrical volume of a rock. Function $F(x_f)$, which expresses an integral law, indicates that part of the entire volume of blasted rock which belongs to the fractions with fragments measuring between 0 and x_f.

According to this definition, $F(x_f)$ can be written as the following relationship:

$$F(x_f) = \frac{U}{U_0}, \qquad (5.51)$$

where U is the volume of rock fragments measuring between 0 and x_f, and U_0 the total volume of the broken rock.

While calculating the granule-size composition of the fragmented rock it is convenient to use not the absolute size of a fragment x_f but its relative size

$$\hat{x}_f = \frac{x_f}{x_{f\,max}}, \qquad (5.52)$$

where $x_{f\,max}$ is the size of the largest fragment determined by expression (5.46).

In the same way we can find the relative sizes $\hat{x}_{r\,max}$ by dividing the right- and left-hand sides of expressions (5.43) and (5.46), (5.44) and (5.47) by one another:

for a cylindrical charge

$$\hat{x}_{r\,max} = \hat{r}^{4/3}; \qquad (5.53)$$

for a spherical charge

$$\hat{x}_{r\,max} = \hat{r}^{2}, \qquad (5.54)$$

where

$$\hat{r} = \frac{r}{W}. \qquad (5.55)$$

While calculating the volume U of crushed rock with fragments measuring between 0 and x_f it is necessary to bear in mind that fragments of size x_f are formed at a distance r_{x_f} for which the quantity $x_{r\,max}$, determined by expression (5.43), will be greater than that for the given size x_f. Therefore, while calculating the volume U it is necessary to use a double integral

$$U = \int_0^{\hat{x}_{f.loc}} \int_{r_{x_f}}^{W} 2\pi r \, dr f_{loc}(x_{f.loc}) \, d\hat{x}_{f.loc}. \qquad (5.56)$$

Since the volume U_0 of a cylinder with radius W and height of unit length is given by

$$U_0 = \pi W^2,$$

therefore, by dividing the right- and left-hand sides of expression (5.56) respectively by U_0 and πW^2, we have

$$F(\hat{x}_f) = \int\limits_0^{\hat{x}_f} \int\limits_{\hat{r}_{x_f}}^1 2\, r\, f_{loc}(\hat{x}_{f.loc})\, d\hat{r}\, d\hat{x}_{f.loc}.$$

Quantity $\hat{x}_{f.loc}$ used in this expression, according to the definition given above, is equal to

$$\hat{x}_{f.loc} = \frac{x_f}{x_{r\,max}}.$$

Dividing the numerator and denominator by $x_{f\,max}$ according to expression (5.46), we obtain:

$$\hat{x}_{f.loc} = \frac{\hat{x}_f}{\hat{x}_{r\,max}},$$

so that

$$d\hat{x}_{f.loc} = \frac{d\hat{x}_{f.loc}}{\hat{x}_{r\,max}}.$$

By substituting these expressions and equation (5.53) in the double integral given above, we obtain the integral law of distribution

$$F(\hat{x}_f) = \int\limits_0^{\hat{x}_f} \int\limits_{\hat{r}_{x_f}}^1 \frac{2}{\hat{r}^{1/3}} f_{loc}\left(\frac{\hat{x}_f}{\hat{x}_{r\,max}}\right) d\hat{r}\, d\hat{x}_f. \qquad (5.57)$$

The lower limit in this integral is calculated from expression (5.53):

$$\hat{r}_{x_f} = \hat{x}_{r\,max}^{3/4}.$$

The differential law of distribution $f(x_f)$ is obtained from the expression (5.57) by considering that $f(\hat{x}_f) = dF(\hat{x}_f)/d\hat{x}_f$.

$$f(\hat{x}_f) = \int\limits_{\hat{x}_f^{3/4}}^1 \frac{1}{\hat{r}^{1/3}} f_{loc}\left(\frac{\hat{x}_f}{\hat{r}^{3/4}}\right) d\hat{r}. \qquad (5.58)$$

This law gives the density of distribution of rock in fractions and shows which part of the crushed rock belongs to the fraction measuring $\hat{x}_f$ per unit volume. The localized law of distribution $f_{loc}(\hat{x}_{f.loc})$ used in expressions (5.57) and (5.58) depends on the physicomechanical properties of the rock and is determined by the nature of distribution of discontinuities and microfissures in the medium which is being fragmented. Since the latter condition is not governed by definite laws and is of an incidental nature, the localized law of distribution can be determined only through experiments. For this purpose, the method of concentric layers of colored concrete is used [56]. The local law of distribution obtained through experiments can be approximated by the Rozin-Rammler distribution:

$$f_{loc}(\hat{x}_{f.loc}) = A_1 \, \hat{x}_{f.loc}^{n_1-1} e^{-b_1 \, (\hat{x}_{f.\,loc} - a_1) \, n_1}$$

or by the following power dependence

$$f_{loc}(\hat{x}_{f.loc}) = A_2 \, \hat{x}_{f.loc}^{n_2},$$

which has a physical significance only when $n_2 > 0$ and $n_2 \neq \frac{1}{2}$.

In these expressions, the coefficients A_1 and A_2 are normalization constants. Their value is determined from a known property of the differential law of distribution:

$$A \int_0^1 \hat{x}_{f.loc} \, f_{loc}(\hat{x}_{f.loc}) \, d\hat{x}_{f.loc} = 1.$$

If in the Rozin-Rammler distribution, we take $n_1 = 2$, then

$$A_1 = \left\{ \frac{1}{2b_1} \, e^{-b_1 \, (1-a_1)^2} \left[1 - e^{-b_1 \, (1-2a_1)} \right] \right.$$

$$\left. + \frac{a_1}{\sqrt{b_1}} \left[\int_0^{\sqrt{b_1} \, (1-a_1)} e^{-z^2} \, dz - \int_0^{-a_1 \sqrt{b_1}} e^{-z^2} \, dz \right] \right\}^{-1}.$$

Integration of the power dependence according to expression (5.57) results in:

$$f(\hat{x}_f) = 3 \, \frac{n_2+1}{2n_2-1} \left[\hat{x}_f^{1/2} - \hat{x}_f^{n_2} \right], \tag{5.59}$$

where it is considered that $A_2 = n_2 + 1$.

For ballistic calculations it is very important to know the expectation value of the fragment size $M(\hat{x}_f)$.

As we know:

$$M\left(\hat{x}_f\right) = \int\limits_0^1 \hat{x}_f f(\hat{x}_f)\, d\hat{x}_f.$$

The calculations show that for the law of distribution

$$f(\hat{x}_f) = 3\,\frac{n_2+1}{2n_2-1}\left[\hat{x}_f^{1/2} - \hat{x}_f^{n_2}\right];$$

the expected value is given by

$$M\left[\hat{x}_f\right] = \frac{3\,(n_2+1)}{2n_2-1}\left[\frac{2}{5} - \frac{1}{n_2+2}\right].$$

When n_2 is taken to be equal to 0; 0.25; 1; 2 and ∞, the expectation values $M\left(\hat{x}_f\right)$ will be 0.3; 0.333; 0.4; 0.45 and 0.6 respectively. It should be noted that for $n_2=0$ the local differential law of distribution $f_{loc}\left(\hat{x}_f\right) = A_2 x_f^{n_2}$ is the law of equal probability. However, because the exponential law $f_{loc}\left(\hat{x}_{f.loc}\right)$ has a physical significance only when $n>0$ and $n\neq\frac{1}{2}$, therefore, the law of equal probability, in fact, does not hold. Numerical integration of the Rozin-Rammler law of distribution for the most probable values $b_1=5$; $a_1=0.5$ and $A_2=2.8$ gives an expectation value of $M\left[\hat{x}_f\right]=0.6$. This means that in case of symmetrical explosions the average size of fragments is 60% of the maximum value of $\hat{x}_{cy\ max}$ for the Rozin-Rammler law of localized distribution.

Calculations show that the integral and differential laws of distribution of fragments for a spherical charge are identical to those for a cylindrical charge.

The following conclusions flow from the analysis of the law of crushing discussed above:

For calculating the differential law of distribution of fragments (5.58), it is necessary to know the localized law which can be found by theoretical calculations. The easiest way of finding the differential law of distribution is the experimental method. For this purpose it is necessary to carry out small-scale blasting in the same rock in which an actual explosion is to be conducted and a geometric similarity of charges and volume of rock to be blasted should be strictly observed. The explosives used in experimental and actual explosions should be the same. It is essential to prevent additional fragmentation of the rock due to kinetic energy accumulated during blasting. The obtained differential law of distribution will conform to the law of distribution of an actual explosion at the final moment of general breaking.

From formulas (5.46) and (5.47), it follows that the size of the largest fragment in the case of an actual explosion will be equal to the model size multiplied by the coefficient of geometric similarity raised to the power 2/3.

8. CRUSHING OPERATIONS IN UNBROKEN ROCK

Let us calculate the efficiency for crushing rock. Here, efficiency implies the ratio between the work done in blasting, which is used directly for rock breaking and the total blast energy.

The work done in breaking rock to fragments measuring between x_f and $x_f + dx_f$ due to blasting of a cylindrical charge, is equal to:

$$dE_1 = 6\,\frac{x_f^2}{k_{loc}^2}\,e\,dN,$$

where dN is the number of fragments belonging to a given fraction in a unit volume; and E_1 the total energy required in crushing a unit volume of rock.

Because the volume of all fragments of this fraction in a unit volume is $dU = f(\hat{x}_f)\,\hat{dx}_f$ and the volume of a single rock fragment is equal to x_f^3/k_{loc}^3 therefore,

$$dN = \frac{f(\hat{x}_f)\,\hat{dx}_f\,k_{loc}^3}{x_f^3}.$$

Substituting this value of dN in the expression for dE_1, we get:

$$dE_1 = \frac{6k_{loc}\,ef(\hat{x}_f)\,\hat{dx}_f}{\hat{x}_f}.$$

In accordance with (5.52)

$$x_f = \hat{x}_f\,x_{f\,max},$$

therefore,

$$dE_1 = \frac{6k_{loc}\,f(\hat{x}_f)\,\hat{dx}_f}{\hat{x}_f\,x_{f\,max}}.$$

By applying the differential law $f(\hat{x}_f) = 3/2\,\hat{x}_f^{1/2}$ which is obtained from expression (5.59) when $n_2 \to \infty$ and integrating for $\hat{x}_f$ within the limits of $\hat{x}_{f\,min}$ to $\hat{x}_f = 1$, we obtain

$$E_1 = \frac{18ek_{loc}}{x_{f\,max}}\left(1 - \hat{x}_{f\,min}^{1/2}\right).$$

Because $\hat{x}_{f\,min} \ll 1$, therefore,

$$E_1 = \frac{18k\,e_{loc}}{\hat{x}_{f\,max}}.$$

Substituting the value of $x_{f\,max}$ in accordance with expression (5.46), we have

$$E_1 = \frac{18k_{loc}\,e}{\psi_{cy}\,\omega_{cy}\,W^{2/3}} \left(\frac{p^{1/2}}{W}\right)^{2/3}. \qquad (5.60)$$

During blasting of a cylindrical charge laid along the axis of the cylindrical volume of the rock, the energy expended per unit volume of the rock is $pe_{exp}/\pi W^2$, where p is the mass of the cylindrical charge per unit length.

The blast efficiency for fragmentation is

$$\eta = \frac{E_1}{pe_{exp}}\,\pi W^2. \qquad (5.61)$$

From equations (5.60) and (5.61), we obtain

$$\eta = \frac{A_1}{W^{2/3}\left(\dfrac{p^{1/2}}{W}\right)^{4/3}}, \qquad (5.62)$$

where

$$A_1 = \frac{18k_{loc}\,\pi e}{\psi_{cy}\,\omega_{cy}\,e_{exp}}.$$

From expression (5.62), it follows that with an increase in W, i.e. with an increase in the scale of blasting, the relative portion of work done in breaking is reduced whereas the charge mass ($p/W^2 = $ const) remains constant.

The reason for this lies in the fact that with an increase in the scale of blasting, the relative area of the re-formed surface decreases because, in accordance with formulas (5.46) and (5.47), the size of the largest fragments increases. An inverse proportional dependence of η on $p^{1/2}/W$ is due to the fact that when $W = $ const, the charge radius decreases with a decrease in its mass. The latter phenomenon, by virtue of expression (5.25), leads to an increase in the velocity gradient in the nearer zone due to which the average size of the rock fragment, as is clear from expression (5.35), is reduced. Consequently, the relative area of the re-formed surface will increase which leads to an increase in crushing energy with a reduction in $p^{1/2}/W$.

Experiments [57, 58] indicate that for a constant specific consumption of explosives, the emergence of small fractions increases with an increase in the diameter of the deep-hole charge. This is easily explained by expression (5.58). In fact, in accordance with formula (5.46), the value of $x_{cy\,max}$ decreases with an increase in diameter of the charge (with increase in parameter p, when $W = $ const.). This leads to an increase in the upper limit $\hat{x}_f$ in the integral (5.57) which follows directly from equation (5.52) if we take into account that the highest value of the size of a fragment x_f of a given fraction does not vary.

The calculations made by formula (5.62) for rocks with a breaking strength of 1000 N/cm² show that where $W > 1$ m and $p^{1/2}W = 1$ kg$^{1/2}$/m$^{3/2}$ the blast efficiency does not exceed 3%, which is closer to the values given in [25]. It should be noted that formula (5.62) is obtained for a charge which has a cylindrical volume. For usual blasts (with one free face), the efficiency will be less.

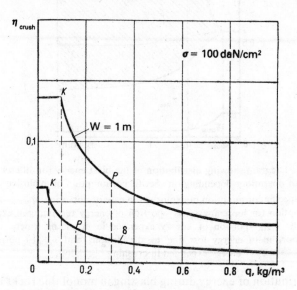

Fig. 67. Dependence of efficiency of rock breakage on specific consumption of explosives.

In the same way we can obtain the formula for determining the efficiency of a spherical charge:

$$\eta_{sp} = \frac{B_2}{W^{2/3}\left(\dfrac{Q^{1/3}}{W}\right)^{5/3}}, \qquad (5.63)$$

where

$$B_2 = \frac{24\pi e}{\psi_{sp}\,\omega_{sp}\,e_{exp}}.$$

Figure 67 shows curves of efficiency η_{crush} for crushing rock through a blast depending on the specific consumption q of explosives and explosion depth. The curves have been plotted on the basis of formula (5.62) given above for monolithic rock with an instantaneous breaking strength $\sigma = 100$ daN/cm².

146

While plotting the curves it was assumed that the specific consumption of explosives during loosening and implosion along two different lines of least resistance are in the same ratio (points P and K respectively).

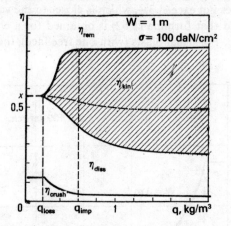

Fig. 68. Diagram showing distribution of the blast energy for different types of operations depending on specific consumption of explosives:

q_{imp}—specific consumption of explosives in case of implosion; q_{loose}—specific consumption for loosening; η_{rem}—portion of energy remaining in explosion products; η_{kin}—portion of energy expended in providing kinetic energy; η_{diss}—portion of energy lost due to dissipation; and η_{crush}—portion of energy expended in crushing operation.

The distribution of energy during blasting in monolithic rocks is illustrated in the diagram (Fig. 68). It can be seen that the values of η_{kin} increases with the increase in the specific consumption of explosives, i.e. with an increase in the mass of explosives per unit volume of broken rock. Loss of energy due to dissipation and the portion of the energy remaining in the explosion products remain constant. This assertion is valid only in the case of monolithic rocks. In highly fissured semirocks the losses due to dissipation increase (dotted line) due to attenuation of the compression wave. The position of point x along the ordinate depends on the coefficient of springing for rock: the greater the coefficient of springing the farther the curve from the origin of coordinates and vice-versa. It should be noted that the diagram given in Fig. 68 does not take into account the further crushing of the rocks when they fall on the free face.

9. DIFFERENTIAL LAW OF DISTRIBUTION AND OPTIMUM PARAMETERS FOR FIRING A SYSTEM OF DEEP-HOLE CHARGES

It is of practical significance to determine whether the fragment size will be maximum during firing of a cylindrical or spherical charge when the specific consumption of explosives is constant. For this purpose it is necessary to express the quantities $p^{1/2}/W$ and $Q^{1/3}/W$ in terms of specific consumption q of the explosives, use them in formulas (5.46) and (5.47) and then find the ratio $x_{\text{cy max}} : x_{\text{sp max}}$.

The specific consumption of explosives for cylindrical charges is, respectively, equal to

$$q = \frac{p}{\pi W^2} \; ;$$

$$q = \frac{Q}{\frac{4}{3}\pi W^3} \, ,$$

from where we obtain:

$$\frac{p^{1/2}}{W} = \sqrt{\pi q};$$

$$\frac{Q^{1/3}}{W} = \sqrt[3]{\frac{4}{3}\pi q}.$$

Using these expressions in formulas (5.46) and (5.47), we find

$$x_{\text{cy max}} = \frac{\psi_{\text{cy}}\,\omega_{\text{cy}}}{\pi^{1/3}} \cdot \frac{W^{2/3}}{q^{1/3}} \; ;$$

$$x_{\text{sp max}} = \frac{\psi_{\text{sp}}\,\omega_{\text{sp}}}{\left(\frac{4}{3}\pi\right)^{4/9}} \cdot \frac{W^{2/3}}{q^{4/9}} \, .$$

Calculations made on the basis of these formulas show that $x_{\text{cy max}}$: $x_{\text{sp max}} = 1$ when $q = 1$ kg/m³ and $\rho_{\text{exp}} = 10^3$ kg/m³. In this manner when $q \leqslant 1$ kg/m³ rock can profitably be crushed to smaller fragments by cylindrical charges; when $q > 1$ kg/m³, rock is broken to smaller fragments by using spherical charges.

The process of breaking rock through blasting mentioned here and the method devised on this basis prove helpful in determining optimum parameters for the blasting of a system of deep-hole charges which are meant for removal or for explosive cutting of rocks. Besides, this process makes it possible to find the differential law of distribution for the most common type of laying of deep-hole charges in the massif to be blasted. This law is very important for solving the problem of rock movement in the air.

First let us find the relation between the maximum size of rock fragment and the coefficient of converging holes. If we assume that rock movement due to firing of several charges is independent, then rock fragments of maxi-

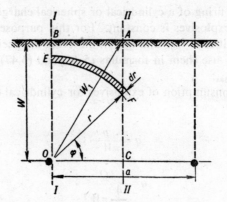

Fig. 69. Calculation for granule-size composition during blasting
of a series of deep-hole charges.

mum size will be formed on the free face in that volume of the rock which is
equidistant from two adjacent charges. By denoting the distance of this
volume from the charge by W_1, from Fig. 69, we obtain:

$$W_1 = \sqrt{W^2 + \left(\frac{a}{2}\right)^2}, \qquad (5.64)$$

where a is the distance between the charge, or

$$W_1 = W \frac{\sqrt{4 + m^2}}{2},$$

where

$$m = \frac{a}{W}.$$

Let us introduce the notation

$$\xi_1 = \frac{4 + m^2}{2}, \qquad (5.65)$$

then

$$W_1 = W \xi_1.$$

Substituting W by W_1 in expression (5.46) to find the size of the largest rock
fragment for a given hole convergence, we have

$$x_{max} = \psi_{cy} \, \omega_{cy} \frac{W^{2/3}}{\left(\frac{p^{1/2}}{W}\right)^{2/3}} \xi_1^{4/3}. \qquad (5.66)$$

For a system of parallel deep-hole charges, the specific consumption of explosives is expressed by the following dependence

$$q = \frac{p}{Wa},$$

where

$$\frac{p^{1/2}}{W} = (qm)^{1/2}. \tag{5.67}$$

From expressions (5.46) and (5.67) we find

$$x_{cy\ max} = \psi_{cy}\ \omega_{cy} f(m) \frac{W^{2/3}}{q^{1/3}}, \tag{5.68}$$

where

$$f(m) = \left[\frac{(4+m^2)^2}{16m}\right]^{1/3}. \tag{5.69}$$

Formulas (5.44) and (5.68) enable us to calculate the maximum size $x_{f\ max}$ of fragment for a given set of explosion parameters. For example, let us calculate $x_{f\ max}$ for the following conditions of explosion: $\sigma = 2 \times 10^6$ daN/m² (breaking strength of solid limestone); $e_{exp} = 4.4 \times 10^6$ J/kg (ammonite No. 6); $\rho_{exp} = 1000$ kg/m³ (ammonite No. 6); $\Delta = 0.5$ (volume of air gaps equal to the volume of charge in the hole); $q = 0.3$ kg/m³; $W = 5$ m; and $m = 1$. Let us consider $\psi_{cy} = 1.8$ and $\psi_{sp} = 1.15$ (see Sec.5). By calculating $x_{f\ max}$ from formula (5.68) we find that $x_{f\ max} \approx 1$ m, which conforms to experimental data. If the selected value of m is not optimal (for example, $m = 0.2$) then $f(m) = 2$. Moreover, $f(m_{opt}) = 1.15$ for $m_{opt} = 1$. Consequently, the maximum size of rock fragment for $m = 0.2$. If during explosion, air gaps are not used, i.e. $\Delta = 1$, then the maximum size of a rock fragment, according to formula (5.68), increases by $\Delta^{-1/3} = 0.5^{-1/3} = 1.26$ times.

From formulas (5.68) and (5.45), it is clear that an increase in the specific consumption of explosives does not have much effect on the crushing of rock. For example, in order to reduce $x_{f\ max}$ by two times, it is necessary to increase the quantity of explosives by eight times. This phenomenon is proved by experimental blasting operations.

The curve of function $f(m)$ and experimental point taken from [59, 60] are given in Fig. 70. The optimal value of m lies between the limits 0.8–1.3 which is also proved in practice.

In this way in breaking of rock by detonating a system of parallel deep-hole charges the ejection of a large fraction increases when the distance between the holes is reduced as also when it is very considerably increased. It must be remembered that the specific consumption of explosives is constant.

The presence of an optimum value for m can be easily explained on the basis of the crushing process stated above. Its essence lies in the fact that the shock shear caused by the velocity gradient dv_r/dr is the main working factor

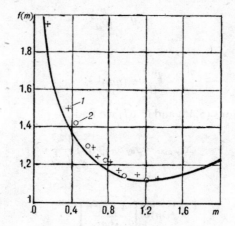

Fig. 70. Curve of $f(m)$ and experimental points
for holes with the following diameters:
1—43 mm; and 2—225 mm.

for crushing. If the given volume moves with a velocity gradient dv_r/dr then, according to formula (5.35), it will be broken to fragments of the following average size:

$$x_f = \frac{B_1}{\left(\dfrac{dv_r}{dr}\right)^{2/3}} .$$

The value of the velocity gradient dv_r/dr for cylindrical and spherical symmetry, in accordance with formula (5.25), will be:

for a cylindrical charge

$$\left|\frac{dv_r}{dr}\right| = \frac{v_0 r_0}{r^2} ; \tag{5.70}$$

for a spherical charge

$$\left|\frac{dv_r}{dr}\right| = \frac{2v_0 r_0^2}{r^3} . \tag{5.71}$$

While writing these expressions it was assumed that the medium is incompressible. In this case the exponent n in formula (5.25) for cylindrical and spherical symmetry is taken to be equal to 1 and 2 respectively. We can obtain more accurate laws of crushing if the compressibility of the medium is taken into account. However, we shall, for the present, restrict ourselves to a simple examination of this problem.

The rate of movement v_0 of the medium at the boundary of the charge depends very little on the density of charging of the hole (up to a certain limit).

Therefore, in future we shall consider v_0 to be constant. If the specific consumption of explosive is considered to be invariable and the distance a between the holes is reduced, the hole radius r_0 will be reduced. Consequently, in accordance with formula (5.70), gradient dv_r/dr will decrease at a given distance $r = W_1$ which, according to formula (5.35) leads to an increase in the maximum size of a rock fragment and a deterioration in crushing [61]. The maximum size of a fragment will decrease with an increase in distance a. However, this decrease in size takes place up to a certain optimum value of a after which the velocity gradient for $r = W_1$ will start diminishing and the maximum size of a fragment will start increasing. Function $f(m)$ also reflects the sequence of events described here.

Formulas (5.45) and (5.68) explain why the air gaps between charges in the holes improve the quality of fragmentation [62, 63]. From these formulas it is evident that a reduction in the relative loading density Δ (formation of air gaps in the holes) leads to a reduction in the value of $x_{f\,max}$. From the viewpoint of the breaking process adopted by us, the formation of air gaps (q and m being constant) will lead to enlargement of the effective radius r_0 of a hole and, consequently, an increase in gradient dv_r/dr at the same distance W. As a result, the fragment size will be reduced in accordance with formulas (5.35) and (5.70). Let us examine the method of finding the differential law of distribution of rock fragments $f(x_f)$ with the given coefficient of hole convergence m. As

$$f(x_f) = \frac{dU}{U_0\,dx_f}, \tag{5.72}$$

where dU is the volume of the elementary circular layer EF between spaces I–I and II–II (see Fig. 69) with fragments measuring $x_{r\,max}$; and U_0 being half of the rock volume to be blasted between neighboring deep-hole charges. Therefore, taking into consideration that in the present case

$$U_0 = \frac{a}{2}\,W;$$

$$dU = \frac{\pi}{3}\,r\delta_1\,dr,$$

we obtain

$$f(x_f) = \frac{\pi r\,dr}{aW\,dx_f}\,\delta_1. \tag{5.73}$$

In these expressions coefficient δ_1 is the ratio between the length of arc EF of radius r within the limits of rectangle $OBAC$ and one-fourth of the circumference of a circle of the same radius (see Fig. 69).

In the present instance rectangle $OBAC$ is under review, therefore, by virtue of explosion symmetry in other volumes of the rock between the holes, the distribution of granule-size composition will be similar. In this manner

$$\delta_1 = \frac{2l_{EF}}{\pi r}.$$

From Fig. 69 it is clear that $\delta_1 = 1$ for $r \leqslant a/2$. If $a/2 < r \leqslant W$, then

$$\delta_1 = 1 - \frac{2\varphi}{\pi}. \tag{5.74}$$

The value of δ_1 can be calculated easily with the help of a graph if $W < r \leqslant W_1$.
The value of $\cos \varphi = a/2r$ within the limits $a/2 < r \leqslant W$.

Since $a = mW$, therefore, taking expression (5.64) into account, we obtain

$$\cos \varphi = \frac{m}{\hat{r}(4+m^2)^{1/2}}, \tag{5.75}$$

where

$$\hat{r} = \frac{r}{W_1}.$$

From expressions (5.74) and (5.75), we have

$$\delta_1 = 1 - \frac{2}{\pi} \cos^{-1} \frac{m}{\hat{r}(4+m^2)^{1/2}}.$$

This relation is true when

$$\frac{m}{(4+m^2)^{1/2}} < \hat{r} \leqslant \frac{2}{(4+m^2)^{1/2}}.$$

As $\hat{x}_f = \hat{r}^{4/3}$ in accordance with Eq. (5.53), therefore,

$$\delta_1 = 1 - \frac{2}{\pi} \cos \frac{m}{\hat{x}_f^{3/4}(4+m^2)^{1/2}}$$

when

$$\left[\frac{m}{(4+m^2)^{1/2}}\right]^{4/3} < \hat{x}_f \leqslant \left[\frac{2}{(4+m^2)^{1/2}}\right]^{4/3}.$$

The differential law (5.73) can be written in a simpler dimensionless form as follows:

$$f(\hat{x}_f) = \frac{4\pi \hat{r} \, d\hat{r}}{(4+m^2) m \, d\hat{x}_f} \delta_1. \tag{5.76}$$

This expression is obtained from Eq. (5.73) by dividing the numerator and denominator by W_1^2 in accordance with relation (5.64). Here

$$\hat{x}_f = \frac{x_f}{x_{f\,max}},$$

where the quantity $x_{f\,max}$ conforms to $r = W_1$, i.e. the distance at which fragments of maximum size are formed.

By determining $\hat{r}$ and $d\hat{r}/d\hat{x}$ from equation (5.53) and substituting them in expression (5.76), we obtain

$$f(\hat{x}_f) = \frac{3\pi\,(4+m^2)\,\delta_1}{16m}\,\hat{x}_f^{1/2}. \qquad (5.77)$$

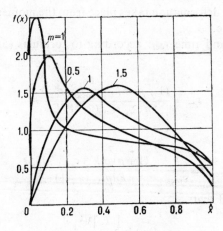

Fig. 71. Differential laws of distribution $f(\hat{x}_f)$ for certain values of coefficient of convergence of holes m.

Figure 71 indicates the differential laws of distribution $f(\hat{x}_f)$ for various values of coefficient m of hole convergence. It is evident that the law considerably depends on quantity m. In the figure the relative size $\hat{x}_f$ of a rock fragment is shown along the abscissa and the quantity $f(\hat{x}_f)$ along the ordinate. In order to reconstruct curve $f(\hat{x}_f)$ on the curve for the exit of broken rock as a percentage of fractions (this curve is often used in practice), it is necessary, for the average size fragment of a given fraction within the range of x_{f1} to x_{f2}, to integrate the differential expression $\int_{\hat{x}_{f1}}^{\hat{x}_{f2}} f(\hat{x}_f)\,d\hat{x}_f$ and multiply by 100%.

Here $x_{f1} = \hat{x}_{f1}x_{f\,max}$; $x_{f2} = \hat{x}_{f2}x_{f\,max}$; and $x_{f\,max}$ is calculated from formula (5.68).

10. VELOCITY OF CRACK PROPAGATION DURING GENERAL FRAGMENTATION OF ROCK

The velocity of crack propagation during fragmentation of rock through blasting can be easily determined for rocks which are brittle.

At a given distance r from the charge the general fragmentation of the rock (appearance of cracks in an elementary volume of a highly fissured system) begins from the moment of detonation of the charge after a time lapse of

$$t = t_w + t_{def},$$

where t_w is the time of arrival of the compressional wave and t_{def} the time during which shear deformation takes place from the moment of arrival of the compressional wave.

As $t_w \approx r/c_{rock}$ and time t_{def}, according to formulas (5.22) and (5.40), is equal to:

$$t_{def} = \frac{(1+\mu)\,\sigma\,k_{w1}\,c_{rock}\,\rho\,\hat{r}_{def}^{n}\,r}{\pi n \rho_{exp}\,e_{exp}\,(\varkappa - 1)\,E},$$

therefore,

$$t = \frac{r}{c_{rock}} + \frac{(1+\mu)\,\sigma\,k_{w1}\,c_{rock}\,\rho\hat{r}_{def}^{n}\,r}{\pi E n \rho_{exp}\,(\varkappa - 1)\,e_{exp}}.$$

Bearing in mind that:

$$c_{rock} = \left[\frac{E}{\rho}\right]^{1/2}, \tag{5.78}$$

we find that

$$t = \frac{r}{c_{rock}} + \frac{(1+\mu)\,\sigma\,k_{w1}\,\hat{r}_{def}^{n}\,r}{\pi n \rho_{exp} e_{exp}\,(\varkappa - 1)\,c_{rock}}, \tag{5.79}$$

where

$$\hat{r}_{def} = \frac{r}{r_0}.$$

The crack propagation velocity is then

$$v_{crack} = \frac{dr}{dt}.$$

After differentiating expression (5.79) with respect to t and solving the obtained equation with respect to $v_{crack} = dr/dt$, we obtain

$$v_{crack} = \frac{c_{rock}}{D_{crack}}, \tag{5.80}$$

where

$$D_{\text{crack}} = 1 + \frac{(1+\mu)\,\sigma\,k_{w1}\,(n+1)}{\pi n \rho_{\text{exp}}\,e_{\text{exp}}\,(\varkappa-1)}\,\hat{r}_{\text{def}}^n.$$

For incompressible mediums in the case of cylindrical ($n=1$) and spherical ($n=2$) symmetry, respectively, we have

$$D_{\text{crack cy}} = 1 + \frac{2\,(1+\mu)\,\sigma\,k_{w1}}{\pi \rho_{\text{exp}}\,e_{\text{exp}}\,(\varkappa-1)}\,\hat{r}_{\text{def}}; \qquad (5.81)$$

$$D_{\text{crack sp}} = 1 + \frac{3\,(1+\mu)\,\sigma\,k_{w1}}{2\pi \rho_{\text{exp}}\,e_{\text{exp}}\,(\varkappa-1)}\,\hat{r}_{\text{def}}^2. \qquad (5.82)$$

From formulas (5.80) to (5.82), it follows that, firstly, the crack propagation velocity v_{crack} during general crushing is less than the velocity of sound c_{rock}; secondly, v_{crack} depends on the elastic (μ, k_{w1}) and tensile (σ) characteristics of the rock as well as on the properties of the explosives (ρ_{exp}, e_{exp}, $\varkappa$) and, thirdly, v_{crack} depends on the relative distance $\hat{r}_{\text{def}}$: the farther the volume of the rock from the charge the lower the velocity v_{crack}.

Let us calculate $D_{\text{crack cy}}$ and $D_{\text{crack sp}}$ for the following parameters: $\sigma = 2 \times 10^6$ daN/m^2 (instantaneous breaking strength of solid limestone); $\rho_{\text{exp}} = 1000$ kg/m^3 (ammonite No. 6); $e_{\text{exp}} = 4.4 \times 10^6$ J/kg (ammonite No. 6); $\mu = 0.33$; $\varkappa = 2$ (when relative loading density $\Delta = 0.7$ to 1.2); and $k_{w1} = 1.25$ (for rocks of average strength). Substituting these in formulas (5.81) and (5.82), we have:

$$D_{\text{crack cy}} = 1 + 4.8 \times 10^{-3}\,\hat{r}_{\text{def}};$$

$$D_{\text{crack sp}} = 1 + 3.6 \times 10^{-3}\,\hat{r}_{\text{def}}^2.$$

As an example, let $\hat{r}_{\text{def}} = 25$, then $D_{\text{crack cy}} = 1.12$ and $D_{\text{crack sp}} = 3.3$. Consequently, the crack propagation velocity for a general crushing of rock at a relative distance $\hat{r}_{\text{def}} = 25$ for a spherical charge is almost three times less than the velocity of sound. For a cylindrical charge, velocity v_{crack} differs considerably from the wave velocity. If $\hat{r}_{\text{def}} \approx 1$, then $v_{\text{crack}} \approx c_{\text{rock}}$ i.e. at small distances for all charges cracks appear immediately after the compressional wave.

Detonation of a spherical (concentrated) charge is distinguished by a high velocity gradient near the charge. This creates a dense network of cracks in the neighboring zone which may, in certain cases, be on spiral surfaces [64]. At great distances the cracks appear some time after the compressional wave, especially in the case of spherical charges. Therefore, in some cases, especially for spherical charges, 'break away' cracks may appear on the free face in addition to the wave of cracks due to general crushing as a result of the charge being detonated. These cracks, however, as explained in [42] do not cause a general fragmentation of rock.

These peculiarities in the pattern of general crushing of rock are confirmed by conducting experimental blasts in transparent monolithic mediums. It should be noted that the formulas obtained for v_{crack} are true only for monolithic mediums which are brittle in nature. In the presence of cracks the reduction in v_{crack} in comparison with c_{rock} will be much greater. Till now it has not been possible to quantitatively evaluate this reduction in view of inadequate knowledge of the process connected with the propagation of the compressional wave in highly fissured rock.

11. LAWS GOVERNING FRAGMENTATION OF FISSURED ROCK

If the massif to be blasted is full of fissures, the compressional wave is rapidly attenuated and the fragmentation process based mainly on shock shear ceases [65]. In this case highly fissured rock will be broken due to an inelastic impact between separate blocks as a result of propagation of the consolidation wave (see Sec. 3).

Let us suppose that the rock massif is formed of separate blocks of average size x_1. On detonating the charge, a condensation wave will pass through the massif. As a result, all fissures will be closed, and when this moving mass of rock comes in contact with individual fragments, deposited ahead of the wave front, an elastic impact will take place. The energy lost during impact will be partly utilized in crushing the blocks into fragments of average size x_f [66, 67].

Let the mass velocity of the medium beyond the condensation wave at a distance r from the charge be equal to v. Then, according to the laws of impact, the energy lost during the impact of mass M beyond the condensation wave with the mass of rock fragment measuring x_1 will be

$$E_{lost} = \frac{1}{2} M_1 \frac{M}{M + M_1} (1 - z_{rest}^2) v^2,$$

where z_{rest} is the coefficient of restoration in case of impact

$$M_1 = x_1^3 \rho.$$

As $M \gg M_1$, therefore, $E_{lost} = \frac{1}{2} M_1 v^2 (1 - z_{rest}^2)$.

Only a part of this energy will be utilized in breaking the rock fragment. Let us denote this part by ξ_b. The quantity ξ_b depends on the shape of the fragment. In the first approximation, we may take $\xi_b \leqslant 0.25$. The other part of the energy, in a certain proportion, is redistributed between masses M_1 and M. In this way, energy $\frac{1}{2} \xi_b (1 - z_{rest}^2) M_1 v^2$ will be utilized in forming new fissures. Let us suppose that a block measuring x_1 is crushed to fragments of average size x_f. The re-formed surface will then be:

$$S_f = 6x_f^2 \frac{N_f}{k_{loc}^2} - S_0,$$

where N_f is the number of re-formed fragments of sizes x_f, and S_0 the initial area of the surface of the block. It can be assumed that

$$S_0 = \frac{6}{k_{loc}^2} x_1^2.$$

Besides, it is obvious that

$$N_f = \frac{x_1^3}{x_f^3}.$$

Taking into account that energy e is utilized in forming a unit surface, the equation of energy balance can be written as follows:

$$\frac{6e}{k_{loc}^2} \left(\frac{x_1^3}{x_f} - x_1^2 \right) = \frac{1}{2} \xi_b (1 - z_{rest}^2) M_1 v^2.$$

However, the fragment measuring x_1 will begin to break into equal parts only at a definite minimum value of absorbed energy. This energy depends on the volume of the rock fragment and strength of the rock. Let us denote this minimal energy by E_{min}. In order to break a rock fragment of size x_1 it is necessary that at least one new fissure with a surface area x_1^2 be formed in it. Therefore, $E_{min} = e x_1^2 / k_{loc}^2$.

Let us modify the equation of energy balance written above as follows taking E_{min} into account:

$$\frac{6e}{k_{loc}^2} \left(\frac{x_1^3}{x_f} - x_1^2 \right) = \frac{1}{2} \xi_b (1 - z_{rest}^2) M_1 v^2 - e \frac{x_1^2}{k_{loc}^2}.$$

Considering that $M_1 = x_1^3 \rho$, we obtain

$$\frac{6e}{k_{loc}^2} \left(\frac{x_1^3}{x_f} - x_1^2 \right) = \frac{1}{2} \xi_b (1 - z_{rest}^2) x_1^3 \rho v^2 - \frac{e x_1^2}{k_{loc}^2},$$

whence

$$x_f = \frac{x_1}{\dfrac{1}{6} \left[\dfrac{x_1 (1 - z_{rest}^2) \xi_b \rho v^2 k_{loc}^2}{2e} - 1 \right] + 1}. \tag{5.83}$$

From this expression it follows that if

$$\frac{x_1 (1 - z_{rest}^2) \xi_b \rho v^2 k_{loc}^2}{2e} = 1, \tag{5.84}$$

the fragment will not be crushed. This means that the energy absorbed during impact is less than E_{min}.

This minimal energy corresponds to the so-called critical impact velocity at which the fragmentation of rock into equal parts begins.

In general the critical impact velocity is determined from the following energy equation:

$$\frac{1}{2} M_f \xi_b (1 - z_{rest}^2) \frac{M_1}{M_1 + M_f} v_{cr}^2 = e \frac{x_f^2}{k_{loc}^2} \cdot$$

By substituting $M_f = x_f^3 \, \rho / k_{loc}^3$; $M_1 = x_1^3 \, \rho / k_{loc}^3$ and e from Eq. (5.38), we obtain

$$v_{cr} = \frac{B_f}{x_f^{1/2}},$$

(5.85)

where

$$B_f = \left[\frac{\sigma^2 (1 + \varkappa_f^3) \, \beta_b}{\rho \xi_b (1 - z_{rest}^2) E} \right]^{1/2};$$

$$\varkappa_f = \frac{x_f}{x_1};$$

σ being the compression strength limit.

While writing the energy equation in [68], it was not taken into account that the critical impact velocity depends on the size of the rock fragment. As a result, the relations given in the cited work do not represent the actual laws.

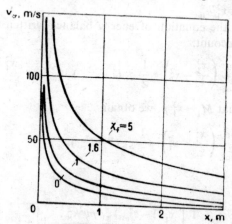

Fig. 72. Dependence of critical impact velocity for solid limestone on size of rock fragment.

Curves describing the relationship between v_{cr} and the size of the fragment for different values of $\varkappa_f$ are shown in Fig. 72. These curves are plotted for solid limestone with the following parameters: $\sigma = 6 \times 10^7$ N/m^2; $\rho = 2700$ kg/m^3; $E = 7.3 \times 10^{10}$ N/m^2; $\xi_b = 0.01$; $z_{rest} = 0$, $\beta_b = 0.1$ m; and $k_{loc} = 0.7$. The value $\varkappa_p = 0$ corresponds to the impact by the fragment on a very large mass ($M_f \ll M_1$). When $\varkappa_f = 1$, rocks of equal mass strike against each other. If $\varkappa_f > 1$, the size of the striking is greater than that of the struck fragment.

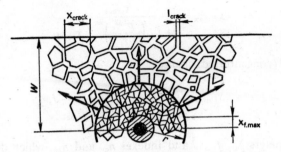

Fig. 73. Diagram showing crushing of fissured rock
by a condensation wave.

In [69] it is mentioned that the compressional wave cannot be propagated in rock with a large number of fissures. Instead, as stated earlier, a condensation wave is produced. The openings of the fissures in width l_{crack} are closed when the condensation wave passes, and the loose mass in the fissure is compressed (Fig. 73). It can be considered that the condensation wave front in highly fissured rock is also the front for general breaking (crack formation) in rock. It is well known that in the same type of rock, the velocity of propagation of the front for general breaking is greater in monolithic rock than in fissured mediums [70]. Fissured rock, especially when the fissures are partly filled with a loose mass, is transformed into a cemented conglomerate after the passage of the condensation wave (sphere of radius r in Fig. 73). In case of impact of this conglomerate with separate fragments, deposited ahead of the condensation wave front, there is an inelastic impact which breaks the rock fragments. Let us make an equation for crushing on the basis of the balance of energy: the work done in the formation of new free faces in the fragment is equal to the energy lost during inelastic impact. Using Carnot's theorem of energy lost during impact and bearing in mind that the mass of the medium beyond the condensation wave front is considerably greater than the mass of the fragment, the following relationship can be written:

$$\frac{1}{2}\zeta_b \frac{x_{crack}^3 \rho}{k_{loc}^3} v^2 = e \left[\int_0^\infty 6 \frac{x_{crack}^3}{k_{loc}^2 x_{f\,max}} f(\hat{x}_f) \frac{d\hat{x}_f}{\hat{x}_f} - 6 \frac{x_{crack}^2}{k_{loc}^2} \right], \qquad (5.86)$$

where x_{crack} is the size of an individual rock depending on the network of fissures; v the mass velocity beyond the condensation wave front; and $x_{f\,max}$ the size of the largest fragment after crushing of an individual fragment measuring x_{crack}.

Let us express mass velocity v by the following expression $v = v_0 \, (r_0/r)^n$. By solving equation (5.86) with respect to $x_{\text{f max}}$ we have:
for a cylindrical (deep-hole) charge:

$$x_{\text{f max}} = \frac{x_{\text{crack}}}{1 + f_{\text{cy}} \left(\dfrac{p^{1/2}}{r}\right)^{2n_{\text{cy}}} x_{\text{crack}}} \; ; \tag{5.87}$$

for spherical (concentrated) charges:

$$x_{\text{f max}} = \frac{x_{\text{crack}}}{1 + f_{\text{sp}} \left(\dfrac{Q^{1/3}}{r}\right)^{2n_{\text{sp}}} x_{\text{crack}}} \, J, \tag{5.88}$$

where parameters f_{cy}, f_{sp}, J and indexes n_{cy} and n_{sp}, which depend on the physicomechanical properties of the explosives and the rock, vary over wide limits. They should, therefore, be determined experimentally. They are expressed as follows:

$$f_{\text{cy}} = \frac{\xi_b \, \rho v_0^2}{12 k_{\text{loc}} \, e \, (\pi \rho_{\text{exp}} \Delta)^{n_{\text{cy}}}} \; ; \quad f_{\text{sp}} = \frac{\xi_b \, \rho v_0^2}{12 k_{\text{loc}} \, e \left(\dfrac{4}{3} \pi \rho_{\text{exp}} \Delta\right)^{\frac{2n_{\text{sp}}}{3}}} \; ;$$

$$J = \int_0^1 f(\hat{x}_{\text{f}}) \, \frac{d\hat{x}_{\text{f}}}{\hat{x}_{\text{f}}} \, .$$

From formulas (5.87) and (5.88) it follows that natural blocks in fissured and very hard rocks (with a high value of σ) at certain distances r from the charge will not be crushed. This will happen if the following conditions hold good:

$$f_{\text{cy}} \left(\frac{p^{1/2}}{r}\right)^{2n_{\text{cy}}} x_{\text{crack}} \ll 1;$$

$$f_{\text{sp}} \left(\frac{Q^{1/3}}{r}\right)^{2n_{\text{sp}}} x_{\text{crack}} \ll 1.$$

In case these inequalities hold good, the probability of obtaining large-size fragments is greatly reduced by increasing the specific consumption and density of explosives [71].

As explained in [72 to 79], the breaking of fissured rock along natural fissures begins only at certain distances from the charge. In this way, during blasting of fissured rock with unfilled fissures for low values of Q, p and a high value of σ there always remains a zone which is not broken. In this case, during ballistic calculations, the largest fragment of the rock massif before blasting is taken as the largest rock fragment.

It was stated earlier that if the fissures in the rock are filled with a solid mass or water, a compressional wave can be propagated in such a medium. However, it is greatly attenuated in comparison with propagation in a monolithic medium. The compressional wave, in passing through individual rock fragments forming a massif, breaks them by the same process as in the case of a monolithic medium, i.e. as a result of shock shear. However, because of a quick attenuation of the compressional wave the exponent n in expression (5.40) will be less than that for monolithic incompressible mediums. Its value can be determined experimentally [65].

12. FURTHER CRUSHING DUE TO COLLISION OF FRAGMENTS AND THEIR FALLING ON A HARD SURFACE

Rock fragmented by blasting is further crushed due to the kinetic energy accumulated during flight and potential energy due to gravity when rock fragments fall on a hard surface. The degree of further crushing depends on a number of factors: specific consumption of explosives, size of flying pieces, hardness of rock, density, etc. Further crushing due to the kinetic energy accumulated during flight can be fully ensured only when a special technique of blasting is used due to which rock fragments strike against one another. Good results are achieved during blasting in an untidy rock mass or retaining wall [80–82]. Further crushing is most effective when the impact is total (impact without damping) during which the kinetic and potential energies are utilized for the crushing operation with a maximum efficiency.

The energy balance equation for further crushing is formed on the basis of the following considerations. During flight, the kinetic energy in a unit volume of a fragment is $E_{1\,kin} = q\,e_{exp}\,\eta$, and the kinetic energy in a fragment measuring $x_{r\,max}^3$ and of volume x^3/k_{loc}^3 will be equal to $E_{kin} = E_{1\,kin}\,x_{r\,max}^3/k_{loc}^3$. The potential energy in a fragment falling from a height H will be $E_{pot} = x^3\,\rho g H/k_{loc}^3$. The total energy $E_{kin} + E_{pot}$ due to the fall, and collision of rock fragments with a hard surface will be partly utilized in the crushing operation:

$$E_{crush} = \xi_i\,(E_{kin} + E_{pot}), \qquad (5.89)$$

where ξ_i is the stiffness coefficient of impact which indicates that part of the energy which is utilized for crushing, and is determined experimentally.

On impact a fragment of size $x_{r\,max}$ will be crushed to fragments of different sizes: from the smallest to a certain largest fragment of size $x_{f\,max}$. In general this diversity in fragmentation is expressed by the differential law of distribution $f(\hat{x}_f)$. This law determines that part of the total volume of these fragments which belongs to the fraction of sizes between x_f and $x_f + dx_f$.

After impact and crushing of rock the re-formed surface is:

$$S = S_\Sigma + S_0,$$

where S_Σ is the total surface area of all the fragments, and S_0 the initial surface area of the fragment of size $x_{r\ max}$.

After impact and crushing of a fragment of size $x_{r\ max}$, we obtain a certain number dN of fragments of size x_f.

$$dN = \frac{x^3_{r\ max}}{x^3_f} f(\hat{x}_f)\, d\hat{x}_f.$$

The surface area of these fragments is

$$dS = 6\, \frac{x^2_f}{k^2_{loc}}\, dN.$$

The total surface area of all fragments is

$$S_{pot} = \int\limits_0^1 6\, \frac{x^3_{r\ max}\, f(\hat{x}_f)\, d\hat{x}_f}{k_{loc}\, x_{f\ max}\, \hat{x}_f},$$

where

$$\hat{x}_f = \frac{x_f}{x_{f\ max}}.$$

The initial surface area is:

$$S_0 = 6\, \frac{x^2_{r\ max}}{k^2_{loc}}.$$

The energy required for crushing the striking fragments into separate pieces of sizes varying between the smallest to $x_{f\ max}$ is expressed by the following formula:

$$E_{crush} = e\,(S_\Sigma - S_0), \tag{5.90}$$

where e is the specific energy required for the formation of a unit area.

From equations (5.89) and (5.90) we get the equation of energy balance for further crushing of a rock fragment:

$$\xi_b \eta\, \frac{x^3_{r\ max}}{k^3_{loc}}\, qe_{exp} + \xi_b\, \frac{x^3_{r\ max}}{k^3_{loc}}\, \rho g H$$

$$= e\left\{ 6\, \frac{x^3_{r\ max}}{k^2_{loc}\, x_{f\ max}} \int\limits_0^1 \frac{f(\hat{x}_f)\, d\hat{x}_f}{\hat{x}_f} - 6\, \frac{x^2_{r\ max}}{k^2_{loc}} \right], \tag{5.91}$$

whence

$$x_{f\ max} = \frac{x_{r\ max}}{k_{crush}} J, \qquad (5.92)$$

where

$$k_{crush} = 1 + \frac{x_{r\ max}}{6k_{loc}\ e} (\eta q e_{exp} + \rho g H);$$

$$J = \int_0^1 \frac{f(\hat{x}_f)\ d\ \hat{x}_f}{\hat{x}_f}. \qquad (5.93)$$

For calculations of estimated values, we can consider $J \approx 1$. As is evident, coefficient k_{crush} shows by how much the initial size of a fragment $x_{r\ max}$ is reduced as a result of further crushing due to impact.

In expression (5.93), the figures given in parentheses represent the sum total of the kinetic E_{kin} and potential E_{pot} energies accumulated up to the moment of impact in a unit volume of rock. Height H, for which $E_{kin} = E_{pot}$, is calculated by the following expression:

$$H = \frac{\eta e_{exp}\ q}{\rho g}.$$

By substituting the mean values of parameters used in this expression ($q = 0.5$ kg/m^3; $e_{exp} = 4 \times 10^7$ J/kg; $\eta = 0.15$ and $\rho = 2300$ kg/m^3), we find that $H = 15$ m. Thus, the dominant role in further crushing is played by potential energy due to gravity if a fragment falls from a height $H = 15$ m.

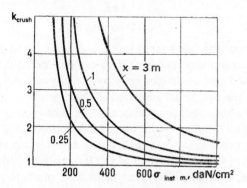

Fig. 74. Dependence of coefficient of further crushing on breaking strength and size of fragment.

Figure 74 shows the curves of coefficient k_{crush} when $E_{kin} \gg E_{pot}$. For calculating coefficient k_{crush} the parameters were: $\xi_b = 0.01$; $E = 5 \times 10^{10}$ N/cm^2; $\beta_b = 0.1$ m; $k_{loc} = 1.3$; $q = 0.5$ kg/m^3; $e_{exp} = 4 \times 10^6$ J/kg and $\eta = 0.15$.

From the curves it is clear that a fragment of size $x=1$ m for rock of average hardness ($\sigma_m = 400$ to 500 daN/cm²) will, on impact, be broken to pieces of sizes $x_{f\,max} < 0.5$ m. If $E_{pot} \gg E_{kin}$ coefficient k_{crush} is chiefly determined, on

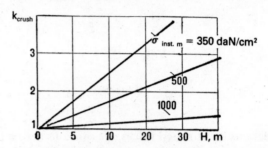

Fig. 75. Dependence of coefficient of further crushing k_{crush} on height from which a fragment falls.

the basis of height H. Figure 75 shows curve k_{crush} depending on height H for a fragment of size $x=1$ m and parametric values being the same as in the previous case. By analyzing the equations (5.92) and (5.93) and the curves of Figs. 74 and 75, it follows that further crushing of rock plays an important role only for rock of which the breaking strength is less than 500–700 daN/cm³. Moreover, these expressions reveal that if

$$\frac{x_{r\,max}}{6k_{loc}\,e}\,\zeta_b\,(\eta e_{exp}\,q + \rho g H) \gg 1,$$

then

$$x_{f\,max} = \frac{6k_{loc}\,e}{\zeta_b\,(\eta e_{exp}\,q + \rho g H)}\,J.$$

This means that for fairly large reserves of kinetic and potential energies the granule-size composition of rock does not depend on the size of the striking fragments and can be defined only by the physicomechanical properties of the rock and the energy accumulated at the moment of impact with a hard surface. This was observed, for example, while constructing the flood-control dam at Medeo and the dam of the Baipazinsk hydroelectric grid. The size of the biggest fragments $x_{f\,max}$ measured at the dam surface at Medeo was 30–40 cm, and about 140 cm at the dam of the Baipazinsk hydroelectric network. Moreover, the sizes of the falling fragments were much larger than what can be observed from the films of the explosion process.

Chapter 6

Design of Wedge-shaped Charges

1. ROCK BLASTING MECHANISM

Wedge-shaped charges are included among slab charges. However, they differ from the latter in one respect that their thickness is variable. In cross section, their shape is trapezoidal in which the longer sides meet at an acute angle.

The practical advantage of wedge-shaped charges is that they enable one-sided directional blasting of the rock when the free face is horizontal. It has been established through experiments that firing of a wedge-shaped charge forms a trench of triangular section and the muck pile is thrown to one side.

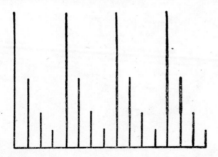

Fig. 76. Wedge-shaped charges in the form of a system
of detonating cord segments.

A number of model and actual explosions were conducted with a view to understanding the blasting process in case of firing of wedge-shaped charges and also to verify the technique developed for designing these charges. In small-scale experimental explosions, wedge-shaped charges were made of several sections of three or four parallel detonating cords. In each section the length of detonating cord was variable (Fig. 76).

The experiment was conducted in sandy and loamy soils. A charge was laid under the massif to be blasted in the following manner. A trench of triangular section was initially made with one side steep, and the other gently sloping. The prepared charge was laid on the gently sloping side and the

165

166

earth dug out of the trench was scattered over it and tamped. The charge was fired by an electrodetonator connected to the main cord on the free face.

Control experiments were conducted simultaneously with the main experiments with a view to checking the effect of scattered soil on directional accuracy and extent of blasting. For these experiments wedge-shaped charges were prepared in the form of a series of parallel inclined holes which were charged nonuniformly with powder explosives. It was found after comparison that the scattered soil does not have a noticeable influence on blast efficiency.

In small-scale experiments four different arrangements were studied for blasting (Fig. 77). In each series of experiments, the angle of inclination α and depth W for laying wedge-shaped charges were kept constant. Only the mass of the charge was varied.

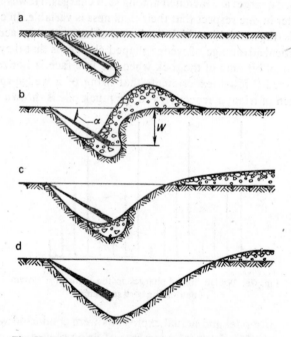

Fig. 77. Different arrangements for firing wedge-shaped charges of varying mass.

An implosion (Fig. 77, a) takes place when the charge mass is relatively small. There is no visible effect on the earth surface in this case. However, a cavity of wedge-shaped profile was formed by this type of explosion. If the mass of the charge is increased by several times without changing the angle of inclination and depth of laying, a one-sided throw can be observed during detonation (Fig. 77, b) and the exploded soil is dumped on the right side of the trench. However, the depth of the pit thus formed is less than the depth of laying of the charge. The charge-laying depth is the distance between the free face and the base of the charge. This happened because the blasted rock was torn away from the right-hand slope and was partly strewn in the pit itself formed by the explosion.

When the mass of the wedge-shaped charge was increased further (in comparison with the previous experiments), the pit had a symmetrical contour (Fig. 77, c) and its depth was equal to the original depth of laying of the charge. One of the experimental explosions conducted with this arrangement is shown in Fig. 78. It can be seen that the blasted earth lies on the right side of the pit. The blasted earth did not fall on the left side. The results of experimental, as well as actual explosions, have proved that a practically cent percent blast directivity can be achieved with the wedge-shaped charge.

Fig. 78. Crater and muck pile created by firing a wedge-shaped charge.

If the mass of the wedge-shaped charge is increased further (in comparison with the previous case) then an asymmetrical pit will be formed due to its explosion (Fig. 77, d). The right-hand slope will be more gentle and the blasted soil will be scattered over a larger area without forming a pile crest. Besides, a surface heave is observed on the left side.

Visual observations of the process of blasting in case of a wedge-shaped charge reveal that the direction of throw from the central zones of the rock mass to be blasted is perpendicular to the plane of the charge. Exceptions to this rule are the zones near the ends of the charge. Figure 79 indicates typical projection velocities. It can be seen that the volume of the rock nearer to the right-hand side of the trench is blasted with gradually decreasing velocities and the direction of throw changes in such a way that it becomes parallel to the right-side slope of the trench at the right edge.

A different situation arises when blasting with a wedge-shaped charge is conducted in hard rock and the charge itself is laid at a small angle to the verticle plane. If this angle is less than a certain limiting value γ_{lim}, then in case of laying of a wedge-shaped charge, for example, in the plane BB (Fig. 80)

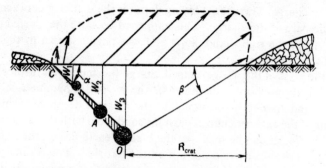

Fig. 79. Diagram showing direction of projection velocity during firing of a wedge-shaped charge.

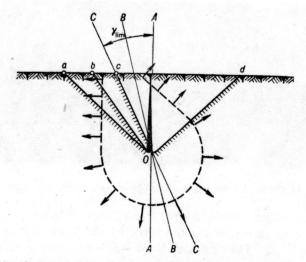

Fig. 80. Shapes of craters formed after firing of a wedge-shaped charge laid at different angles to the vertical plane.

an asymmetrical trench will be formed within the limits of Obd and its left slope Ob will be further left of plane BB. If the wedge-shaped charge is laid in plane CC which is at an angle γ_{lim} to the vertical plane then the left-hand slope formed by blasting will coincide with the same plane CC in which the charge is laid. It is characteristic that the left-hand slope Od of the trench will be approximately in the same position as in the previous case. The calculations show that the value of the limiting angle of inclination γ_{lim} of the charge depends on the physicomechanical properties of the rock: for hard rock $\gamma_{lim} \approx 10°$, and for cohesive soil $\gamma_{lim} = 15$ to $20°$.

That situation in which the slope of the crater formed by firing of wedge-shaped charges laid at angles $\gamma > \gamma_{lim}$ coincides with the plane of charges, can be explained thus: the compressional wave propagated from the wedge-shaped charge toward the left side (Fig. 80, dotted outline) is perpendicular to the free face and, consequently, at these angles the wave is not deflected and the 'breakaway' phenomenon does not occur. The situation is different on the right side of the charge. As is evident from Fig. 80, the condensation wave propagated to the right approaches the free face at a certain angle and as a result of its deflection from the face a rarefaction wave is produced which causes the 'breakaway' phenomenon and fragmentation of the rock. The broken rock is projected beyond the crater limits when the explosion products expand further.

From the above analysis it is clear that a one-sided throw due to blasting of a wedge-shaped charge can be achieved only when the slope of the charge is $\gamma > \gamma_{\lim}$.

2. THREE-CHARGE LAYOUT

If the angle of inclination α of the wedge-shaped charge to the horizontal is between 45–80°, then an ideal wedge-shaped charge can be replaced with a system of three linear charges which is equally effective, and the axes of these charges are in the plane coinciding with the plane of the slope of the trench formed by the blast (see Fig. 79). In this case the mass of a unit length p_3 of the lower linear charge is equal to the mass of section AO of the wedge-shaped charge. In the same way the mass of a unit length of the second and the first linear charges is equal to the mass of sections BA and CB respectively of the wedge-shaped charge. Considering the lengths of these sections to be equal, we can find the mass of a unit length of the respective linear charges from the following ratio:

$$P_1 : P_2 : P_3 = 1 : 3 : 5. \tag{6.1}$$

This ratio can be achieved if we calculate the areas equal in height to the sections of the triangle, and find their ratio.

The value of p_3 of the linear charge is calculated from the following hypothesis: if the explosion is conducted with one linear charge of mass p_3 at a depth W_3, a symmetrical trench of triangular section is formed with an angle of slope β to the free face. The formula for calculating the value of p_3 can be found by bearing in mind that the linear charge forms a crater of roughly the same dimensions as is formed by a series of concentrated charges laid at the following distance from each other [34]:

$$a_3 = W_3 \frac{n_e + 1}{2}, \tag{6.2}$$

where the index of blast effect n_e is the ratio between the half-width R_{crat} of the crater and its depth W_3:

$$n_e = \frac{R_{crat}}{W_3} = \cot \beta.$$

Obviously, the average mass per unit length of a series of concentrated charges is $p_3 = Q_3/a_3$.

The mass of the concentrated charge is determined from the following known formula:

$$Q_3 = k_e k_w W_3^3 f(n_e), \tag{6.3}$$

where k_e is the specific consumption of explosives in the case of normal detonations

$$k_w = 1 + \frac{W_3}{50}.$$

Correction of k_w for explosion depth in this formula takes into account the effect of gravity on the dimensions of the crater formed by blasting.

Function $f(n_e)$ in expression (6.3) is, according to the Boreskov formula, written as follows:

$$f(n_e) = 0.4 + 0.6 n_e^3.$$

From the dependences given above we find that

$$p_3 = k_e k_w W_3^2 f'(n_e),$$

where

$$f'(n_e) = 2 \frac{0.4 + 0.6 n_e^3}{n_e + 1}.$$

For the function $f'(n_e)$, when n_e varies between 1 and 3, the following simple approximate expression can be selected:

$$f'(n_e) = n_e^2.$$

Taking into account that $n_e = \cot \beta$, we get the following formula from the relations given above:

$$p_3 = k_e k_w W_3^2 n_e^2. \tag{6.4}$$

However, a linear charge p_3, when blasted together with charges p_2 and p_1, causes a one-sided muck throw from the crater formed due to explosion. Therefore, the mass per unit length of p_3 will be approximately half of that calculated by formula (6.4) which is true for a symmetrical two-sided blast. Consequently, the mass per unit length of the third linear charge component of the wedge-shaped charge assembled according to the three-charge system will be:

$$p_3 = \frac{1}{2} k_e k_w W_3^2 \cot^2 \beta. \tag{6.5}$$

By using relation (6.1) we find

$$p_2 = \frac{3}{10} k_e k_w \, W_3^2 \cot^2 \beta; \tag{6.6}$$

$$p_1 = \frac{1}{10} k_e k_w \, W_3^2 \cot^2 \beta. \tag{6.7}$$

The depth of laying of charges p_2 and p_1 is:

$$W_2 = \frac{2}{3} \, W_3; \tag{6.8}$$

$$W_1 = \frac{1}{3} \, W_3. \tag{6.9}$$

The mass of the wedge-shaped charge p per unit length of the trench formed by the blast is equal to $p_1 + p_2 + p_3$. Therefore,

$$\rho = \zeta_e \, k_w \, W_3^2 \cot^2 \beta, \tag{6.10}$$

where coefficient ζ_e, which depends on the physicomechanical properties of the rock to be blasted, should be determined through experiments. In the first approximation, it can be calculated by the following formula:

$$\zeta_e = 0.9 k_e.$$

It should be noted that these calculation formulas are fully applicable for $\beta > 48$ to $50°$. In case of higher values of β, especially in soft rock, the right-hand slope of the trench collapses (see Fig. 77, b).

It was stated earlier that a linear charge with mass p_3 can be replaced with an equally effective row of concentrated (chamber or sprung-hole) charges which are laid at distances calculated by expression (6.2). The calculations show that quantity $\frac{1}{2}(n_e + 1)$ can be approximated by the following expression with adequate accuracy within the limits $n_e = 1$–3:

$$\frac{n_e + 1}{2} \, n_e^{2/3}. \tag{6.11}$$

Therefore, the distances between the concentrated charges a_1, a_2 and a_3 in the first, second and third rows:

$$a = W n_e^{2/3}. \tag{6.12}$$

The index of explosion n_e is calculated by the following expression

$$p = k_e k_w \, W^2 n_e^2,$$

whence

$$n_e = \left[\frac{p}{k_e k_w \, W^2} \right]^{1/2}. \tag{6.13}$$

From expressions (6.5) to (6.13) we find

$$a_1 = 0.32 \ W_3 \cot^{2/3} \beta; \qquad (6.14)$$

$$a_2 = 0.58 \ W_3 \cot^{2/3} \beta; \qquad (6.15)$$

$$a_3 = 0.79 \ W_3 \cot^{2/3} \beta. \qquad (6.16)$$

Masses of concentrated charges Q_1, Q_2, Q_3 in the first, second and third rows are, respectively

$$Q_1 = p_1 a_1; \quad Q_2 = p_2 a_2; \quad Q_3 = p_3 a_3.$$

Replacement of an ideal wedge-shaped charge with a system of three linear charges or rows of concentrated charges has a disadvantage, namely, the upper charge forms a small muck pile on the left side of the trench (Fig. 79).

This disadvantage can be eliminated if the wedge-shaped charge is made in the form of blast-hole (deep-hole) charges and they are laid in one inclined plane coinciding with the plane of the trench slope. In this case the system of charges consists of sections, each one of which has several deep-hole charges. If a section is formed of four deep-hole charges the first hole is charged along its full length L; the second hole $3/4 \ L$; the third hole $\frac{1}{2} L$ and the fourth hole $\frac{1}{4} L$. A similar charging arrangement is repeated in all other sections. In order to calculate the depth by which each hole in the section is charged, draw a straight line through the upper and lower ends of the first holes of two adjoining sections. The point of intersection of this line with the axis of each hole gives the depth up to which it is to be charged (Fig. 81).

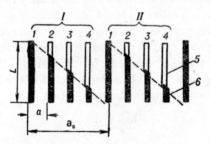

Fig. 81. Layout for charging holes to form a wedge-shaped charge:
I and II—section numbers; 1 to 4—number of charges in one section;
5—stemming; and 6—charge.

For ensuring satisfactory formation of the left-side slope of the trench, along which the section of deep-hole charges is laid, it is necessary to adjust the distance a_s between the sections in such a way that the gas chambers, which are formed during detonation of the first deep-hole charges of the two adjoining sections, come in contact with each other. Let us denote the diameter of the gas chamber by $D_{g.c.}$. Now the condition for contact between gas chambers will be the equation $a_{sec} = D_{g.c.}$.

Diameter $D_{g.c}$ is determined by assuming that the gas chamber has a cylindrical shape. The volume for a 1 m length is equal to $\pi D_{g.c}/4$. The volume of gases in such chambers is determined by the product $p_h u$, where p_h is the mass of explosives in a hole of 1 m length and u the specific volume of the explosion products on their adiabatic expansion down to atmospheric pressure. Therefore,

$$p_h u = \frac{\pi}{4} D_{g.c}^2 .$$

From the two equations given above we observe that

$$a_s = f_{sec} \sqrt{p_h}, \tag{6.17}$$

where

$$f_{sec} = \sqrt{\frac{4u}{\pi}} . \tag{6.18}$$

The mass of explosive charge Q_{sec} is equal to $p a_{sec}$ in all holes of each section. Using expression (6.10) we obtain

$$Q_{sec} = \zeta_e k_w W_3^2 a_s \cot^2 \beta. \tag{6.19}$$

If the holes in all sections are charged as shown in Fig. 81, the total length L_{ch} of the charged areas of holes in one section will be

$$L_{ch} = \frac{1}{2} (N_{sec} + 1) L, \tag{6.20}$$

where N_{sec} is the number of holes in one section; and L the hole length.
By multiplying length L_{ch} by p_h we obtain the mass:

$$Q_{sec} = L_3 p_h. \tag{6.21}$$

Bearing in mind that

$$L = \frac{W_3}{\sin \alpha}; \tag{6.22}$$

$$p_h = \frac{1}{4} \pi d^2 \rho_{exp} \Delta, \tag{6.23}$$

where d is the hole diameter and Δ the relative loading density (ratio between the volume of explosives and the volume of charge area of the hole). From expressions (6.17) to (6.23), we obtain:

$$d = W_3 \frac{4 \zeta_e k_w f_{sec} \cot^2 \beta \sin \alpha}{(N_{sec} + 1)(\pi \rho_{exp} \Delta)^{1/2}} . \tag{6.24}$$

From Fig. 81 it is clear that the distance between the holes is

$$a = \frac{a_s}{N_{sec}} .$$

Fig. 82. Various stages of two-sided directional blasting of wedge-shaped
charges in a four-charge layout.

By substituting from expressions (6.17) and (6.23) we have:

$$a = \frac{df_{sec}}{2N_{sec}} (\pi \rho_{exp} \Delta)^{1/2}. \tag{6.25}$$

In accordance with formula (6.18) we can take $f_{sec} \approx 1$ kg$^{1/2}$/m$^{-3/2}$ for calculation; besides, it should be borne in mind that the minimum number of holes in one section is $N_{sec} = 2$. By varying the number N_{sec}, we can vary the diameter d, which is closer to the diameter of the available drilling bit.

3. MULTICHARGE LAYOUT

For angles in the range $0 < \alpha < 45°$, especially for blasting soft rocks with a relatively low specific consumption of explosives, the three-charge layout does not ensure a complete removal of the rock to the right side of the trench. In this case, it is necessary to use a multicharge layout. Follow the procedure given below for designing the charges. The specific consumption q of explosives should ensure that the rock is projected to a distance D calculated by the following formula:

$$D = 2W_{des} [\cot \alpha + \cot \beta],$$

where W_{des} is the depth of the design pit (depth at which the next row of charges is laid).

The specific consumption of explosives is calculated by formula (1.20). In the case of soft rock we should bear in mind that the efficiency is calculated by the expression (1.22).

Depth W_1 of laying the first linear charge or the first row of concentrated charges is:

$$W_1 = \frac{1}{4} W_{des}.$$

Distance b_1 between the first and second rows of charges is calculated by the following formula:

$$b_1 = W_1 \sqrt{\frac{q}{k_e}}.$$

In the same manner we can calculate the distance between the ith and $(i+1)$th row of charges:

$$b_i = W_i \sqrt{\frac{q}{k_e}},$$

where

$$W_i = W_{i-1} + b_{i-1} \tan \beta.$$

The distance between charges in the ith row:

$$a_i = W_i \sqrt{\frac{q}{k_e}} \, .$$

The mass of concentrated charges in the first row:

$$Q_1 = q W_1 a_1 b_1.$$

For the ith row

$$Q_1 = \frac{1}{2} q W_i a_i (b_i + b_{i-1}).$$

For the last row

$$Q_{des} = q W_{des} \, a_{des} \, b_{des}.$$

If linear charges are used in place of rows of concentrated charges, the mass per unit length of the first, ith and last linear charge is determined by the following expressions:

$$p_1 = q W_1 b_1;$$

$$p_i = \frac{1}{2} q W_i (b_i + b_{i-1});$$

$$p_{des} = q W_{des} b_{des}.$$

Angle β of the right-hand slope of the designed pit is calculated by the following formulas:

$$\cot \beta = \sqrt{\frac{p_{des} + p_{des+1}}{W_{des}^2 \, k_e}} \; ;$$

$$\cot \beta = \sqrt{\frac{2 (Q_{des} + Q_{des-1})}{k_e W_{des}^2 (Q_{des} + Q_{des-1})}}$$

Wedge-shaped charges can be used for making trenches of triangular section. Besides, an embankment can be formed by using a two-sided directional blasting procedure. Figure 82 shows three slides which illustrate the progress of explosion. This explosion was conducted with a four-charge layout with $W = 3$ m. The first slide shows the beginning of the explosion, the second shows that the earth blown away by explosion moves from two sides to the center of the embankment. The third slide shows the position of the forward fronts of the soil at the moment of their mutual impact.

Chapter 7

Subcritical Short-delay Firing of Slab-charge Systems

1. A NEW TECHNIQUE OF DIRECTIONAL BLASTING

A new phenomenon came to light during experimental researches conducted by the Union of Industrial Explosives in 1967 in which the present author also participated. It was observed that if in short-delay firing of a row of charges, the time delay is less than a certain critical value then muck is thrown not toward the side of the charge blasted earlier, as is customarily thought, but at the side of the charges blasted later.

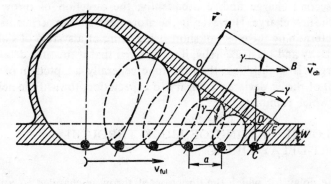

Fig. 83. Diagram of one-sided blasting of rock during short-delay firing of a horizontally laid system of charges with a subcritical time delay.

Figure 83 shows a diagram of trial blasts in which a practically 100% one-side directional blast was achieved when the free face was horizontal. In these trials elongated charges were laid in a horizontal, parallel to the free face. The whole system of charges was fulminated from the left side and the time delay between successive blasts was less than the critical value for the given conditions. As is evident, the exploded rock is thrown in the direction of v_{ful} which coincides with the direction of fulmination of the charge system. In this way, the muck is thrown on the side of the charges which detonate later in

177

such a way that the entire rock deposited above the charges is moved in the given direction.

Analysis of the results obtained from trials proves that the practical utilization of this phenomenon opens new possibilities for improving the effectiveness of directional blasting.

At present in practical blasting operations, the short-delay firing of a system of chamber, as also vertical or inclined deep-hole charges laid in one plane, is being used extensively for achieving a directed explosion. A fully worked-out system of short-delay firing is used for concentrating the blasted rock in the required direction. In these cases, the following principle is borne in mind: if in a given row chamber or deep-hole charges are fired one after another with a predetermined time delay, the direction of throw from each charge deviates from its line of least resistance toward the neighboring charge which was detonated earlier by the assigned delay period. In other words, the direction of throw deviates from the line of least resistance toward that side from which began the short-delay firing of the entire system of charges.

This principle of determining the new direction of throw in case of short-delay firing of two or more charges is reflected in a separate excavation explosion of two chamber charges. In this case the charge which exploded earlier formed a new uncovered plane. As a result, the line of least resistance for the second charge and, consequently, the direction of throw deviated toward the first charge. However, in the aforementioned experiments conducted for determining the effect of short-delay firing of a system of slab charges on directivity and blast accuracy, it was proved that if the time delay between the charges is slightly less than the time generally adopted in practice the direction of throw deviates, contrary to expectation, toward the neighboring charges which fulminate later.

2. EXPLOSIVE CASTING OF MUCK THROUGH SHORT-DELAY FIRING

The time delay for which the direction of throw is changed to an opposite direction will be known as the *critical* time delay.

In order to explain the phenomenon, let us analyze the aforementioned process of interaction of charges during short-delay firing. Figure 84, a, shows a system of five inclined deep-hole charges which are used for directional blasting with a maximum concentration of the blasted rock along the axis of the muck pile. According to established practice for achieving a particular result, we use a short-delay blasting technique and that too, in such a way that the central charge is detonated first and each pair laid symmetrically with respect to the central charge is fired after a certain delay. Such a method of firing will be called *short-delay firing with diverging detonations*. In Fig. 84, a, b and c, the direction of charge detonation is indicated by two arrows origi-

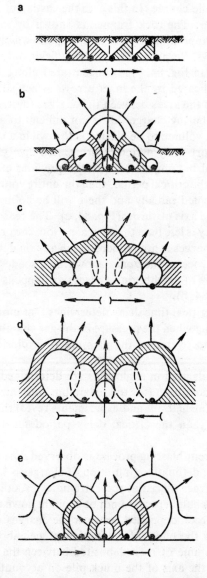

Fig. 84. Schematic diagram of rock blasting during short-delay firing
with different time delays and directed detonations:

a—before blasting; b—time delay greater than critical value with
diverging detonations; c—time delay less than critical value with
diverging detonations; d—time delay less than critical value with
coinciding detonations (new method of directional blasting); e—
time delay greater than critical value with coinciding detonations.

nating from the middle charge. In this case the casting of muck progresses in the following manner: The rock fragments shown by unshaded areas (see Fig. 84, a) move in approximately radial directions which are shown in Fig. 84, b, by outer arrows. Because of short-delay firing only the rock to be exploded (shaded zone in Fig. 84, b) is concentrated along the axis of the blast material which is indicated by the inner arrows which show the direction of rock movement from the areas between the charges toward the gas chambers formed by the neighboring charges. It is not difficult to see that in this case only half of the rock volume to be blasted is thrown in a concentrated form.

If in case of a short-delay period of firing with diverging detonation, the time delay is reduced in comparison with the previous example and is made equal to or less than the critical period, then the entire volume of rock between charges will be projected radially and there will be no noticeable concentration of rock along the axis of throw (Fig. 84, c). The reason for this is that if the blasting time delay is less than the critical period, the pressure in gas chambers of the blasted charges is not reduced much and on detonation of the next pair of charges, the rock in the zones shown as a shaded area cannot move into the gas chambers of neighboring charges as happens in case of a longer time delay (see Fig. 84, b).

In this way, the critical time delay determines that moment in the process of short-delay blasting when the pressure in the gas chamber is still sufficiently high and is able to overcome the pressure of explosion products of the neighboring charge.

The delay intervals used in explosions are determined experimentally for the given explosion condition (for the maximum concentration of rock along the axis of throw). Calculations and experiments reveal that these time delays are always greater than the critical delay period for the given explosion conditions.

A basically different blasting process is observed during a subcritical time delay and coinciding detonation (initiation of charges is begun simultaneously from the two charges placed at the extremities). As the pressure of explosion products in gas chambers could not be brought down considerably during the time delay, the rock deposited between the charges does not move into gas chambers under the effect of detonation of neighboring charges. As a result, the entire volume of rock deposited between the two end-charges is concentrated along the axis of the muck pile on account of a change in the direction of throw (Fig. 84, d).

If in case of a coinciding detonation the time delay is greater than the critical period, the fragments from the shaded areas shown in Fig. 84, a, are projected toward the gas chambers of neighboring charges (Fig. 84, c). As a consequence, this volume of rock in contrast to the explosion diagram shown in Fig. 84, b, is thrown out toward the longer sides of the axis of the muck pile and a considerable part of the rock remains near the point of blast.

From analysis of the throw process under examination it follows, from the viewpoint of maximum accuracy, correct directivity and complete throw, that the short-delay firing of charges with subcritical delay and coinciding detonation (see Fig. 84, d) is most effective because the entire rock mass deposited above the charge is projected in the direction of the axis of the muck pile. If the short-delay firing is conducted with a time delay greater than the critical period then only half of the entire blasted rock volume will be blown out in a concentrated form and the other half will be projected in radial directions (see Fig. 84, b and e). Besides, in this case the entire rock mass will not be blasted but a part of it will remain near the point of blast. This method of short-delay blasting was checked experimentally, and the theoretical and experimental data did not greatly differ (see Sec. 7).

A question arises: why was the technique of blasting with subcritical time delay not discovered so far in practical operations? The answer, in our opinion, lies in the following. Firstly, calculations show that in the majority of cases the delays used in practice are greater than the critical time delays. Secondly, the explosion diagram shown in Fig. 84, d is rarely found in practice. Moreover, it is only during blasting by this method that the effect of the critical time delay is distinctly revealed when the entire rock mass is concentrated along the axis of the blasting material.

3. CRITICAL TIME DELAY

Let us study the short-delay firing of two charges of equal mass laid at a depth W (Fig. 85). The left-side charge is detonated earlier. Therefore, during the delay period t_{del} the gas chamber of the left-side charge expands and

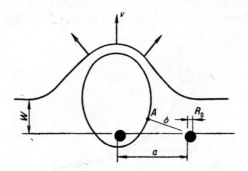

Fig. 85. Diagram for calculating critical time delay.

the pressure of explosion products in it drops to Σ. During expansion of the gas chamber of the left-side charge the right-side charge explodes. The compressional wave produced by it is attenuated after repeated deflection from the face and the gas chamber of the left-side charge. Thereafter, a quasi-

182

stationary process of expansion of the gas chamber of the right-side charge begins. The pressure at all points in the medium will fall with an increase in the distance from the center of the charge. Pressure Σ_A at point A, nearest to the right-side charge and located on the surface of the gas chamber of the left-side charge, at the initial moment after detonation of the right-side charge will be:

$$\Sigma_A = \Sigma_0 \left(\frac{R_0}{b} \right)^a, \tag{7.1}$$

where R_0 is the radius of the gas chamber when the pressure is Σ_0, b the distance between the center of the charge and point A, and Σ_0 the initial pressure.

The value of exponent a depends on the type of charge symmetry. The condition, under which the rock is not moved into the gas chamber of the left-side charge due to detonation of the right-side charge, is the following inequality:

$$\Sigma \geqslant \frac{\Sigma_A}{k_{dy}}, \tag{7.2}$$

where Σ is the pressure in the gas chamber of the left-side charge at the moment of detonation of the right-side charge, k_{dy} is a coefficient greater than unity which takes into account the dynamic head and elastic properties of the medium. This inequality determines the critical time delay during which the pressure in the gas chamber of the left-side charge as well as the dynamic head of the moving medium are capable of preventing the projection of rock into the gas chamber from the right side. The pressure in the gas chamber decreases from a certain initial value with an increase in time. It is not difficult to appreciate that the time during which the pressure in the gas chamber of the left-side charge drops to Σ_A/k_{dy} is the *critical time delay* (Fig. 86).

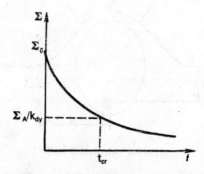

Fig. 86. Graphic method of determining critical time delay.

Let us find the expression for pressure depending on the time. Research [83] shows that dependence of pressure on a specific volume u of the explosion products is expressed by the following isentropic relation:

$$\Sigma \mu^{\varkappa} = C. \tag{7.3}$$

Here $\varkappa = 3$, for $\Sigma > \Sigma_{cr}$ and $\varkappa = 1.25$ for $\Sigma < \Sigma_{cr}$, where critical pressure $\Sigma_{cr} \approx 2 \times 10^8$ N/m².

Constant C is determined from the condition that when $\Sigma = \Sigma_{cr}$,

$$C = \Sigma_{cr} u_{cr}^{\varkappa},$$

where u_{cr} is the specific volume of explosion products at pressure Σ_{cr}.

The value of u_{cr} is found from the isentropic equation for pressure $\Sigma > \Sigma_{cr}$, for which $\varkappa = 3$. By denoting the pressure and specific volume of products of explosion at the moment of detonation by Σ_{det} and u_{det} we obtain the following equation:

$$\Sigma_{cr} u_{cr}^3 = \Sigma_{det} u_{det}^3,$$

whence

$$u_{cr} = u_{det} \left(\frac{\Sigma_{det}}{\Sigma_{cr}} \right)^{1/3}.$$

Therefore

$$C = \Sigma_{det} \left(\frac{\Sigma_{det}}{\Sigma_{cr}} \right)^{\frac{\varkappa}{3}} u_{det}^{\varkappa}.$$

In this way, in the general form for conjugated isentropy we obtain

$$\Sigma = \Sigma_0 \left(\frac{u_{det}}{u} \right)^{\varkappa}, \tag{7.4}$$

where

$$\Sigma_0 = \Sigma_{cr} \left(\frac{\Sigma_{det}}{\Sigma_{cr}} \right)^{\frac{\varkappa}{3}}.$$

It must be borne in mind that $\varkappa = 1.25$ for $\Sigma \leqslant \Sigma_{cr}$ and $\varkappa = 3$ for $\Sigma > \Sigma_{cr}$ [17].

The specific volume u and radius R of a gas chamber with charge mass Q and time t are related by the following equation:

$$Qu = \beta_{cr} R^z; \tag{7.5}$$

$$R = R_0 + c_{cr} t^n, \tag{7.6}$$

where the constant factor c_{cr} depends on the conditions of explosion, but coefficient β_{cr} and exponents z and n depend on the shape of the charge.

By taking the derivative of R in time, we get the rate of expansion of the chamber in time t.

$$v = \frac{dR}{dt} = c_{cr} n t^{n-1}. \tag{7.7}$$

The gas chamber stops transferring its energy to the moving rock at the moment of maximum expansion. It may be considered that in this period of blasting the rate of movement of the boundary point between the gas chamber and the rock is proportional to the initial state of blasting:

$$v = a_v A_v \left(\frac{Q^{1/z}}{W} \right)^{m_1},$$

(7.8)

where W is the explosion depth; A_v, m_1 values which depend on the properties of the rock being blasted, and a_v the proportionality factor.

From expression (7.6), we find the period at the end of which the expansion process of the gas chamber comes to an end:

$$t_{cr} = [(R_{cr} - R_0) \, c_{cr}^{-1}]^{\frac{1}{n}}.$$

(7.9)

During period t_{cr} the gas chamber radius R_{cr} becomes maximum and the velocity of projection is determined by formula (7.8).

Radius R_{cr} is expressed by the following expression:

$$R_{cr} = a_r \, (Qu\beta_{cr}^{-1})^{\frac{1}{z}},$$

(7.10)

where the coefficient a_r takes into account the deviation of the gas chamber shape from spherical and cylindrical symmetry; and u is the specific volume of explosion products at the moment of maximum expansion of the gas chamber.

The constant c_{cr} is calculated from the expressions (7.7) to (7.10):

$$c_{cr} = c_0 \left[\frac{Q^{1/z}}{W} \right]^{m_1 n} Q^{\frac{1-n}{z}},$$

where

$$c_0 = \left[A_v \, a_v \, \beta^{\frac{n-1}{nz}} \, n^{-1} \, u^{\frac{1-n}{zn}} \, a_r^{\frac{1-n}{n}} \right]^n.$$

By substituting the expression for c_{cr} in Eq. (7.6) and ignoring R_0 in comparison with R, we find

$$R = c_0 \, Q^{\frac{1}{z}} \left(\frac{Q^{1/z}}{W} \right)^{m_1 n} \left(\frac{t}{Q^{1/z}} \right)^n.$$

(7.11)

From expressions (7.3), (7.5) and (7.11) we find the expression for pressure Σ on time,

$$\Sigma = \Sigma_0 \left(\frac{u_{cr}}{\beta_{cr}} \right)^{\varkappa} Q^{\varkappa} \left[c_0 \left(\frac{Q^{\frac{1}{z}}}{W} \right)^{m_1 n} Q^{\frac{1}{z}} \left(\frac{t}{Q^{\frac{1}{z}}} \right)^n \right]^{-z\varkappa}.$$

(7.12)

Let us assume that the distance b is proportional to the distance a between the charges, i.e.

$$b = \alpha_e \, b,$$

where α_e is the proportionality factor. From expressions (7.1), (7.3) and (7.12), we get

$$\hat{t}_{cr} = \frac{t_{cr}}{Q^{\frac{1}{z}}} = B_{ch} \hat{W}^{m_1} \hat{a}^{d_{cr}},$$ (7.13)

where

$$B_{ch} = (\alpha_e \, k_{dy})^{\frac{\varkappa}{nz\varkappa}} u_{cr}^{\frac{z\varkappa - x}{nz^2\varkappa}} nu^{\frac{n-1}{n}} a_r^{\frac{n-1}{n}} \left(a_v \, A_v \, \beta_{cr}^{\frac{nz\varkappa - x}{nz^2\varkappa}} \right)^{-1};$$

$$\hat{t}_{cr} = \frac{t_{cr}}{Q^{\frac{1}{z}}}; \quad \hat{a} = \frac{a}{Q^{\frac{1}{z}}}; \quad \hat{W} = \frac{W}{Q^{\frac{1}{z}}}; \quad d_{cr} = \frac{x}{nz\gamma}.$$ (7.14)

Table 7 contains the values of all parameters for spherical and cylindrical symmetries in regard to certain rocks. In the calculations $\varkappa = 1.25$.

TABLE 7

Symmetry	Rock	A_v	$m,$	B_{ch}, ms	x	β_{cr}	d_{cr}	z	n
	Loess	8	3	150					
Spherical	Loam	16	2	90	2	$4/3\,\pi$	5/3	3	1/3
	Hard rock	45	1.5	30					
	Loess	8	2	120					
Cylindrical	Loam	16	1.5	60	1	π	5/6	2	1/2
	Hand rock	45	1	20					

Coefficient B_{ch} for loamy soils and a cylindrical symmetry has been determined on the basis of experimental data (see Sec. 7). For other soils, the coefficient B_{ch} of spherical and cylindrical symmetries shown in Table 7 is calculated using formula (7.14). These should, therefore, be considered as approximate values.

While calculating $\hat{t}_{cr}$ by formula (7.13) it must be borne in mind that for concentrated charges (spherical symmetry) Q implies the mass of the charge and for a cylindrical charge it means the mass per unit length. Therefore, for spherical (concentrated) charges

$$t_{cr} = B_{ch} \, W^{m_1} \, a^{\frac{5}{3}} \, Q^{\frac{1}{3}\left(m_1 + \frac{2}{3}\right)};$$

for cylindrical (linear) charges

$$t_{cr} = B_{ch} \, W^{m_1} a^{\frac{5}{6}} \, p^{-\frac{6m_1-1}{12}} .$$

The values of coefficient B_{ch} in Table 7 are for the following dimensions: W is measured in meters, a in meters, Q in kilograms (spherical charge) and p in kilogram per meter (cylindrical charge). Hence t_{cr} is measured in meter second.

It is not possible to substitute any value of the derivatives of $\hat{a}$ and $\hat{W}$ in Eq. (7.13) because for the given explosion conditions there are definite maximum values of $\hat{a}_{lim}$ and $\hat{W}_{lim}$, which are interrelated.

For spherical (concentrated) charges, when $\hat{W}_{lim} \geqslant 1$, throwing out of muck weakens and the explosion gradually changes to an implosion. Moreover, with an increase in $\hat{W}$, the limiting value of the relative distance $\hat{a}_{lim}$ between charges decreases because the charges cease to have an effect on each other. In practical blasting operations [34] the limiting value of the distance between concentrated charges is calculated by the formula $a_{lim} = \frac{1}{2} W (n_e + 1) \approx W n_e^{2/3}$, where n_e is the index of blast effect. It is approximately calculated by the following expression:

$$n_e \approx \frac{Q^{1/3}}{W} .$$

From this we obtain the following relationship:

$$\hat{a}_{lim} = \hat{W}_{lim}^{\frac{1}{3}} . \tag{7.15}$$

In the same way we can find the relation between $\hat{a}$ and $\hat{W}$ for cylindrical charges. The corresponding calculations give the following result:

$$\hat{a}_{lim} = \hat{W}_{lim}^{\frac{3}{7}} . \tag{7.16}$$

Let us calculate the maximum value of $\hat{t}_{cr \, max}$ for which in formula (7.13), we substitute the expressions obtained for the limiting values of $\hat{a}$.

By calculating, we find that:
for concentrated charges:

$$\hat{t}_{cr \, max} = B_{ch} \, \hat{W}_{lim}^{\frac{3m_1 + d_{cr}}{3}} ; \tag{7.17}$$

for cylindrical charges:

$$\hat{t}_{cr \, max} = B_{ch} \, \hat{W}_{lim}^{\frac{7m_1 + 3d_{cr}}{7}} . \tag{7.18}$$

It is observed that the blasting technique made use of is implosion in case of soils when $\hat{W}_{\lim} \approx 1$ and in case of hard rock when $\hat{W}_{\lim} \approx 1.5$, therefore, the maximum value of t_{cr} is as follows:

for soils

$$t_{cr\,max} = B_{ch};$$

for hard rock

$$\hat{t}_{cr\,max} = B_{ch} \cdot 1.5^{\frac{7m_1 + d_{cr}}{7}}.$$

In practical utilization of short-delay firing with a subcritical time delay, it should be borne in mind that the minimum distance between the charges should be a certain safe value a_{min} which guarantees the safety of the firing circuit of neighboring charges. Experimental investigations conducted by us for this type of blasting reveal that, in general, the safe distance between deep-hole charges is determined by the following formula:

$$a_{min} = k_{safe}\,Wq,$$

where k_{safe} is the coefficient which depends on the physicomechanical properties of the rock; and q is the specific consumption of explosives.

Experimental investigations reveal that for elastoplastic clays the coefficient $k_{safe} \approx 0.9$ m^3/kg. Because clay soils have high springing indexes, therefore, the area of elastoplastic deformation spreads to greater distances. As a consequence the value of k_{safe} for clay is high. For other rocks, especially hard rock, one can expect that coefficient k_{safe} will be considerably lower. For the given conditions of explosion, the coefficient k_{safe} should be determined experimentally.

4. BALLISTIC ANALYSIS

Let us study a short-delay firing of a system of parallel cylindrical charges laid in one horizontal plane at a depth W from the free face. If the time delay is subcritical and fulmination of the charges is begun from the extreme left, the blast front will be rotated at an angle γ to the horizontal as shown in Fig. 83. From the right-angled triangle OAB formed on the vectors of the velocity of projection v and the mean velocity of short-delay firing v_{ch} we obtain:

$$\sin \gamma = \frac{v}{v_{ch}}.$$

In this expression, velocity v is the projection velocity of a rock fragment along a new line of least resistance, which is at an angle γ to the calculated line of least resistance W, and v_{ch} is the average initial velocity imparted by the charges.

As

$$v_{ch} = \frac{a}{t_{ch}},$$

where t is the time delay, therefore,

$$\sin \gamma = \frac{v}{a} t_{ch}. \tag{7.19}$$

Because the distance a between charges is bounded by the values a_{min} and a_{lim} and the time delay cannot be greater than the critical, therefore, the angle of rotation of the thrown-out material γ for the given conditions of blasting can be varied between zero ($t_{ch}=0$) to a certain limiting value γ_{llm} ($t_{ch}=t_{cr}$). Obviously, γ_{lim} can be found if t_{ch} is substituted by t_{lim} in formula (7.19), i.e. the maximum possible time delay at which the rock will still be projected from the area between the charges inside the gas chambers without being broken up.

For given values of v, a and t_{ch} we can calculate the angle of rotation of the blast front from expressions (7.8) and (7.19):

$$f(\gamma) = G_2 a^{d_{cr}-1}, \tag{7.20}$$

where

$$f(\gamma) = \frac{\sin \gamma}{\cos^{m_1} \gamma};$$

$$G_2 = A_v B_{ch}.$$

Limiting value γ_{lim} is found by substituting relations (7.15) and (7.16) in expression (7.19):

for spherical charges

$$f(\gamma_{lim}) = G_2 \hat{W}_{lim};$$

for cylindrical charges

$$f(\gamma_{lim}) = G_2 \hat{W}_{lim}^{\frac{3(d_{cr}-1)}{7}}.$$

It should be noted that in Eq. (7.20) we cannot substitute a value of $\hat{a}$ which is greater than the value obtained from formulas (7.15) and (7.16).

Figure 87 shows curves of the function $f(\gamma)$ for various values of exponent m_1. The value of γ_{lim} can be found by calculating the product $G_2 \hat{a}^{d_{cr}-1}$ which is used for determining angle γ with the help of curves $f(\gamma)$ for a given value of m_1 (see Table 7). According to the data given in Table 7, the quantity $G_2 = A_v B_{ch}$ is equal to 1.2 for spherical charges and 0.96 for cylindrical charges. For these values of G_2, the values of γ_{lim} for spherical and cylindrical symmetries varies, according to Fig. 87, between 40 and 60°.

In case of short-delay firing of an inclined slab charge the blast front, as in the previous case, will be deflected by an angle γ. The vector of initial velocity of projection will also be deflected by this angle. For calculating the flight of rock it is necessary to know the corresponding angle of rotation of the projection of the vector describing the initial velocity of projection in the horizontal plane.

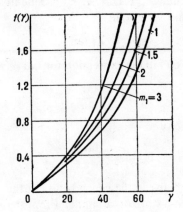

Fig. 87. Curves of $f(\gamma)$ versus γ for different values of parameter m_1.

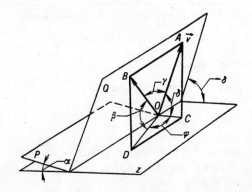

Fig. 88. Diagram for determining angle of rotation φ.

Figure 88 represents a slab-charge system laid in a plane P with an angle of inclination α and plane Q perpendicular to the slab charge. In plane Q, the vector describing the initial velocity of projection $\vec{v}$ rotates through an angle γ. Here, the projection of $\vec{v}$ on the horizontal plane is rotated at an angle φ. Because $\angle OCD$ in $\triangle DOC$ is a right angle, therefore:

$$\tan \varphi = \frac{CD}{OC}.$$

From the right-angled triangle AOC, we find

$$OC = OA \cos \delta.$$

Because line BA lies in a plane parallel to plane z, therefore, $AB = DC$.

Using a similar argument, $BA \perp OA$. Consequently, $\triangle BAO$ is a right-angled triangle. From this we obtain

$$DC = OA \tan \gamma.$$

By substituting DC and OC in the expression for $\tan \varphi$, we have

$$\tan \varphi = \frac{\tan \gamma}{\cos \delta}.$$

Because $\delta = 90° - \alpha$, therefore,

$$\tan \varphi = \frac{\tan \gamma}{\sin \alpha}. \qquad (7.21)$$

Let us calculate angle β between the new position of the initial velocity vector and its projection OD on a horizontal plane. From the right-angled triangle OBD we find

$$\sin \beta = \frac{BD}{OB}.$$

Because $BD = AC = OA \sin \delta$, and $OB = AO/\cos \gamma$, therefore,

$$\sin \beta = \sin \delta \cos \gamma,$$

or, taking into account that $\delta = 90° - \alpha$,

$$\sin \beta = \cos \alpha \cos \gamma. \qquad (7.22)$$

5. THE MUCK PILE

After short-delay firing of a slab-charge system with coinciding detonations and subcritical time delay, the muck pile is formed in the shape of symmetrically arranged lobes (Fig. 89).

The two lateral lobes are made by that volume of blasted rock which is primed for a slab charge at the ends (see Fig. 89, shaded areas). The axis of the lateral lobes is at about 45° to the direction of the main blast. The thickness of the muck pile in the lateral lobes is negligible because a comparatively small volume of rock is spread over a wide area. The lateral lobes become visible only in case of large-scale blasts. The rear lobe of the dump is formed by rock which is primed for a slab charge from the upper (end) face (see shaded area). It appears only in case of a relatively small angle of inclination of the slab charge (less than 30°). The rear lobe of the muck pile is not formed if the angles are large and the specific consumption q of explosives is less than 3 kg/m³. The main dump (area $ABCD$ in Fig. 89) in case of coinciding short-delay blasting, is formed by rock deposited exactly above the slab charge (unshaded rectangular area $ABCD$) and also because of preloading from the lower (end) face of the slab charge. The main muck pile outline is in the form of an elongated trapezium the longer sides of which are at an angle φ to the main direction of throw. Angle φ is calculated by formula (7.21). If the time delay is increased and is brought closer to the critical period, the muck pile contour will be in the shape of a wedge, and its base will touch the lower (end) face of the slab charge. If short-delay firing is conducted with a one-sided initiation (see Fig. 83), the main muck pile will be oval-shaped and the right-side lobe (along the direction of throw) will merge with the main pile (area $ABEFG$ in Fig. 89).

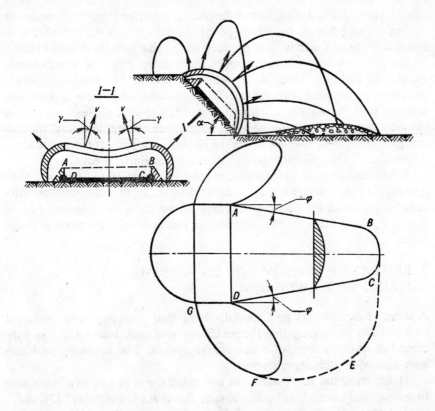

Fig. 89. Diagram of explosive casting and the resultant muck pile in case of slab-charge system with subcritical short-delay firing and coinciding detonations.

6. PRACTICAL APPLICATION OF THE NEW
TECHNIQUE OF DIRECTIONAL BLASTING

A pin-pointed directional blasting can be achieved by firing a number of chamber charges on a steep slope in accordance with the layout shown in Fig. 84, d. Such a blasting technique is effective especially in case of erection of an embankment, dam or dike by blasting the side of a steep slope of monolithic rock in which the blast efficiency is sufficiently high in comparison with soils. This factor is helpful for a concentrated throw of rock fragments to greater distances with a relatively low expenditure of explosives. A dam or dike can be made by directional blasting by such a layout of deep-hole charges.

If the free face is horizontal, it is very difficult to conduct directional blasting by the usual methods. Here, directional blasting implies a rock throw mainly in a predetermined direction. Short-delay firing of a slab-charge system with a one-sided initiation deflects the blast front at the required angle which ensures a one-sided throw of the blasted rock (see Fig. 83). A negligible part of the rock flies to the side opposite the main throw because of a symmetrical expansion of explosion products of the charge at the extreme ends.

Explosive casting of rock through short-delay firing of a slab-charge system can find practical application in strip mining. The bulk of the ore is deposited under the horizontal free face. In order to achieve a maximum range of rock throw, it is necessary to ensure an optimal angle of rotation of the blast front γ_{opt}, the value of which depends on raising (lowering) the slab charge in relation to the face where the blasted rock falls. If the difference in height is zero, the optimal angle of rotation of the throw is equal to $45°$.

It has been proved through experiments that short-delay firing of a horizontally laid slab charge ensures a directed blast. In this case the directivity factor depends on the linear dimensions of the slab-charge system (in the direction of the main throw). The former increases with an increase in the latter.

7. RESULTS OF EXPERIMENTS ON SHORT-DELAY
FIRING OF SLAB CHARGES

A series of experiments by a specially developed procedure was conducted with the aim of investigating the peculiarities of rock blasting during subcritical short-delay firing of a slab-charge system. The following problems were studied in these experiments:

1) Experimental confirmation of a directed throw in case of a subcritical time delay and calculation of the constant factor B_{ch} in formula (7.13); and

2) Experimental exploration of the possibility of a directed throw in subcritical short-delay firing of a horizontally laid system of deep-hole charges.

Experimental investigations were carried out by the 'Soyuzvzryvprom'

Laboratory Trust* at the Afanas'ev open-pit mine in loamy soil with a density of 1750 kg/m³. The author participated in these experiments.

The slab charge was prepared by laying a system of cylindrical charges on the prepared inclined surface with equal distances between them. Cartridged ammonite No. 7 was used as explosive for small-scale blasting and ammonite No. 6 for large-scale blasting. The charges were covered with a layer of soil of a calculated thickness with the help of dump trucks and bulldozers. The delay between successive firings of charges was effected with short-delay detonating cord or detonating loops. For comparison, slab charges were detonated without any time delay in addition to subcritical short-delay firing.

As stated earlier, rock fragments are projected along the direction of firing of the charge in the case of short-delay firing at subcritical time delays. If the time delay is greater than a critical period, the rock is thrown in an opposite direction, i.e. in the direction from which the charges are fired. With a view to confirming this effect, the time delay was varied in a wide range and that time delay was recorded for which the direction of rock blasting had changed.

Table 8 contains the results of experiments carried out at the Afanas'ev open-pit mine. The trial blasts, marked with an asterisk in the table, were carried out with a coinciding detonation and the remaining with a one-sided detonation.

In all experiments, except experiments No. 11 and 12, the fragments were thrown in a direction parallel to the direction of firing of charges. However, in experiments 11 and 12, in which the time delay was greater than the critical period, the direction of blast was changed to an opposite direction. The blast direction was noted visually as well as on the basis of the layout of the pile contours. In the case of a one-sided detonation the muck pile was within the limits of $ABCEFG$ (see Fig. 89) with a clearly noticeable angle φ between the axis of symmetry of the slab charge and the left edge (along the direction of blasting) of the muck pile.

In case of coincident detonation of a slab charge with a subcritical time delay (experiments No. 5, 9, 10 and 13), rock was blasted according to the desired pattern (see Fig. 84, d). This was the case of a pin-pointed throw with a soil concentration along the axis of symmetry of the slab charge, as a result of which the muck pile attained the shape of an elongated narrow stub and was located within the limits $ABCD$ (see Fig. 89). The thickness of the soil heap was maximum along the axis of the throw and it gradually decreased toward the pile edges.

Table 8 presents calculated values of γ and φ. Angle γ could not be found from the experiment in view of the rapidity of the blasting process. However, angle φ depends on angle γ, therefore, the value of γ was indirectly checked

*Trust—in the USSR, a group of industrial or commercial enterprises with centralized direction—The Oxford Russian-English Dictionary. Ed. B.O. Unbegaun. Clarendon Press. 1972.

TABLE 8

No. of blast	$L \times B$, m	α, deg	W, m	q, kg/m³	p, kg/m	a, m	t_{ch}, ms	γ, deg	φ, deg	B, ms
1	2.5×1.5	26	0.5	2.0	0.5	0.5	2	7.5	17	6.3
2	2.5×1.5	30	0.5	2.0	0.5	0.5	4	14	27	12.7
3	2.5×1.5	37	0.5	2.0	0.5	0.5	4	14	23	12.7
4	2.5×1.5	25	0.5	2.0	0.5	0.5	10	40	63	32
5*	3.75×1.5	36	0.75	1.5	0.9	0.75	10	20	31	36
6	3.75×1.5	31	0.75	1.6	0.9	0.75	10	20	35	36
7	2.5×1.5	30	0.5	2.0	0.5	0.5	10	14	27	32
8	2.5×1.5	36	0.5	2.0	0.5	0.5	20	90	60	64
9*	28.5×6	24	1.8	2.2	14.5	3.5	20	12	28	17.5
10*	24.1×6	29	2.0	2.06	14.5	3.5	55	33	58	4.5
11	2.5×1.5	24	0.5	2.0	0.5	0.5	70	—	—	—
12	2.5×1.5	27	0.5	2.0	0.5	0.5	105	—	—	—
13*	3.75×1.5	30	0.75	1.4	0.8	0.75	20	37	56	50
14	2.5×1.5	33	0.5	2.0	0.5	0.5	20	90	62	64

*Coincident detonation.

on the basis of the value of φ in accordance with formula (7.21). Angles observed during blasts were sufficiently close to the calculated values given in Table 8. Angle φ in the case of a coincident detonation of the charge characterizes the degree of soil concentration along the axis of throw. The greater the angle φ, the narrower the loop of the pile and the greater the height of the crown.

The value of coefficient B_{ch} used in formula (7.13) and necessary for calculating the critical time delay was determined in the experiments for short-delay firing of slab charges.

From expression (7.13), we find that

$$B_{ch} = \frac{\hat{t}_{cr}}{\hat{W}^{m_1} \hat{a}^{d_f}}.$$

For determining the value of B_{ch} for a period t_{cr}, it was necessary to find the maximum value of the time delay $\hat{t}_{ch} = t_{ch}/p^{1/2}$, at which the rock is thrown along the direction of firing of a slab-charge system. Obviously, if coefficient B_{ch} is calculated for all trial blasts in which rock is thrown along the direction of firing then according to the formula given above, the maximum value of B_{ch} will correspond to the critical time delay. From the last column of Table 8, it is evident that this value is 64 and is found in experiments No. 14 and 8. The value of coefficient B_{ch} depends on the physicomechanical properties of the soil and the explosive. Therefore, the value of B_{ch} determined from these experiments, strictly speaking, cannot cover all types of cohesive soils. However, as is evident from formula (7.13), the value of B_{ch} is proportional to $A_v^{-1} \alpha_e^{\frac{x}{nzx}}$ which, for a majority of soils, varies negligibly. For solid rock the value of B_{ch} should be determined through experiments according to the procedure given above. The figures given in Table 8 cannot be used as approximate values for hard rock.

Several experiments were conducted for determining the maximum possible distance between cylindrical (hole) charges after blasting of which no crown is formed in the space between them and the throw is fairly complete. In these experiments the distance a between charges was varied while specific consumption q of the explosives, the line of least resistance and W were kept constant. As a result of these experiments it was established that formula (1.5) for the maximum distance between linear charges gives satisfactory results when the specific consumption of explosives is up to 10 kg/m³.

Chapter 8

Practical Recommendations

1. SELECTION OF A BLAST LAYOUT

A blast layout is selected on the basis of those specific conditions in which blasting is to be conducted and the objective that is to be fulfilled. The following factors should be borne in mind while selecting the most appropriate layout for the given conditions.

If the blasting operation is to be conducted on level ground with the intention of moving some volume of rock in order to strip a mineral or create a sufficiently large trench or crater and it is necessary that the exploded rock should fall on one side of the trench, then the most effective method will be short delay-blasting with subcritical time delay (see Chapter 7). This method of blasting (see Fig. 83) is especially suitable for rocks with a relatively high index of springing. This factor is helpful in laying the charge under the massif to be blasted by a prior springing of the pits and their subsequent charging with explosives.

While selecting and calculating for the given scheme, it is necessary to bear in mind that a small volume of rock will fly to the right side (i.e. in a direction opposite to the main directed throw) which is shown as a shaded area in Fig. 90. If the width of the crater B is considerably greater than depth W, i.e. when $B/W \gg 1$, the specific consumption of explosives should gradually increase from the left to the right along the width of the pit with a view to economizing on the quantity of explosives being expended. In Fig. 90 this is shown by a gradual increase in the mass of the charges toward the right side of the trench. In such calculations, the total mass of the charges should be reduced in comparison with cases where the specific consumption of explosives is considered to be constant.

Blasting in accordance with the layout shown in Fig. 90 can also be undertaken for constructing embankments, dikes and coffer-dams, for laying railway lines and building the roads which are necessary for open-pit mining in hilly areas. For a better concentration of rock during blasting, it is advisable to use peripheral charges (see Chapter 1).

For erecting embankments, a very good effect of throw directivity can be

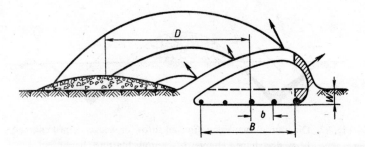

Fig. 90. Layout for a one-sided directed throw during subcritical short-delay
firing of a horizontally laid slab-charge system with variable
specific consumption of explosives.

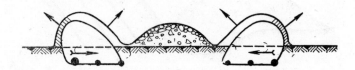

Fig. 91. Layout for two-sided directed throw with subcritical time delay
and variable consumption of explosives.

achieved by using the aforementioned layout for blasting in a two-sided ver-
sion (Fig. 91). Laying of the charges on two sides of the embankment to be
made, leads to a piling up of rock along its axis which results in the maximum
possible density of rock in the embankment. This method, in contrast with the
usual layout, excludes the possibility of an irregular throw of rock fragments
on both sides. Blasting in hard rock according to the layout shown in Fig. 91,
is complicated because the method of concentrated charges for laying charges
under the massif to be blasted, becomes difficult due to the poor springing
property of these rocks. Since in this case, it is advisable to use the method
of deep-hole charges, a directional blasting of wedge-shaped charges (see
Chapter 6) may be conducted for erecting an embankment by laying charges
on both sides of the proposed embankment (Fig. 92). With a view to economizing on explosive, and also for eliminating fish garths and destruction of
slopes at the embankment foundation, it is advisable to lay an additional series
of converging deep-hole charges for contour blasting [84 to 88]. In this case
the additional charges, which are blasted earlier than the wedge-shaped
charges, form a continuous spalling cavity and thereby reduce the work involved in blasting, in the case of wedge-shaped charges.

It is advisable to use a multirow short-delay firing of inclined deep-hole
charges (Fig. 93) for stripping minerals by moving the burden to a predetermined area.

198

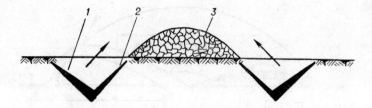

Fig. 92. Diagram of two-sided directed throw of wedge-shaped charges:
1—wedge-shaped charge; 2—contour blasting charge;
3—profile of muck pile.

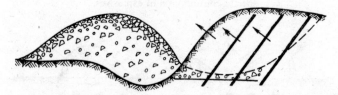

Fig. 93. Diagram of short-delay firing of a system of rows of deep-hole
charges on a steep slope.

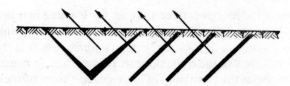

Fig. 94. Diagram of directional blasting of a system of wedge-shaped charges
and rows of deep-hole charges for making a trapezium-shaped trench.

Excavation in hard rock, for which the method of sprung-hole charges is
most difficult, can be easily carried out by using short-delay firing for a multirow system of deep-hole charges (Fig. 94). In order to make the blasting of
subsequent rows of charges more effective, it is advisable to lay the first row
in the form of a wedge-shaped charge (see Chapter 6). With a view to ensuring
the strength and stability of the sides of the formed pit, the technique of
contour blasting may be used for laying the first and last rows of charges.

In practice, one may come across such instances when during blasting on
a slope, it is necessary to create an embankment or dike which must intersect
a depression situated at a level lower than the slope. Directional blasting with
a pin-pointed throw (see Chapter 7) is very effective for these purposes. It can
be used for erecting embankments or dikes, for laying railway lines or constructing roads for open-pit mines.

Considerable practical complications arise when an embankment is built in the bed of a water reservoir by blasting. Of all the known methods, blasting under a gas chamber (Fig. 95) is the most reliable method. The essence of this method is that, initially, a system of cylindrical charges laid in the reservoir bed is blasted with a view to throwing a water layer to the left and right. These charges are blasted one after another beginning with the middle one. The blasting is conducted with a subcritical time delay at which the water layer is thrown at an optimum angle of 45°. The main charges are blasted simultaneously with the firing of the end charges in the reservoir bed and, as a result, an embankment body is formed.

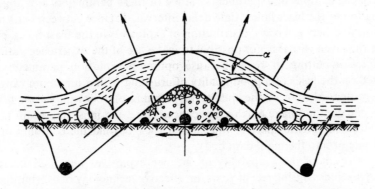

Fig. 95. Diagram of underwater directional blasting with the formation of a gas chamber for making a dam from a reservoir bed.

The method of blasting under a gas chamber has a number of advantages. Firstly, with the blasting of the first series of charges, the reservoir bed is made compact which provides stability of the embankment being made. Secondly, the layer of water thrown aside does not hinder the blasting of rock when the main charges are fired. Thirdly, conditions of a directed throw are created when the main charges are fired, because these charges detonate at that moment when the pressure of explosion products of the end charges in the reservoir bed, and the hydrostatic pressure of the water hinders the symmetry of rock flight when the main charges are exploded. This factor creates a very high priming on the main charges for throwing the rock to the side opposite the embankment being made. As a result of this, the direction of rock throw by the main charges is changed toward the gas chamber in which the pressure approaches atmospheric pressure at the moment of firing of the main charges. Finally, a directed explosion under a gas chamber creates conditions for collision of the blasted rock and maximum consolidation of the fragments in the embankment body. In this case, the harmful damping effect of water is totally excluded.

The technique of blasting under review is, in addition to the advantages stated above, highly reliable for achieving the desired dimensions of the proposed embankment and, what is more significant, reduces the specific consumption of explosives by several times.

2. OPTIMUM BLASTING PARAMETERS

Blast energy is utilized most economically for stripping minerals, constructing dams, dikes, coffer dams and other hydraulic installations where a maximum range of rock blasting is achieved with the same specific consumption of explosives. With this aim the optimum explosion parameters are selected on the basis of calculations or experiments. Some of these parameters are: angle of inclination of the blast holes, time delay interval, distance between blast holes in a row and between rows, distribution of explosives in the blast holes, etc.

In the given case it is not possible to make use of the experience gathered in explosion-cutting of rocks because optimum explosion parameters are selected on the basis of a good quality of crushing with a minimum range of rock scatter. Moreover, the quality of crushing of rocks moved to form a muck pile does not have much significance. Here, the most important factor is to move the greater part of the blasted rock beyond the contours of the muck pile or into the main dam body.

As a result of the availability of cheaper explosives and high-efficiency boring machines during recent years, an effective technology of overburden removal is being used in Soviet and foreign open-pit mine fields for mining without using a means of transportation. This technology is based on an optimum combination of explosion and excavation for moving the rock to form a muck pile [89, 90]. The bulk of overburden removal operations is based on a most suitable channeling of the blast energy. This factor is characterized by a parameter of the overburden removal operation such as the index of displacement by explosion i_e which is the ratio between the volume of the rock U_e moved by explosion and the volume of blasted rock U.

$$i_e = \frac{U_e}{U}. \tag{8.1}$$

It is obvious that the greater the value of the index of displacement by explosion with the same specific consumption of explosives, the larger the part of overburden removal operation based on explosion. This index can also characterize the economic effectiveness of the technique of directional blasting during excavations for dams and other constructions.

In this manner, the optimum value of a given explosion parameter can be determined on the basis of the maximum throw range or index of displacement by explosion with a minimum specific consumption of explosives. These criteria will be borne in mind in future research for optimization of the explosion parameters.

3. ANGLE OF INCLINATION OF DEEP HOLES

During recent years the method of inclined deep-hole charges [91 to 94] has been increasingly used for breaking of rocks by explosives. This method improves the crushing of rock and making of the base of an open-pit bench. Under certain conditions, it is highly effective even for overburden removal by explosion to a predetermined area.

It is well known that in most practical cases, the maximum projection effect is achieved when inclined deep-hole charges laid at 45° to the horizontal are exploded. For example, let us compare the range of rock scatter in two different cases with the specific consumption of the explosives being constant: a system of vertical deep-hole charges, and deep-hole charges laid at 45° to the horizontal (Fig. 96) are exploded.

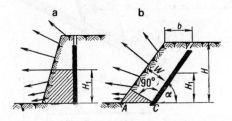

Fig. 96. Diagram of rock scatter during explosion of a) vertical and b) inclined deep-hole charges.

Let us study an elementary volume of rock deposited at height H_1 from the deep-hole base. H_1 is approximately half the length of the deep-hole. The ratio between the desired ranges can be found on the basis of the data given in [8]. The maximum range D_{45} of the rock blasted away by explosion of a slab charge laid at 45° to the horizontal is calculated by the following expression:

$$D_{45} = \frac{2e_{\exp}\eta}{\rho g}\, q.$$

If the vector describing the initial velocity of projection is parallel to the horizontal, the body projected with this velocity from a height H_1 will fall at a distance:

$$D_0 = \sqrt{\frac{4e_{\exp}\eta}{\rho g}}\, H_1\, q\, .$$

By dividing the left- and the right-hand sides of these expressions we find

$$\frac{D_{45}}{D_0} = \sqrt{\frac{e_{\exp}\eta q}{\rho g H_1}}\, .$$

202

For hard rock the average values are: $\rho = 2500$ kg/m³; $\eta = 0.3$; and $e_{exp} = 4.3 \times 10^6$ J/kg. Taking $H_1 = 10$ m and the specific consumption of the explosives $q = 0.3$ to 1 kg/m³, we find the relation $D_{45}/D_0 = 1.25$ to 2.3. Hence it is evident that for typical open-pit mining conditions, the throw range of the inclined holes, in comparison with vertical holes, is greater by approximately 2 times whereas the specific consumption of the explosives is the same in both cases. Consequently, the index of displacement by explosion will be increased when for the given conditions of explosion, the optimal angle of inclination of deep-holes is equal to 45°.

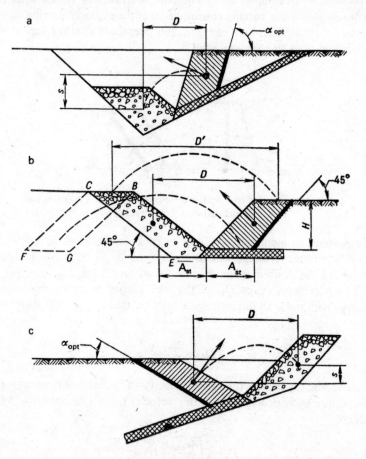

Fig. 97. Ballistics of flyrock for different methods of ore mining:
a—for a depression; b—for the same horizontal level; c—for a higher plane.

The major disadvantage of vertical deep-holes is that the rock in the lower half of the ledge scatters at negative angles to the horizontal (see Fig. 96, a).

Therefore, the rock is thrown near the zone of explosion and does not form a muck pile. A different picture is obtained in the case of explosion of inclined deep-hole charges. The rock in the lower half of the ledge, except for a small volume near the base, is projected at positive angles (see Fig. 96, b). Consequently, practically the entire volume of exploded rock is moved to form a muck pile.

The optimum angle of inclination α_{opt} of deep holes depends on topographic conditions in which the mining of minerals is carried out without a means of transportation. Figure 97 shows typical layouts for mining in different topographic conditions. The value α_{opt} depends only on the relation between the relative altitude s and range D by which the center of gravity of the blasted rock moves. This factor is numerically characterized by the coefficient of relative altitude ζ which is equal to the ratio between the relative height s and range D (see Chapter 1).

$$\zeta = \mp \frac{h}{D}.$$

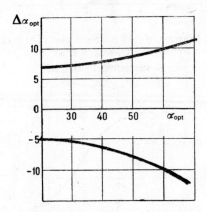

Fig. 98. Relationship between permissible deviations and optimum angle of inclination of a slab charge.

Figure 98 shows the limits of $\pm \Delta\alpha_{opt}$ in which the deviation from α_{opt} causes a reduction in index i_e by about 10%. The value of i_e decreases sharply when α_{opt} is greater than the values shown in the chart.

4. LOADING BLAST HOLES WITH EXPLOSIVES

By comparing the blast layouts shown in Fig. 99 it becomes clear that in case of a partial charging of the blast holes (diagram a), the entire volume of the overburden deposited above the level of charges in the blast holes, i.e. above line A–A, receives an impulse mainly along the vertical. As a result, the overburden is not blasted beyond the boundary of the pile. In case of a complete charging of blast holes (over their entire depth) the bulk of the rock removed by explosion to form a muck pile increases sharply (diagram b).

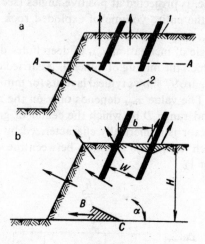

Fig. 99. Diagram of rock scatter in case of a) partial and b) complete loading of blast holes:

1—stemming; 2—charge.

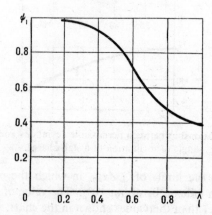

Fig. 100. Relationship between relative loss of charge energy and the ratio between the lengths of the charge and the blast hole.

However, the absence of stemming or its short length leads to a considerable loss of blast energy due to a nonproductive ejection of gases through the blast hole. Figure 100 shows the curve of explosive energy losses through the blast hole after ejection of the stemming. This curve is plotted on the basis of calculations given in [13]. The coefficient of energy losses ψ_1 (ratio between blast efficiency of a deep-hole charge after ejection of stemming and maximum possible efficiency) is shown along the ordinate and the coefficient of charging k_{ch} (ratio between the length of the charge H_{ch} and length of the hole H_h) along the abscissa. It is clear from the graph that in case of a complete charging of the deep hole ($k_{ch} = 1$) more than 50% of the total explosive energy is dissipated due to a flow of explosion products. A short length of stemming (less than 30–40% of the hole depth) also does not improve the utilization of explosive energy: about 50% of the charge energy is wasted due to the explosion products escaping from the blast hole. This can be explained by the fact that due to the tremendous pressure of the explosion products at the moment of detonation of the charge, the stemming material, in view of its low density, gathers a high momentum of several hundred meters per second in a very short period. This is the period in which the stemming flies out and opens the blast hole and the explosion products transfer about 30–50% of the energy to the medium being exploded. After ejection of the stemming the energy remaining in the gas chamber is almost fully carried away by the explosion products.

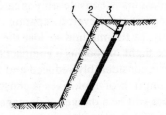

Fig. 101. Diagram of placement of self-wedging stemming
for a deep-hole charge:

1—main charge; 2—blocking charge; 3—stemming section.

In order to avoid this wastage, special attention should be paid to stemming. In case of a short hole length, the stemming should be of the self-wedging type. The experiments mentioned in Chapter 2 show that stemming possesses self-wedging properties in those cases where so-called blocking charges (Fig. 101) are laid along its entire length. Self-wedging stemming should simultaneously satisfy two conditions: 1) it should not contain more than one, or two, blocking charges; and 2) its length l_{ch} should not be less than half the distance b between the rows of charges.

If the second condition is not fulfilled a part of the massif being exploded may break off together with the self-wedging stemming due to the effect of explosion of the main charge, and the explosion products may fly out of the slot thus formed.

5. SEQUENCE OF FIRING AND TIME DELAY

In case of a multirow explosion of a system of deep-hole charges, the most expedient layout will be that in which the explosion of the row fired earlier does not affect the progress of blasting in the next row. This condition is fulfilled by a row-wise short-delay firing with successive detonations of charges beginning with the first row.

Experimental investigation [95] shows that in case of instantaneous explosion of rows of deep-hole charges, the muck pile width increases in comparison with explosion with a certain optimum time delay. This can be explained in the following manner: in case of instantaneous explosion the subsequent rows of charges transfer their energy by propagating compressional waves to that volume of rock which is blasted by explosion of the first row of charges. In this case the range of overburden blasted by explosion of the first row of charges will increase, but will gradually decrease for subsequent rows. Such a distribution of blast energy between rows of charges during removal of the overburden on blasting is highly undesirable because the index of displacement by explosion (8.1) will be reduced. From this point of view a separate explosion of each row of charges would have been ideal (when the time delay between successive explosions is considerably greater than the time taken by the entire explosion process). However, a separate explosion is not desirable for many reasons. Firstly, by this method we lose the main advantage of short-delay firing, namely, the flight of rock as a compact mass due to which the retarding effect of the air is considerably reduced and the flight range increases while the specific consumption of explosives is constant. Secondly, in case of separate explosions there will be a danger of damage to the detonating circuit of the rows of charges which explode later. As a result there may be failures in explosion. Finally, separate explosions will lead to increased fissures in the rock which will sharply reduce the explosive efficiency of subsequent rows of charges.

The time delay t_{ch} between successive explosions of rows of charges should be of such duration as is sufficient for the completion of the process of expansion of explosion products of a previous row of charges to atmospheric pressure. Denoting the time for expansion of explosion products by t_{exp} we get:

$$t_{ch} = t_{exp}. \qquad (8.2)$$

The value of t_{exp} is calculated by the following formula:

$$t_{exp} = \frac{c_{inst}\, H_{g.c}}{v}, \qquad (8.3)$$

where $H_{g.c}$ is the thickness of the gas chamber along the normal to the slab-charge system in case of adiabatic expansion of explosion products to atmospheric pressure; v the initial velocity of projection of rock fragment and c_{inst}

a coefficient which takes into account that the moving rock does not achieve velocity v instantaneously but gradually over the whole process of expansion of the gas chamber [96].

Because

$$H_{g.c} = qWu;$$

$$v = \sqrt{\frac{2e_{exp}\,\eta}{\rho}}\,q\,,$$

therefore

$$t_{ch} = c_{inst}\,Wu\,\sqrt{\frac{\rho q}{2e_{exp}\,\eta}}\,. \tag{8.4}$$

Let us compare the time delays for two cases: in the first case the medium is loosened by explosion; in the second case the same volume of rock is moved by blasting to form a muck pile. Here, all explosion parameters used in expression (8.4) are identical and their values are: $W = 5$ m; $\rho = 2500$ kg/m^3; $q = 0.8$ kg/m^3; $e_{exp} = 4.3 \times 10^6$ J/kg; $u = 1$ m^3/kg; $c_{inst} = 1.5$; and $\eta = 0.3$. According to formula (8.4) t_{ch} is 210 ms for the second case. If the explosion is conducted for loosening the rock, then, on the basis of the data for these conditions, $t_{ch} = 25$ ms. Evidently, if a maximum projection is to be obtained from the explosion the time delay between successive explosions of the rows of deep-hole charges in comparison with the explosion for loosening should be increased by about 10 times. Only after such a sharp rise in the value of t_{ch} will the explosion of the previous row of charges not hinder the blasting process due to explosion of the next row.

However, the time delay given in formula (8.4) should be improved in the course of preliminary experimental explosions because, as mentioned earlier, in case of an excessively high value of t_{ch} the rarefaction wave, propagated deeper into the massif after explosion of the previous row of charges, will open the fissures through which the explosion products will issue forth after detonation of the next rows of charges and thereby, the energy transfer for the propellant effect will be reduced sharply.

If the explosion circuit is made up of detonating cords, the probability of breakage of the cord increases at such high values of t_{ch}. This leads to a failure in explosion and, as a consequence, in an inferior blasting of charges. In order to prevent this undesirable occurrence it is advisable to use an electric detonator for a short-delay firing.

6. DISTANCES BETWEEN ROWS OF CHARGES AND BETWEEN CONSECUTIVE CHARGES

In Fig. 99 b, the shaded area shows the volume of rock which, because of an end effect, will be projected to a zone nearer the point of explosion. It is very easy to obtain from this figure an expression for the maximum possible value of index i_e calculated by formula (8.1):

$$i_e = 1 - \frac{\sin 2\alpha}{4} \cdot \frac{b}{H},$$

whence

$$\frac{b}{H} = \frac{4(1 - i_e)}{\sin 2\alpha}, \qquad (8.5)$$

where b is the distance between rows; and H the height of the stope being blasted.

While deriving this formula, it was assumed that the rock projected to the zone nearer the point of explosion is below the plane BC, passing perpendicularly to the hole axis through its base.

Using a practically acceptable value of $i_e = 0.9$ in expression (8.5), we obtain the corresponding optimum distance between rows of blast holes:

$$\frac{b}{H} \leqslant \frac{0.4}{\sin 2\alpha},$$

whence

$$b \leqslant H \frac{0.4}{\sin 2\alpha}. \qquad (8.6)$$

While moving the overburden by explosion the distance between charges in a row should not exceed the value calculated by formula (1.10), otherwise, a part of the rock being exploded, which lies between the charges, will not be blown out of the explosion zone, and thus the coefficient of displacement by explosion (8.1) will be reduced.

7. WIDTH OF STOPE AND SPECIFIC CONSUMPTION OF EXPLOSIVES

Let us determine the specific consumption of explosives for displacing the overburden to a distance D (see Fig. 97) with the aforementioned optimum explosion parameters.

Let us calculate the value of criterion (1.12) for typical open-pit mining conditions. The value of B in this case is equal to the depth of the blast hole L_{hole}. Since $L_{hole} = H/\sin \alpha$; $W = b \sin \alpha$, and $b/H \leqslant 0.4/\sin 2\alpha$, substituting the expressions for these parameters in (1.12) we find

$$\chi = \frac{L_{hole}}{qu\,W} = \frac{5}{\tan \alpha\, qu}. \qquad (8.7)$$

As according to condition (1.12)

$$\frac{L_{\text{hole}}}{qu\,W}=\frac{5}{qu\tan\alpha}\geqslant\sqrt{\frac{16e_{\text{exp}}\,(c_{\text{s}}-1)\,ik^{2}\,q}{c_{\text{s}}^{2}\,\rho\,gu\,W}}\;,$$

therefore, from this equation we get the value of specific consumption of explosive at which a plane-parallel throw is possible with an index $i_{\text{e}}\approx90\%$:

$$q\leqslant c_{q}\sqrt{W}, \tag{8.8}$$

where

$$c_{q}=\sqrt{\frac{1.57\,c_{\text{s}}^{2}\,\rho g\,W}{e_{\text{exp}}\,(c_{\text{s}}-1)\,ik^{2}\,u\tan^{2}\alpha}}\;. \tag{8.9}$$

Using the following average values $c_{\text{s}}=1.15$; $e_{\text{exp}}=4.3\times10^{6}$ J/kg; $u=1$ m^3/kg; $\alpha=45°$; $k=0.9$; $\rho=2500$ kg/m^3 (solid rock); $\rho=1800$ kg/m^3 (cohesive soil); $i=0.33$ m^3/kg (for hard rock when $q\leqslant1.5$ kg/m^3); and $i=0.04$ m^3/kg (for cohesive soil) we find:

for hard rock

$$q\leqslant0.67\sqrt[3]{W}; \tag{8.10}$$

for cohesive soil

$$q\leqslant1.3\sqrt[3]{W}. \tag{8.11}$$

The values of specific consumption of explosives calculated by these equations help in finding approximate upper limits of q at which it is still possible to displace the rock as a compact mass (without a breaking of the blast dome into fragments). Exact values of these limits can be obtained only in special experiments conducted for this purpose. It should be borne in mind that the calculated values of specific consumption of explosives should fulfill the conditions (1.12) and (1.13).

Let us calculate the specific consumption of explosives which ensures a range D when inequalities (1.12), (1.13) and (8.6) are satisfied.

According to formula (1.20)

$$q=\frac{f(\alpha,\zeta)}{c_{1}}\,D, \tag{8.12}$$

where

$$c_{1}=\frac{4\eta e_{\text{exp}}}{\rho g}. \tag{8.13}$$

The value of function $f(\alpha,\zeta)$ for optimal angles (see Fig. 9) can be found from the graph shown in Fig. 10.

As for soft and hard ($q<1.5$ kg/m^3) rock $\eta=iq$, therefore, from expressions (1.20) and (1.22) we find

$$q=\sqrt{\frac{f(\alpha,\zeta)\,\rho g}{4ie_{\text{exp}}}}\,D. \tag{8.14}$$

It should be borne in mind that the proportionality factor i for the given type of rock should be determined through experiments (see Chapter 2, Sec. 4). For approximate calculations, we may take

$i=0.25$ m³/kg (hard rock, $q<1.5$ kg/m³) and

$i\approx0.04$ m³/kg (cohesive soil).

Let us transform formula (8.14) for a system of mineral mining that does not involve transportation, but removes the overburden to a predetermined area by explosion. For this purpose, let us express the range of moving the rock through a width of stope A_s and height H. For calculation purposes let us examine the blast layout shown in Fig. 97, b. Here, it should be borne in mind that the specific consumption of explosives should ensure movement of the upper part of the ledge to a distance D'. Let us denote the slope angles of the muck pile and the ledge to be exploded by φ_0 and φ_y, respectively. If the density of the mineral is considerably lower than the density of the enclosing layer, it is easy to obtain the following expression from Fig. 97, b:

$$D' = A_s + H\,(\cot\varphi_0 + \cot\varphi_y).\tag{8.15}$$

By substituting it in formulas (8.12) and (8.14) in place of range D, we have

$$q = \frac{f(\alpha,\,\zeta)}{c_1}\,[A_s + H\,(\cot\varphi_0 + \cot\varphi_y)];\tag{8.16}$$

$$q = \sqrt{\frac{f(\alpha,\,\xi)\,\rho g}{4ie_{exp}}}\,[A_s = H\,(\cot\varphi_0 + \cot\varphi_y)].\tag{8.17}$$

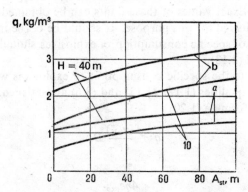

Fig. 102. Relationship between specific consumption of explosives and width of stope and height of ledge for a) soil and b) hard rock.

Figure 102 shows curves for the specific consumption of explosives calculated by formulas (8.16) and (8.17) for hard rock and loose soil. The following values have been used in the calculations: $f(\alpha,\,\zeta)=2$; $\varphi_y=45°$; $\varphi_0=35°$, $\rho=2300$ kg/m³ (hard rock); $\rho=1800$ kg/m³ (soils); $i=0.2$ m³/kg (hard rock); $i=0.04$ kg/m³ (soils); and $e_{exp}=4.3\times10^6$ J/kg.

From a comparison of these curves it follows that the specific consumption of explosives sufficient for removing the overburden to a muck pile in case of loose soil is two times more than in case of hard rock. However, it should be mentioned that utilization of an explosive effect in air chambers [62, 63] sharply increases efficiency and leads to a reduction in the specific consumption of explosives.

8. DISTRIBUTION OF EXPLOSIVES IN A BLAST HOLE

Formulas (8.16) and (8.17) have been obtained with the condition that the entire exploded mass attains an approximately uniform initial velocity of projection which ensures movement of the center of gravity of that mass to a distance D' (see Fig. 97, b). If the dump slope CE were not in the way of movement of the rock then the rock would have been in the area $CBGF$. Hence, it is clear that the lower part of the mass being exploded receives considerable energy, capable of displacing the rock to a distance which is roughly three times greater than that required. Therefore, in an ideal situation, the distribution of blast energy in the ledge volume should be proportional to the required range of the trajectory which, as shown in Fig. 97, b, is increased with an increase in the height of the rock above the minerals.

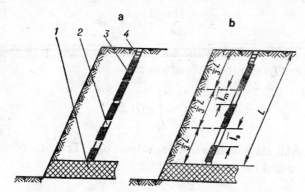

Fig. 103. a) Multi-section and b) three-section charging of blast holes with variable mass of charge along holes:
1—mineral; 2—inert tamping material; 3—charge;
4—self-wedging stemming.

A nonuniform distribution of energy in the rock to be exploded can be achieved by using a variable loading density of explosive along the blast hole. This can be easily effected if the explosive is charged in the blast hole in such a way that the quantity of inert material between dispersed parts of the charge is gradually decreased from the base toward the mouth of the blast hole (Fig. 103, a). However, in practice such nonuniform charging is a labor-consuming operation. It is simpler to divide a blast-hole charge into three sections and place the required quantity of inert material between them (Fig. 103, b). For detonation, the charge sections are connected to the main detonating cord. Let us study the design of three sections of a deep-hole charge for an instance in which the mineral is deposited horizontally.

According to the energy principle [97] the mass of charge per meter of blast hole for any of its points is calculated by the following formula:

$$p = qab \sin \alpha, \tag{8.18}$$

where the distance a between the blast holes in a row is determined by formula (1.10).

In a general form, the specific consumption of explosives for any point of the blast hole is expressed by the following expression:

$$q = \frac{f(\alpha, \zeta)}{c_1} \left[\frac{l}{L} (D' - A_s) - A_s \right], \tag{8.19}$$

where l is the distance from the bottom of the blast hole, and L its total depth.

The mean specific consumption q_m of explosives for the entire rock to be blasted will, obviously, correspond to distance $l = \frac{1}{2} L$. Substituting this value in formula (6.19) we find

$$q_m = \frac{f(\alpha, \zeta)}{2c_1} (D' + A_s). \tag{8.20}$$

The mass of explosives at any point in the blast hole, with ends placed at distances l_1 and l_2 from the bottom, will be:

$$Q = \int_{L_1}^{L_2} p(l) \, dl.$$

Let us divide the hole into three equal sections. The mass of charge in the upper, middle and lower sections will be:

$$Q_{up} = \int_{\frac{2}{3}L}^{L} p(l) \, dl = B_s (5D' + A_s); \tag{8.21}$$

$$Q_{mid} = \int_{\frac{1}{3}L}^{\frac{2}{3}L} p(l) \, dl = B_s (3D' + 3A_s); \tag{8.22}$$

$$Q_{low} = \int_{0}^{\frac{1}{3}L} p(l) \, dl = B_s (D' + 5A_s), \tag{8.23}$$

where

$$B_s = \frac{f(\alpha, \zeta)}{18c_1} Lab \sin \alpha. \tag{8.24}$$

The entire charge mass in the blast hole Q will be equal to $Q_{up} + Q_{mid} + Q_{low}$. By summation we get

$$Q = B_s \, 9 \, (D' + A_s). \tag{8.25}$$

From expressions (8.21) to (8.23) we obtain

$$Q_{up} : Q_{mid} : Q_{low} = (5D' + A_s) : (3D' + 3A_s) : (D' + 5A_s). \tag{8.26}$$

Let us calculate diameter d of the blast hole with the condition that one-third of the upper section of the blast hole should contain a charge Q_{up} and self-wedging stemming.

Because the stemming length $l_{stem} = 0.5 \, b$; $b = 0.4 \, H/\sin 2\alpha = 0.4 \times L \sin \alpha / \sin 2\alpha$, therefore, for an optimal angle $\alpha = 45°$, we have

$$l_{stem} = 0.14L. \tag{8.27}$$

On this basis we obtain the following equation:

$$\frac{\pi d^2}{4} \left(\frac{1}{3} L - 0.14L \right) \rho_{exp} \, \Delta = Q_{up}, \tag{8.28}$$

where ρ_{exp} is the bulk density of explosives and Δ the relative loading density.

From relation (8.28) we get:

$$d = 2.6 \sqrt{\frac{Q_{up}}{L \rho_{exp} \, \Delta}} \, , \, \text{m}, \tag{8.29}$$

where Q_{up} is measured in kilograms, L in meters and ρ_{exp} in kilograms per cubic meter.

In the same way we can find the formula for the diameter of a blast hole for continuous charging:

$$d = 1.23 \sqrt{\frac{Q}{L \rho_{exp} \, \Delta}} \, , \tag{8.30}$$

where Q is the mass of the charge in a blast hole calculated by the following formula:

$$Q = q \, Lab \sin \alpha. \tag{8.31}$$

Substituting the use of Q from the expression (8.31) in formula (8.30) we obtain the formula for calculating a blast-hole diameter when charges are placed in a dispersed manner:

$$d = 0.87 \sqrt{\frac{f(\alpha, \zeta) \, ab \sin \alpha}{18c_1 \, \rho_{exp} \, \Delta} (5D' + A_s)}. \tag{8.32}$$

Let us calculate the length of inert material in the middle and lower sections of a blast hole (Fig. 103, b). In the middle section

$$l_{mid} = \frac{1}{3}L - l_{ch}.$$ (8.33)

where l_{ch} is the length of charge in this section of the blast hole.

The value of l_{ch} is calculated by the following formula:

$$l_{ch} = \frac{Q_{mid}}{\frac{1}{4}\rho_{exp}\,\Delta\pi d^2}.$$ (8.34)

From expressions (8.22) to (8.34), we find

$$l_{mid} = L\left[\frac{1}{3} - 0.19\frac{3D' + 3A_s}{5D' + A_s}\right].$$ (8.35)

In the same way we find the formula for calculating the length of inert material in the lower section of a blast hole:

$$l_{low} = L\left[\frac{1}{3} - 0.19\frac{D' + 5A_s}{5D' + A_s}\right].$$ (8.36)

Charging of a blast hole with a variable mass per unit length economizes on the specific consumption of explosives which, in this case, is equal to the specific consumption of explosives for the middle section of the blast hole:

$$q = \frac{f(\alpha, \zeta)}{c_1} \cdot \frac{A_s + D'}{2}.$$ (8.37)

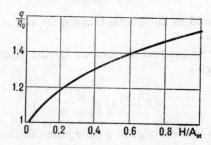

Fig. 104. Dependence of ratio between specific consumption
for continuous and dispersed charging of hole on ratio
between ledge height and width of stope.

By comparing the obtained expression with formula (8.16) for specific consumption of explosives in continuous charging of a blast hole and $\varphi_0 = \varphi_y = 45°$ it becomes clear that considerable economy in specific consumption of explosives is achieved when $H/A_s > 1$. It is evident from Fig. 104 in which the ratio curve q/q_0 is depicted between specific consumption of explosives for dispersed and continuous charging of blast holes against the value of the ratio H/A_s.

In case of breaking of rock by explosion the blast hole is charged in such a way that the center of gravity of the charge is shifted toward the lower half of the blast hole. This becomes necessary in order to process the ledge base properly. For the same purpose redrilling of blast holes is carried out.

The charging method, examined here, with variable mass of the charge along the blast hole, however, shifts the center of gravity to the upper half of the blast hole. Such a distribution of the charge for breaking of rock by explosion leads, as a rule, to a poor processing of the ledge base. In the present example this undesirable occurrence is avoided because stable bonds between the minerals and covering rock are, as a rule, weaker than in the case of monolithic rock.

Displacement of the center of gravity of the charge to the upper section of the blast hole is also beneficial from another aspect: it reduces crushing of the minerals and prevents their being thrown into the muck pile. A safety cushion made of an inert material and laid in the lower section of the blast hole also serves this purpose.

9. ECONOMIC EFFICIENCY OF REMOVING THE OVERBURDEN TO MUCK PILE BY EXPLOSIVE CASTING

It has been proved that under certain favorable geologic conditions (relatively short distance of muck piles from the faces, and a fairly high efficiency of explosive excavation) the movement of the overburden directly to the muck pile by blast energy is more economical than transporting the muck by road and rail.

Let us calculate the expenditure incurred on transporting one ton of spoils by different means of transport and compare these with the cost involved in case of removing it by explosive casting. Let us carry out calculations for a situation when the overburden is raised by blasting upward from the base of the pit. Suppose the depth of the open-pit mine is H. Then expenditure C_{exp} in case of movement of muck by explosive excavation is expressed by the formula:

$$C_{exp} = C_{exp_1} q,$$

where C_{exp_1} is the price of 1 kg of explosives; and q the specific consumption.

In the given case we take into account only the cost of removing the overburden by explosive casting to the shortest distance without taking into account the costs on drilling and charging the holes (these comprise a negligible part of the total costs involved C_{exp}). According to expression (1.20)

$$q = f_{opt}(\alpha, \zeta) \frac{\rho g}{4\eta e_{exp}} D,$$

where function $f(\alpha, \zeta)$ can be approximated by the following relationship

$$f(\alpha, \zeta) = 2 + 3\zeta.$$

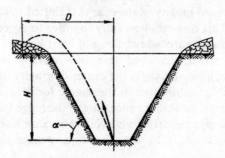

Fig. 105. Diagram showing removal of overburden from deep open-pit mine to muck pile by blasting.

From Fig. 105 and the definition of coefficient ζ, it follows that

$$\zeta = \frac{H}{D} = \tan \alpha,$$

therefore,

$$f(\alpha, \zeta) = 2 + 3 \tan \alpha.$$

Consequently,

$$q = (2 + 3 \tan \alpha) \frac{\rho g}{4 \eta e_{exp}} D.$$

Considering that

$$D = \frac{H}{\tan \alpha} = \frac{H}{\zeta},$$

we determine

$$q = \frac{2 + 3 \tan \alpha}{\tan \alpha} \cdot \frac{\rho g}{4 \eta e_{exp}} H.$$

If the specific consumption of explosives is expressed in terms of consumption per ton of burden rather than per cubic meter, then

$$q = \frac{2 + 3 \tan \alpha}{\tan \alpha} \cdot \frac{g}{4 \eta e_{exp}} H.$$

In this way the cost involved in the case of movement of the overburden to the muck pile by explosive excavation is

$$C_{exp} = \frac{2 + 3 \tan \alpha}{\tan \alpha} \cdot \frac{g C_{exp_1}}{4 \eta e_{exp}} H.$$

While calculating, we assume that: $\alpha = 60°$; $C_{exp_1} = 8$ kopecks/kg (igdanites); $\eta = 0.5$; and $e_{exp} = 4.3 \times 10^6$ J/kg. Hence $C_{exp} = 4 \times 10^{-2} \times H$ kopecks/ton.

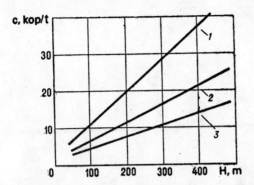

Fig. 106. Costs involved in removing overburden to muck pile
according to different methods of transportation:
1—by road; 2—by rail; 3—by blasting.

Figure 106 indicates the costs incurred in removing one ton of overburden by explosive excavation as a function of depth H of an open-pit mine. For the sake of comparison the same figure shows the costs incurred on transportation by road and rail [98]. Evidently, the cheapest method of moving the overburden is the explosive casting method.

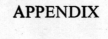

APPENDIX

Ballistic Tables

In the following pages are given the ballistic tables of trajectories of movement of rock fragments in an orthogonal system of coordinates. Air drag has been taken into account while preparing these tables. For plotting the trajectory, the origin of the coordinate system coincides with the center of gravity of the rock fragment before explosion. The orthogonal system of coordinates is selected in such a way that the x axis is along the horizontal and y axis vertically upward.

Three values have been used in the tables: air drag factor b_d; initial velocity of projection v, and angle of departure θ (the angle between the horizontal and the vector of the initial velocity of projection).

The trajectory of rock movement has been calculated to a point which is 200 m below the point of explosion. For plotting the entire trajectory, the tables contain ten points, each one of which is determined by the corresponding x and y coordinates.

If our task is to determine where a rock fragment falls on the free face, then only the last section of the trajectory is plotted, and the point of intersection with the free face determines the point where the fragment falls.

The ballistic tables cover parameters b_d, v and θ in the following ranges: $b_d = 10^{-4}$ to 1.1×10^{-1} m^{-1}; $v = 14$ to 248 m/s and $\theta = +85$ to $-45°$.

For plotting the trajectory the air drag factor b_d is calculated by the method given in Chapter 4 and having selected coordinates x and y for the given b_d, v and θ, the entire trajectory is plotted.

If the tables do not contain the values of x and y for a given value of the air drag factor toward lower values of velocity v, the trajectory can be plotted by the usual methods taking air drag into account.

The tables have been compiled in such a way that by rounding off the true value of b_d to the nearest value given in the table the error thus created does not exceed 15% of the true values of coordinates x and y. It should be borne in mind that if the value of the velocity is rounded off toward the higher side then the air drag factor b_d should also be rounded off toward the higher side and conversely. As a result, the error caused by rounding off is compensated to a certain extent. The values of θ and v should be interpolated for plotting

221

the trajectory accurately.

The ballistic tables are limited by the coordinate $y = -200$ m. For $y < -200$ m, the ballistic tables given in [18] should be consulted.

While using the tables it should be borne in mind that for angles $\theta \leqslant 0$, coordinate y is negative (the trajectory is below the horizontal).

$b_d = 10^{-4}$ m^{-1}; $v = 150$ m/s

85°		75°		60°		45°		30°	
x, m	y, m	x, m	y, m	x, m	y, m	x, m	y, m	x, m	y, m
10.77	119.7	31.18	113.2	54.99	92.58	65.14	63.28	57.94	32.47
53.84	528.8	155.9	499.6	274.9	407.9	325.7	278.0	289.7	142.3
96.91	819.7	280.6	773.8	494.9	630.2	586.3	428.3	521.5	218.9
140.0	985.3	405.3	928.8	714.8	753.5	846.8	510.1	695.3	253.4
172.3	1025	467.7	690.1	879.8	777.8	1042	524.2	869.1	267.6
215.4	962.0	592.4	927.5	1045	741.5	1238	495.9	1101	253.5
247.7	827.5	717.1	765.4	1210	642.4	1433	423.6	1333	200.0
280.0	615.7	810.6	855.6	1375	478.0	1628	305.3	1564	105.4
312.3	323.3	904.2	266.2	1595	151.8	1889	72.34	1796	−32.70
355.3	−200.8	1018	−203.0	1771	−204.0	2113	−200.5	2011	−201.1

15°		0°		−15°		−30°		−45°	
x, m	y, m	x, m	y, m	x, m	y, m	x, m	y, m	x, m	y, m
34.14	8.875	57.47	0.7228	30.90	8.504	18.05	10.52	11.10	11.16
170.7	38.85	172.4	6.555	92.71	26.86	54.15	32.12	33.31	33.80
307.3	59.80	287.3	18.35	154.5	47.04	90.25	54.49	55.52	56.87
443.8	71.47	344.8	26.53	185.4	57.82	108.3	65.97	66.63	68.57
546.3	73.98	459.8	47.53	247.2	80.78	144.4	89.50	88.83	92.30
751.1	62.45	517.2	60.39	309.0	105.6	162.5	101.6	99.94	104.3
956.0	28.10	632.2	90.93	370.9	132.4	198.6	126.3	122.1	128.7
1161	−30.04	747.1	128.0	401.8	146.4	234.7	151.8	144.4	153.6
1366	−113.0	862.0	171.8	463.6	176.1	270.8	178.1	166.6	178.9
1534	−200.6	931.0	201.4	511.0	200.2	300.8	200.6	185.1	200.3

$b_d = 10^{-4}\ \text{m}^{-1};\ v = 210\ \text{m/s}$

30°		45°		60°		75°		85°	
x, m	y, m	x, m	y, m	x, m	y, m	x, m	y, m	x, m	y, m
110.7	62.07	121.6	118.3	100.7	169.8	56.31	204.8	19.37	215.8
553.4	272.1	608.3	521.2	402.6	621.6	281.6	909.7	96.85	958.4
996.2	416.4	973.2	742.9	805.2	1076	506.8	1411	174.3	1490
1328	477.5	1338	889.0	1208	1341	675.8	1639	251.8	1785
1660	495.2	1825	955.1	1510	1405	844.7	1732	309.9	1841
1993	466.5	2190	899.9	1812	1344	1014	1684	368.0	1754
2435	350.3	2555	747.2	2114	1150	1183	1490	426.1	1518
2767	200.0	2920	488.8	2416	813.9	1352	1141	484.2	1127
3099	−8.763	3285	114.9	2718	321.2	1520	625.8	542.3	567.6
3343	−201.9	3528	−204.8	2962	−206.4	1716	−203.2	602.4	−208.5

−45°		−30°		−15°		0°		15°	
x, m	y, m	x, m	y, m	x, m	y, m	x, m	y, m	x, m	y, m
11.51	11.54	19.21	11.14	35.46	9.652	80.46	0.7239	66.43	12.27
34.53	34.79	57.62	33.76	106.4	29.87	241.4	6.586	332.2	75.55
57.55	58.29	96.03	56.82	177.3	51.30	402.3	18.49	597.9	115.8
69.06	70.12	115.2	68.52	248.2	73.98	563.2	36.64	797.2	133.5
92.07	93.97	153.6	92.25	283.7	85.80	643.7	48.12	996.5	140.2
103.6	106.0	172.9	104.3	354.6	110.4	804.6	76.03	1262	130.9
126.6	130.2	211.3	128.7	390.1	123.2	885.1	92.51	1594	88.58
149.6	154.7	249.7	153.6	461.0	149.7	1046	130.7	1927	9.953
172.6	179.4	288.1	178.9	531.9	177.5	1126	152.4	2192	−80.82
191.8	200.2	320.1	200.4	588.7	200.8	1288	201.3	2463	−200.1

$b_d = 1.4 \cdot 10^{-4}$ m⁻¹; v = 126.8 m/s

85°		75°		60°		45°		30°	
x, m	y, m	x, m	y, m	x, m	y, m	x, m	y, m	x, m	y, m
7.691	85.49	22,27	80.83	39.28	66.13	46.53	45.20	41.39	23.19
38.46	377.7	111.4	356.9	196.4	291.4	232.6	198.6	206.9	101.7
69.22	585.5	200.4	552.7	353.5	450.2	418.8	305.9	372.5	156.3
99.99	703.8	289.5	663.5	510.6	538.2	604.9	364.4	538.1	186.0
123.1	731.9	356.3	688.5	628.4	555.6	744.5	374.4	662.2	191.2
146.1	707.2	423.2	662.6	746.3	529.6	884.0	354.2	827.8	174.2
176.9	591.0	512.2	546.7	903.3	425.0	1024	302.5	1035	112.8
207.7	376.7	601.3	334.4	1060	236.2	1210	182.4	1200	29.92
238.4	58.02	690.4	19.05	1178	35.77	1396	0.0278	1366	−84.81
257.9	−202.1	739.4	−201.8	1288	−200.0	1549	−200.5	1498	−200.7

15°		0°		−15°		−30°		−45°	
x, m	y, m	x, m	y, m	x, m	y, m	x, m	y, m	x, m	y, m
24.39	6.339	48.57	0.7233	28.39	7.871	17.28	10.10	10.81	10.88
122.0	27.75	145.7	6.569	85.16	25.21	51.83	31.03	32.43	33.08
219.5	42.71	242.9	18.42	141.9	44.71	86.39	52.94	54.06	55.85
317.0	51.05	291.4	26.64	170.3	55.29	103.7	64.27	64.87	67.46
390.2	52.84	388.6	47.80	227.1	78.11	138.2	87.70	86.49	91.11
536.5	44.61	437.1	60.78	255.5	90.35	155.5	99.79	97.30	103.2
682.9	20.07	534.3	91.66	312.3	116.6	180.1	124.7	118.9	127.7
853.6	−30.08	631.4	129.2	369.0	145.1	224.6	150.7	140.5	152.8
1024	−104.6	728.6	173.8	425.8	176.0	259.2	177.8	162.2	178.6
1188	−200.4	780.4	200.4	467.5	200.2	286.8	200.2	180.2	200.5

$b_d = 1.4 \cdot 10^{-4}$ m^{-1}; $v = 177.5$ m/s

85°		75°		60°		45°		30°	
x, m	y, m	x, m	y, m	x, m	y, m	x, m	y, m	x, m	y, m
13.84	154.1	40.22	146.3	71.89	121.3	86.89	84.51	79.06	44.34
69.18	685.3	201.1	649.8	287.6	443.8	347.6	308.2	316.2	161.1
124.5	1065	362.0	1008	575.1	768.3	608.3	483.7	632.5	276.3
179.9	1275	482.7	1171	862.7	957.9	955.8	635.0	869.7	329.9
221.4	1315	603.4	1237	1078	1003	1303	682.2	1186	353.7
262.9	1253	724.0	1203	1294	959.9	1564	642.8	1423	333.2
304.4	1084	844.7	1064	1510	821.6	1825	533.7	1740	250.2
359.7	685.6	1006	706.2	1725	581.3	2085	349.2	1977	142.9
401.2	243.3	1126	295.6	1941	229.4	2346	82.05	2214	−6.259
434.0	−204.0	1238	−205.8	2138	−200.1	2558	−201.5	2451	−200.7

15°		0°		−15°		−30°		−45°	
x, m	y, m	x, m	y, m	x, m	y, m	x, m	y, m	x, m	y, m
47.45	12.34	68.00	0.7246	33.28	9.104	18.69	10.86	11.33	11.37
189.8	44.74	204.0	6.605	99.85	28.44	56.08	33.03	34.00	34.36
379.6	76.77	340.0	18.59	166.4	49.29	93.46	55.79	56.67	57.67
522.0	92.03	408.0	26.94	199.7	60.30	112.1	67.39	68.00	69.45
711.8	100.1	544.0	48.52	266.3	83.49	149.5	91.05	90.67	93.26
949.1	89.44	612.0	61.81	299.6	95.69	168.2	103.1	113.3	117.4
1186	54.17	748.0	93.56	366.2	121.3	205.6	127.7	124.7	129.6
1424	−7.394	816.0	112.1	432.7	148.5	242.9	152.9	147.3	154.2
1661	−97.13	952.0	154.6	499.3	177.5	280.4	178.7	170.0	179.2
1863	−200.2	1079	201.1	550.3	200.8	311.5	200.7	188.9	200.3

$b_d = 1.4 \cdot 10^{-4}$ m^{-1}; $v = 248.5$ m/s

85°		75°		60°		45°		30°	
x, m	y, m	x, m	y, m	x, m	y, m	x, m	y, m	x, m	y, m
23.56	263.3	68.90	251.4	125.5	212.3	156.4	152.5	147.8	82.96
94.26	974.6	275.6	929.2	627.6	946.7	782.2	674.5	591.1	301.8
188.5	1705	551.2	1622	1130	1461	1252	958.7	103.4	470.5
282.8	2136	826.8	2025	1506	1672	1721	1134	1478	581.8
353.5	2240	1034	2110	1883	1712	2180	1183	1921	627.9
424.2	2145	1240	1995	2260	1562	2659	1089	2364	600.2
494.8	1844	1447	1664	2636	1197	3129	831.0	2808	488.9
565.6	1317	1654	1093	2887	817.8	3442	554.4	3251	282.3
636.2	528.3	1860	234.6	3138	313.2	3754	182.3	3694	−33.45
684.9	−200.4	1943	−205.6	3343	−206.8	4010	−202.0	3882	−203.8

15°		0°		−15°		−30°		−45°	
x, m	y, m	x, m	y, m	x, m	y, m	x, m	y, m	x, m	y, m
91.72	23.85	95.20	0.7264	37.38	10.14	19.61	11.36	11.64	11.66
458.6	104.2	285.2	6.656	112.2	31.13	58.83	34.33	34.92	35.12
733.8	147.4	476.0	18.83	149.5	42.00	98.05	57.63	46.56	46.91
1009	174.6	571.2	27.36	224.3	64.48	117.6	69.41	69.85	70.63
1284	184.8	761.6	49.54	261.7	76.10	156.9	93.22	81.49	82.55
1651	169.1	856.8	63.28	336.4	100.1	176.5	105.2	104.8	106.5
2018	116.8	1047	96.31	411.2	125.2	215.7	129.6	128.0	130.7
2385	23.91	1142	115.7	486.0	151.3	254.9	154.2	151.3	155.0
2660	−75.12	1333	160.5	560.8	178.5	294.1	179.3	174.6	179.6
2932	−200.2	1479	200.5	618.1	200.2	326.8	200.4	194.0	200.2

228

$b_d = 1.96 \cdot 10^{-4}$ m^{-1}; $v = 107$ m/s

85°		75°		60°		45°		30°	
x, m	y, m	x, m	y, m	x, m	y, m	x, m	y, m	x, m	y, m
5.494	61.06	15.91	57.73	28.05	47.23	33.23	32.28	29.56	16.57
27.47	269.8	79.54	254.9	140.3	208.1	166.2	141.8	147.8	72.61
49.44	418.2	143.2	394.8	252.5	321.5	299.1	218.5	266.1	111.7
71.42	502.7	206.8	473.9	364.7	384.4	432.0	260.2	384.3	132.9
87.90	522.8	254.5	491.8	448.9	396.8	531.8	267.4	473.0	136.5
109.9	490.8	318.1	458.8	561.1	365.1	664.7	243.2	620.8	118.2
131.8	390.6	381.8	359.2	673.3	275.6	797.6	178.5	168.6	67.86
153.8	219.4	445.4	189.7	785.5	125.1	930.6	70.87	886.9	3.063
175.8	−27.88	493.1	13.61	869.7	−30.62	1064	−82.76	1005	−84.77
188.1	−201.8	539.8	−202.4	944.5	−202.2	1144	−200.3	1126	−200.1

15°		0°		−15°		−30°		−45°	
x, m	y, m	x, m	y, m	x, m	y, m	x, m	y, m	x, m	y, m
17.42	4.528	41.05	0.7239	25.81	7.222	16.38	9.607	10.45	10.54
87.10	19.82	123.1	6.586	77.43	23.52	49.12	29.75	31.34	32.18
156.8	30.51	205.2	18.49	129.0	42.34	81.88	51.14	52.23	54.58
226.4	36.46	246.3	26.78	154.8	52.71	98.25	62.31	62.68	66.07
278.7	37.74	328.4	48.13	206.5	75.41	131.0	85.61	83.57	89.63
418.0	27.34	369.4	61.26	323.3	87.76	147.4	97.74	94.02	101.7
557.4	−4.057	451.5	92.54	283.9	114.5	180.1	123.0	114.9	126.4
696.8	−57.66	492.6	110.8	335.9	143.9	212.9	149.6	135.0	152.0
818.7	−123.8	574.7	152.5	387.1	176.2	245.6	177.5	156.7	178.3
929.0	−200.1	656.8	201.5	423.3	200.5	271.8	200.9	174.1	200.9

$b_d = 1.96 \cdot 10^{-4}$ m^{-1}; $v = 150$ m/s

85°		75°		60°		45°		30°	
x, m	y, m	x, m	y, m	x, m	y, m	x, m	y, m	x, m	y, m
9.882	110.1	28.73	104.5	51.35	86.61	62.07	60.37	56.47	31.67
49.41	489.5	143.6	464.1	256.8	383.4	310.3	265.9	282.4	138.8
88.94	760.4	258.6	720.0	462.2	592.0	558.6	408.4	508.2	212.4
128.5	910.7	344.8	836.4	616.2	684.2	744.8	469.3	677.6	243.6
158.1	939.5	430.9	883.8	770.3	716.7	931.0	487.3	847.1	252.7
187.8	894.7	517.2	859.2	924.3	685.6	1117	459.1	1073	227.6
217.4	774.4	603.4	760.1	1078	586.8	1303	381.2	1242	178.7
247.1	574.9	718.3	504.5	1284	340.1	1552	192.6	1468	69.98
286.6	173.8	804.5	211.2	1438	60.74	1738	−19.17	1694	−93.29
313.6	−201.3	894.5	−201.4	1551	−203.4	1864	−200.9	1811	−201.0

15°		0°		−15°		−30°		−45°	
x, m	y, m	x, m	y, m	x, m	y, m	x, m	y, m	x, m	y, m
33.90	8.812	57.47	0.7254	30.90	8.505	18.05	10.52	11.10	11.16
169.5	38.54	172.4	6.628	92.71	26.88	54.15	32.12	33.31	33.80
305.0	59.08	287.3	18.70	154.5	47.10	90.25	54.51	55.52	56.88
406.7	68.14	344.8	27.13	185.4	57.92	108.3	66.00	77.73	80.40
508.4	71.51	459.8	48.98	247.2	81.02	144.4	89.57	88.83	92.33
711.8	60.29	517.2	62.48	278.1	93.31	162.4	101.6	111.0	116.5
915.2	23.43	632.2	94.82	339.9	119.4	198.6	126.4	122.1	128.8
1118	−41.23	689.6	113.7	401.8	147.5	234.7	152.1	144.1	153.7
1288	−118.1	804.6	157.3	463.6	177.8	270.8	178.5	166.6	179.0
1430	−200.0	904.2	201.5	506.8	200.2	299.6	200.3	185.1	200.5

$b_d = 1.96 \cdot 10^{-4}$ m^{-1}; $v = 210$ m/s

85° x, m	85° y, m	75° x, m	75° y, m	60° x, m	60° y, m	45° x, m	45° y, m	30° x, m	30° y, m
16.83	188.1	49.22	179.5	89.66	151.6	111.7	108.9	105.6	59.26
84.16	844.6	246.1	804.9	448.3	676.2	447.0	398.7	422.2	245.6
151.5	1316	442.9	1251	717.3	968.3	782.2	625.2	738.9	336.0
202.0	1526	590.6	1446	986.3	1157	1229	809.9	1056	415.6
252.5	1600	738.3	1507	1255	1228	1564	845.3	1372	448.5
303.0	1532	836.7	1469	1435	1203	1788	812.8	1689	428.8
358.5	1317	984.3	1286	1704	1047	2011	730.5	2006	349.2
404.0	940.8	1132	937.5	1973	729.9	2347	502.8	2322	201.7
454.5	377.4	1280	−397.6	2242	223.7	2682	130.2	2639	−23.90
492.6	−200.2	1398	−203.7	2409	−207.1	2898	−200.9	2829	−201.9

15° x, m	15° y, m	0° x, m	0° y, m	−15° x, m	−15° y, m	−30° x, m	−30° y, m	−45° x, m	−45° y, m
65.52	17.04	80.46	0.7276	35.46	9.652	19.21	11.14	11.51	11.54
262.1	61.73	241.4	6.689	106.4	29.87	57.62	33.76	34.53	34.79
458.6	96.19	402.3	18.99	177.3	51.35	96.03	56.83	57.55	58.29
720.7	124.8	482.7	27.64	212.8	62.57	115.2	68.53	69.06	70.13
917.2	132.0	643.7	50.22	283.7	85.99	153.6	92.29	92.08	93.99
1442	83.43	885.0	98.16	390.1	123.7	211.3	128.8	126.6	130.2
1703	17.08	965.5	118.1	461.0	150.5	249.7	153.8	149.6	154.8
1965	−81.54	1126	164.5	531.9	178.8	288.1	179.2	172.7	179.5
2202	−201.0	1234	200.4	583.9	200.6	320.1	200.8	191.8	200.3

$b_d = 2.74 \cdot 10^{-4}$ m^{-1}; $v = 90.55$ m/s

85°		75°		60°		45°		30°	
x, m	y, m	x, m	y, m	x, m	y, m	x, m	y, m	x, m	y, m
3.924	43.62	11.36	41.24	20.04	33.74	23.74	23.06	21.12	11.84
19.62	182.7	56.81	182.1	100.2	148.7	118.7	101.3	105.6	51.87
35.32	298.8	102.3	282.0	180.3	229.7	213.7	156.1	190.0	79.76
51.01	359.1	147.7	338.5	240.5	266.9	308.6	185.9	274.5	94.90
62.79	373.4	181.8	351.3	320.6	283.5	379.8	191.0	337.9	97.53
78.48	350.6	227.2	327.7	400.8	260.8	474.8	173.7	443.4	84.46
94.18	279.0	272.2	256.6	480.9	169.9	569.8	127.6	549.0	48.47
109.9	156.7	318.1	135.5	561.1	89.36	664.7	50.62	633.5	2.18
125.6	−19.92	363.6	−39.27	641.2	−65.13	759.6	−59.11	739.1	−78.92
138.0	−201.9	396.2	−200.6	696.0	−200.1	852.3	−200.3	855.2	−200.2

15°		0°		−15°		−30°		−45°	
x, m	y, m	x, m	y, m	x, m	y, m	x, m	y, m	x, m	y, m
12.44	3.234	34.69	0.7246	23.24	6.575	15.36	9.054	10.00	10.12
62.21	14.16	104.1	6.605	69.73	21.84	46.07	28.31	30.01	31.10
112.0	21.80	138.8	11.82	116.2	39.99	76.78	49.11	50.02	53.05
161.7	26.05	208.2	26.94	162.7	61.11	107.5	71.49	60.03	64.41
199.1	26.96	242.9	36.91	185.9	72.80	122.8	83.29	80.04	87.86
298.6	19.53	312.2	61.83	232.4	98.52	153.6	108.1	90.04	99.97
398.2	−2.898	346.9	76.84	255.7	112.6	168.9	121.2	110.0	125.0
510.2	−47.13	416.3	112.22	302.1	143.1	199.6	148.5	130.1	151.0
622.1	−112.8	485.7	154.8	348.6	177.0	230.3	177.5	150.1	178.2
732.4	−200.1	550.5	201.5	378.1	200.2	253.9	201.0	166.1	200.6

$b_d = 2.74 \cdot 10^{-4}$ m^{-1}; $v = 126.8$ m/s

30° x, m	30° y, m	45° x, m	45° y, m	60° x, m	60° y, m	75° x, m	75° y, m	85° x, m	85° y, m
40.34	22.62	44.33	43.12	36.68	61.68	20.52	74.64	7.059	78.63
201.7	99.17	221.7	189.9	183.4	273.8	102.6	331.5	35.29	349.7
363.1	151.7	399.0	291.7	330.1	422.9	184.7	514.3	63.53	543.2
484.0	174.0	532.0	335.2	440.2	488.7	246.3	597.4	91.77	650.5
605.1	180.5	665.0	348.1	550.2	511.9	307.8	631.3	112.9	671.1
766.4	162.5	798.0	327.9	660.2	689.7	369.4	613.7	134.1	639.0
927.7	111.7	931.0	272.3	770.3	419.2	431.0	542.9	155.3	553.1
1089	24.65	1064	178.2	880.3	296.6	513.1	360.3	176.5	410.6
1250	−102.4	1197	41.86	990.4	117.0	595.2	65.74	204.7	124.1
1349	−201.2	1367	−201.12	1129	−200.2	649.9	−202.6	227.8	−203.8

−45° x, m	−45° y, m	−30° x, m	−30° y, m	−15° x, m	−15° y, m	0° x, m	0° y, m	15° x, m	15° y, m
10.81	10.88	17.28	10.10	28.39	7.871	48.57	0.7264	24.21	6.295
32.43	33.08	51.83	31.03	85.16	25.23	145.7	6.656	121.1	27.53
54.06	55.86	86.39	52.97	141.9	44.80	242.9	18.83	217.9	42.20
75.68	79.24	120.9	75.94	198.7	66.67	291.4	27.36	290.5	48.67
97.30	103.2	138.2	87.81	227.1	78.48	388.6	49.55	363.2	51.08
108.1	115.5	172.8	112.4	283.9	103.9	437.2	63.30	508.4	43.06
129.7	140.4	190.1	125.0	312.3	117.6	534.3	96.36	653.7	16.73
151.4	165.9	224.6	151.2	369.0	146.8	582.8	115.8	799.0	−29.45
162.2	178.9	259.2	178.6	425.8	178.6	680.0	160.7	944.2	−97.24
179.5	200.1	285.7	200.3	461.8	200.2	754.5	200.8	1106	−200.3

$b_d = 2.74 \cdot 10^{-4}$ m^{-1}; $v = 177.5$ m/s

85° x, m	85° y, m	75° x, m	75° y, m	60° x, m	60° y, m	45° x, m	45° y, m	30° x, m	30° y, m
12.02	134.3	35.16	128.2	64.05	108.3	79.82	77.79	75.40	42.33
60.11	603.2	175.8	574.9	320.2	483.0	399.1	344.1	301.6	154.0
108.2	940.2	316.4	893.9	512.4	691.6	638.5	489.5	527.8	240.0
144.3	1090	421.9	1033	704.5	826.4	878.0	578.5	754.0	296.8
180.3	1143	527.3	1077	896.6	877.1	1117	603.8	980.1	320.4
216.4	1094	632.8	1018	1089	834.0	1277	580.5	1206	306.2
252.5	940.6	738.2	849.3	1281	685.8	1437	521.8	1432	249.4
288.5	672.0	843.7	557.6	1473	417.2	1676	359.1	1659	144.1
324.6	269.6	949.2	119.7	1665	3.790	1916	93.02	1885	−17.07
355.5	−205.9	1008	−201.1	1740	−204.3	2104	−204.5	2073	−200.5

15° x, m	15° y, m	0° x, m	0° y, m	−15° x, m	−15° y, m	−30° x, m	−30° y, m	−45° x, m	−45° y, m
46.80	12.17	68.00	0.7290	33.28	9.104	18.69	10.86	11.33	11.37
234.0	53.14	204.0	6.729	99.85	28.45	56.07	33.03	34.00	34.36
374.4	75.18	340.0	19.17	166.4	49.36	93.46	55.80	56.67	51.68
514.8	89.11	408.0	27.97	199.7	60.43	130.8	79.19	79.34	81.33
655.2	94.26	544.0	51.04	266.3	83.80	149.5	91.12	90.67	93.29
842.3	86.27	612.0	65.45	299.6	96.14	186.9	115.5	113.3	117.4
1030	59.60	748.0	100.4	366.1	122.1	205.6	127.9	124.7	129.7
1217	12.20	816.0	121.1	432.7	149.9	243.0	153.2	147.3	154.4
1451	−79.75	952.0	189.5	499.3	179.7	280.4	179.2	170.0	179.4
1666	−200.8	1029	201.2	543.6	200.6	310.3	200.5	188.9	200.6

$b_d = 2.74 \cdot 10^{-4}$ m^{-1}; $v = 248$ m/s

85°		75°		60°		45°		30°	
x, m	y, m	x, m	y, m	x, m	y, m	x, m	y, m	x, m	y, m
19.10	214.3	56.24	206.0	104.9	178.1	136.1	133.1	135.9	76.46
95.52	975.1	281.2	935.1	419.6	661.4	544.4	490.1	543.7	278.7
152.8	1412	449.9	1351	734.2	1048	952.7	767.6	951.4	431.5
210.1	1701	618.7	1620	1049	1307	1361	940.9	1359	521.5
267.5	1814	787.4	1713	1364	1406	1633	984.4	1631	538.7
324.8	1138	956.1	1611	1678	1315	1905	959.3	1903	515.2
382.1	1461	1069	1422	1888	1128	2178	854.2	2175	444.4
439.4	946.8	1181	1122	2098	821.7	2586	514.6	2583	231.7
477.6	435.4	1294	689.5	2308	367.7	2558	135.1	2854	3.219
513.3	−208.3	1449	−201.9	2496	−205.9	3040	−206.0	3040	−201.6

15°		0°		−15°		−30°		−45°	
x, m	y, m	x, m	y, m	x, m	y, m	x, m	y, m	x, m	y, m
89.33	23.24	95.20	0.7327	37.38	10.14	19.61	11.37	11.64	11.66
357.3	84.10	285.6	6.832	112.2	31.14	58.83	34.33	34.92	35.12
625.3	130.0	380.8	12.37	186.9	53.17	98.05	57.65	58.21	58.75
893.3	158.4	511.2	28.86	224.3	64.57	117.6	69.43	81.49	82.57
1161	166.7	666.4	40.01	299.1	88.20	156.9	93.26	93.13	94.54
1429	151.7	856.8	68.66	336.5	100.4	176.5	105.3	116.4	118.6
1697	109.5	952.0	86.39	411.2	125.8	215.7	129.7	128.1	130.7
1965	35.88	1142	129.3	486.0	152.3	254.9	154.4	151.4	155.1
2233	−14.38	1238	154.7	560.8	180.1	294.1	179.6	174.6	179.7
2460	−200.7	1390	201.5	613.1	200.4	326.8	200.8	194.0	200.3

235

$b_d = 3.842 \cdot 10^{-4}$ m^{-1}; $v = 76.53$ m/s

85°		75°		60°		45°		30°	
x, m	y, m	x, m	y, m	x, m	y, m	x, m	y, m	x, m	y, m
2.803	31.16	8.116	29.46	14.31	24.10	16.96	16.47	15.08	8.453
14.02	137.6	40.58	130.1	71.57	106.2	84.78	72.38	75.41	37.05
25.23	213.4	73.04	201.4	128.8	164.1	152.6	111.5	135.7	56.97
36.44	256.5	105.5	241.8	186.1	196.1	220.4	132.8	196.1	67.79
44.85	266.7	129.9	250.9	229.0	202.5	271.3	136.4	241.3	69.66
56.06	250.4	162.3	234.1	286.3	186.3	339.1	124.1	331.8	56.53
67.27	199.3	194.8	183.3	343.5	140.6	407.0	91.09	422.3	19.52
78.48	111.9	227.2	96.78	400.8	63.83	474.8	36.16	497.8	−30.91
89.70	−14.22	259.7	−28.05	458.0	−46.52	559.6	−65.70	513.2	−100.5
101.9	−201.0	292.9	−200.0	517.7	−200.8	640.4	−200.6	655.6	−200.4

15°		0°		−15°		−30°		−45°	
x, m	y, m	x, m	y, m	x, m	y, m	x, m	y, m	x, m	y, m
8.887	2.310	29.32	0.7254	20.15	5.948	14.24	8.451	9.488	9.639
44.44	10.11	87.96	6.628	62.25	20.22	42.73	26.74	28.46	29.83
79.99	15.57	146.6	18.70	103.7	37.73	71.22	46.91	47.44	51.27
115.5	18.60	175.9	27.13	124.5	47.75	85.47	51.71	56.93	62.47
142.2	19.26	234.6	49.00	166.0	70.36	114.0	80.81	75.90	85.83
231.1	10.97	263.9	62.52	186.7	82.99	128.2	93.11	85.39	98.00
319.9	−14.28	322.6	94.92	228.2	111.0	156.7	119.3	104.4	123.3
408.8	−57.78	351.9	113.9	269.7	142.7	185.2	147.5	123.4	150.1
497.7	−120.9	410.5	157.7	311.3	178.4	213.7	177.9	142.3	178.2
581.5	−200.1	459.4	200.4	334.7	200.4	233.6	200.6	156.9	200.7

$b_d = 3.842 \cdot 10^{-4}$ m⁻¹; $v = 107.1$ m/s

85°		75°		60°		45°		30°	
x, m	y, m	x, m	y, m	x, m	y, m	x, m	y, m	x, m	y, m
5.042	56.16	14.66	53.31	26.20	44.19	31.67	30.80	28.81	16.16
25.21	249.7	73.30	236.8	131.0	195.6	158.3	135.7	144.1	70.84
45.38	388.0	131.9	367.3	235.8	302.1	285.0	208.3	259.3	108.4
65.55	464.6	175.9	426.7	314.4	349.1	380.0	239.4	345.7	124.3
80.67	479.3	219.9	450.9	393.0	365.7	475.0	248.6	432.2	128.9
100.8	440.4	278.5	425.8	497.8	336.9	601.7	223.9	547.4	116.1
121.0	331.9	337.2	331.9	602.6	245.4	728.3	152.9	662.7	79.76
136.1	201.4	395.8	162.9	681.2	131.2	823.3	65.88	777.9	17.60
151.3	23.88	439.8	−19.70	759.8	−26.92	918.3	−53.31	893.2	−73.13
166.1	−201.8	414.5	−202.8	827.0	−201.8	1009	−201.0	1012	−200.0

15°		0°		−15°		−30°		−45°	
x, m	y, m	x, m	y, m	x, m	y, m	x, m	y, m	x, m	y, m
17.29	44.96	41.05	0.7276	25.81	7.223	16.38	9.607	10.45	10.54
86.47	19.66	123.1	6.689	77.43	23.55	49.12	29.76	31.34	32.19
155.6	30.14	205.2	18.99	129.0	42.47	81.88	51.18	52.23	54.61
207.5	34.76	246.3	27.64	180.7	64.11	114.6	73.93	73.12	77.82
259.4	36.49	328.4	50.23	206.5	75.98	131.0	85.81	83.57	89.72
380.4	28.56	369.4	64.29	258.1	101.9	163.8	110.6	104.5	114.2
501.5	2.591	451.5	98.24	283.9	116.0	180.1	123.5	114.9	126.7
622.6	−43.23	492.6	118.3	335.5	146.5	212.9	150.5	135.8	152.4
743.6	−110.9	574.7	164.8	387.1	180.2	245.6	178.9	156.7	178.9
861.2	−200.2	629.4	200.9	416.4	200.9	269.6	200.8	173.4	200.9

$b_d = 3.842 \cdot 10^{-4}\ m^{-1}; v = 150\ m/s$

85° x, m	85° y, m	75° x, m	75° y, m	60° x, m	60° y, m	45° x, m	45° y, m	30° x, m	30° y, m
8.588	95.96	25.11	91.60	45.75	77.36	57.01	55.56	53.85	30.24
42.94	430.9	125.6	410.7	182.9	284.9	285.1	245.8	215.4	110.0
77.29	671.6	226.0	638.5	320.2	449.8	456.1	349.4	377.0	171.4
103.1	778.6	301.3	738.0	503.2	590.3·	627.1	413.2	538.5	212.0
128.8	816.3	376.7	769.1	540.5	626.5	798.2	431.3	700.1	228.8
154.6	781.7	452.0	727.1	777.7	595.7	968.2	397.0	861.7	218.8
180.3	671.9	527.3	606.6	914.9	489.8	1140	302.8	1023	178.2
206.1	480.0	602.6	398.3	1052	298.0	1311	138.9	1185	102.9
231.9	192.5	678.0	85.50	1189	2.707	1425	−15.94	1346	−12.19
257.1	−204.1	729.1	−200.3	1261	−201.9	1534	−202.8	1531	−201.4

15° x, m	15° y, m	0° x, m	0° y, m	−15° x, m	−15° y, m	−30° x, m	−30° y, m	−45° x, m	−45° y, m
33.43	8.693	57.47	0.7307	30.90	8.506	18.05	10.52	11.10	11.16
167.1	37.95	172.4	6.776	92.71	26.90	54.15	32.13	33.31	33.80
267.4	53.70	287.3	19.40	154.5	47.22	90.25	54.54	55.52	56.89
367.7	63.65	344.8	28.37	216.3	69.56	126.4	77.77	66.62	68.61
468.0	67.33	459.8	52.04	247.2	81.52	144.4	89.70	88.83	92.39
635.1	58.16	517.2	66.91	309.0	107.1	180.5	114.2	99.94	104.4
802.2	27.59	632.2	103.2	339.9	120.7	198.6	126.8	122.1	129.0
969.4	−27.38	689.6	124.8	401.8	149.8	234.7	152.7	144.4	153.9
1271	−110.3	804.6	175.7	463.6	181.3	270.8	179.5	166.6	179.4
1136	−200.4	854.4	201.0	498.6	200.4	298.4	200.7	184.3	200.2

$b_d = 3.842 \cdot 10^{-4}$ m^{-1}; $v = 210$ m/s

85° x, m	85° y, m	75° x, m	75° y, m	60° x, m	60° y, m	45° x, m	45° y, m	30° x, m	30° y, m
13.65	153.1	40.17	147.1	74.92	127.2	97.22	95.04	97.09	54.61
54.58	572.6	200.9	667.9	299.7	472.4	291.6	270.6	388.3	199.1
95.52	915.5	321.4	965.1	524.4	748.4	583.3	489.6	679.6	308.2
136.4	1159	441.9	1157	749.2	933.3	875.0	640.3	970.9	372.5
191.0	1296	562.4	1224	973.9	1005	1167	703.2	1165	384.8
231.9	1242	642.8	1191	1124	977.7	1361	685.2	1359	368.0
372.9	1044	723.1	1092	1274	882.2	1555	610.2	1553	317.4
313.8	676.2	843.6	801.3	1423	708.1	1750	468.0	1748	227.4
354.8	81.52	964.1	294.9	1573	439.6	2042	96.46	1942	90.87
369.3	−206.6	1043	−202.7	1798	−201.2	2197	−203.1	22.17	−201.4

15° x, m	15° y, m	0° x, m	0° y, m	−15° x, m	−15° y, m	−30° x, m	−30° y, m	−45° x, m	−45° y, m
63.81	16.60	80.46	0.7351	35.46	9.653	19.20	11.14	11.51	11.54
255.2	60.07	241.4	6.900	106.4	29.89	57.62	33.76	34.53	34.80
446.6	92.83	402.3	20.01	177.3	51.44	96.85	56.85	57.55	58.30
638.1	113.2	482.7	29.45	248.2	74.37	115.2	68.57	69.06	70.14
765.7	118.8	643.6	54.73	283.7	86.37	153.6	92.37	92.07	94.02
951.1	113.9	724.1	70.85	354.6	111.5	172.8	104.5	103.6	106.1
1148	90.58	885.0	110.8	390.1	124.7	211.3	129.0	126.6	130.3
1340	45.86	965.5	135.0	460.9	152.3	249.7	154.1	149.6	154.9
1595	−53.15	1126	192.8	531.9	181.6	288.1	179.8	172.6	179.7
1842	−200.7	1148	201.5	574.5	200.1	318.8	200.7	191.8	200.6

$b_d = 5.378 \cdot 10^{-4}$ m⁻¹; $v = 64.68$ m/s

85°		75°		60°		45°		30°	
x, m	y, m	x, m	y, m	x, m	y, m	x, m	y, m	x, m	y, m
2.002	22.25	5.797	21.04	10.22	17.21	12.11	11.77	10.77	6.038
10.01	98.32	28.99	92.90	51.12	75.85	60.56	57.70	53.87	26.46
18.02	152.4	52.18	143.9	92.01	117.2	109.0	79.64	96.96	40.69
26.03	183.2	75.36	172.7	132.9	140.1	157.5	94.85	140.1	48.42
32.03	190.5	92.75	179.2	163.6	144.6	193.8	97.46	172.4	49.76
42.06	172.1	121.7	160.5	214.7	126.9	254.3	84.17	247.8	37.19
50.05	129.2	144.9	117.9	255.6	88.88	314.9	47.47	312.4	7.788
58.06	60.12	168.1	49.50	296.5	28.22	375.5	−14.52	377.1	−40.26
66.07	−37.33	191.3	−46.91	337.45	−56.95	436.0	−104.3	441.7	−108.6
75.88	−200.1	218.6	−200.3	388.2	−200.9	485.3	−200.2	506.7	−200.2

15°		0°		−15°		−30°		−45°	
x, m	y, m	x, m	y, m	x, m	y, m	x, m	y, m	x, m	y, m
6.348	1.650	24.78	0.7264	18.38	5.352	13.07	7.816	8.903	9.089
31.74	7.224	74.34	6.657	55.14	18.68	39.22	25.09	26.71	28.40
57.13	11.12	99.12	11.94	91.90	35.61	65.36	44.60	44.51	49.27
82.53	13.29	148.7	27.37	110.3	45.49	78.44	55.23	62.32	71.73
101.6	13.76	198.2	48.58	147.0	68.16	104.6	78.27	71.22	83.58
177.7	5.225	223.0	63.36	165.4	81.00	130.7	103.8	89.03	10.85
253.9	−21.01	272.6	96.54	183.8	94.88	156.9	131.8	97.93	121.7
330.1	−66.58	321.2	137.6	202.2	109.8	169.9	146.9	115.7	149.2
399.9	−127.0	346.9	161.3	257.3	161.3	196.1	179.0	133.5	178.6
464.3	−200.0	383.3	200.1	292.9	200.1	212.7	200.8	146.0	200.3

$b_d = 5.378 \cdot 10^{-4}$ m^{-1}; $v = 90.6$ m/s

85°		75°		60°		45°		30°	
x, m	y, m	x, m	y, m	x, m	y, m	x, m	y, m	x, m	y, m
3.601	40.12	10.47	38.08	18.71	31.56	22.62	22.00	20.58	11.54
18.01	178.4	52.35	169.1	93.57	139.7	113.1	96.91	102.9	50.60
32.41	277.1	83.76	242.9	149.7	200.0	181.0	138.1	164.6	71.93
46.82	331.9	115.2	293.4	205.9	240.5	248.8	165.3	226.4	85.87
57.62	342.4	157.1	322.1	280.7	261.2	339.3	177.6	308.7	92.08
72.03	314.5	198.9	304.2	355.6	240.7	429.8	159.8	411.6	78.08
82.83	261.3	230.4	258.7	411.7	196.1	520.2	109.2	493.9	47.67
97.24	143.9	282.7	116.4	486.6	93.68	588.1	47.06	596.8	−17.14
108.0	17.06	314.1	−14.07	542.7	−19.23	678.6	−72.05	679.1	−93.04
121.9	−202.0	348.3	−201.9	609.5	−200.7	750.2	−200.1	766.9	−200.9

15°		0°		−15°		−30°		−45°	
x, m	y, m	x, m	y, m	x, m	y, m	x, m	y, m	x, m	y, m
12.35	3.312	34.69	0.7290	23.24	6.577	15.36	9.005	10.00	10.12
61.76	14.05	104.1	6.729	69.73	21.88	46.07	28.32	30.01	33.11
98.82	19.98	173.5	18.18	92.97	30.65	76.78	49.18	50.02	53.09
135.9	23.96	208.2	27.98	139.5	50.51	92.13	60.23	60.03	64.47
185.3	26.06	277.5	51.07	162.7	61.66	122.8	83.61	80.04	88.03
284.1	18.57	312.2	65.50	209.2	86.47	153.6	108.8	100.0	112.7
382.9	−5.993	381.6	100.6	255.7	114.8	184.3	135.6	120.1	138.4
481.7	−49.61	416.3	121.4	302.1	146.9	199.6	150.0	130.1	151.8
580.6	−114.6	485.7	170.0	325.4	154.4	230.3	179.9	150.1	179.3
675.7	−200.2	522.7	200.1	370.3	201.3	250.8	200.9	164.7	200.3

$b_d = 5.378 \cdot 10^{-4}$ m^{-1}; $v = 126.8$ m/s

30°		45°		60°		75°		85°	
x, m	y, m	x, m	y, m	x, m	y, m	x, m	y, m	x, m	y, m
38.47	21.60	40.72	39.69	32.68	55.26	17.94	65.43	6.134	68.54
153.9	78.56	203.6	175.6	163.4	246.4	89.68	293.3	30.67	307.8
269.3	122.5	325.8	249.6	261.4	352.9	143.5	422.3	49.07	443.8
384.7	151.4	447.9	295.2	359.4	421.6	197.3	508.7	67.47	536.0
500.1	163.5	570.1	308.0	457.5	447.5	269.0	549.4	92.01	583.1
653.9	159.1	692.3	283.6	555.5	425.5	322.9	519.4	110.4	558.3
769.3	112.3	814.5	216.3	653.5	349.9	376.7	433.3	128.8	479.9
884.7	49.46	936.6	99.20	751.6	212.9	430.5	284.5	153.4	282.6
1000	−43.21	1058	−77.90	849.6	1.934	484.3	61.07	171.8	51.26
1139	−201.3	1124	−202.0	918.2	−200.7	529.7	−201.1	186.5	−200.1

−45°		−30°		−15°		0°		15°	
x, m	y, m	x, m	y, m	x, m	y, m	x, m	y, m	x, m	y, m
10.81	10.88	17.28	10.10	28.39	7.873	48.57	0.7327	23.88	6.209
32.43	33.09	51.83	31.04	85.16	25.27	145.7	6.832	119.4	27.11
54.06	55.89	86.39	53.03	141.9	44.99	194.3	12.37	191.0	38.36
64.87	67.52	120.9	76.09	170.3	55.76	291.4	28.86	262.6	45.46
86.49	91.26	138.2	88.05	224.1	79.26	340.0	40.02	334.3	48.09
97.30	103.4	172.8	112.8	255.5	92.02	437.1	68.70	501.4	34.76
118.9	128.1	207.3	138.8	312.3	119.7	485.7	86.47	620.8	6.225
140.5	153.5	224.6	152.3	340.7	134.6	582.8	130.0	740.2	−40.69
162.2	179.7	259.2	180.2	397.4	166.9	631.4	155.1	859.5	−108.7
178.7	200.2	283.4	200.6	450.4	200.1	705.9	200.0	978.1	−200.6

$b_d = 5.378 \cdot 10^{-4}$ m^{-1}; $v = 177.5$ m/s

85°		75°		60°		45°		30°	
x, m	y, m	x, m	y, m	x, m	y, m	x, m	y, m	x, m	y, m
9.747	109.4	28.69	105.1	53.51	90.83	69.44	67.88	69.35	39.01
48.74	497.5	143.5	477.1	214.1	337.5	277.8	250.1	277.4	142.2
77.98	720.7	229.6	689.3	374.6	534.5	486.1	391.6	485.4	220.1
107.2	867.6	315.6	826.4	535.1	666.7	694.4	480.0	693.5	266.1
136.5	925.4	401.7	874.2	695.7	717.6	833.3	502.3	832.2	274.8
165.7	886.9	487.8	821.8	856.2	670.8	972.2	489.4	1040	248.1
194.9	745.4	545.2	725.3	963.3	575.6	1111	435.8	1179	198.4
224.2	483.0	602.6	572.3	1070	419.2	1319	262.5	1318	118.2
253.4	58.23	688.7	210.7	1177	187.6	1458	68.90	1456	1.642
266.4	−204.9	752.7	−202.7	1300	−204.4	1595	−203.2	1625	−200.9

15°		0°		−15°		−30°		−45°	
x, m	y, m	x, m	y, m	x, m	y, m	x, m	y, m	x, m	y, m
45.58	11.86	68.00	0.7379	33.28	9.106	18.69	10.86	11.33	11.37
182.3	42.91	204.0	6.981	99.85	28.48	56.07	33.04	34.00	34.37
319.0	66.31	272.0	12.73	166.4	49.51	93.46	55.84	56.67	57.70
455.8	80.83	340.0	20.41	199.7	60.69	130.8	79.29	68.00	69.49
592.5	85.07	476.0	42.17	266.3	84.45	149.5	91.27	90.67	93.35
774.8	71.84	544.0	56.58	299.6	97.07	186.9	115.8	113.3	117.6
911.5	45.24	680.0	93.39	366.1	123.9	205.6	128.3	136.0	142.2
1048	1.777	748.0	116.2	399.4	138.1	243.0	153.9	147.3	154.6
1231	−87.40	816.0	142.3	466.0	168.4	280.4	180.3	170.0	179.8
1390	−200.9	942.9	200.7	530.3	200.0	307.8	200.2	188.1	200.3

$b_d = 5.378 \cdot 10^{-4}$ m^{-1}; $v = 248.5$ m/s

85°		75°		60°		45°		30°	
x, m	y, m	x, m	y, m	x, m	y, m	x, m	y, m	x, m	y, m
14.37	162.0	42.60	156.7	81.40	138.7	110.5	108.4	118.5	66.84
71.87	748.1	170.4	592.1	325.6	521.1	442.0	402.8	473.9	244.4
115.0	1090	298.2	951.5	569.8	829.1	773.5	629.9	710.8	337.4
158.1	1305	426.0	1197	814.0	1025	994.5	726.5	947.8	402.5
201.2	1362	553.8	1292	976.8	1070	1215	763.8	1185	431.2
244.4	1244	681.6	1203	1140	1032	1436	725.6	1422	412.9
273.1	1055	766.8	1025	1302	892.2	1657	590.7	1659	333.8
301.9	755.2	852.0	726.4	1465	623.5	1768	477.7	1777	265.7
330.6	299.9	937.2	255.6	1628	171.2	1989	129.9	2014	56.66
352.2	−201.8	994.0	−208.5	1720	−210.6	2129	−207.3	2204	−202.6

15°		0°		−15°		−30°		−45°	
x, m	y, m	x, m	y, m	x, m	y, m	x, m	y, m	x, m	y, m
85.08	22.16	95.20	0.7452	37.38	10.14	19.61	11.36	11.64	11.66
340.3	79.96	190.4	3.087	112.2	31.17	58.83	34.34	34.92	35.12
510.5	109.8	285.6	7.198	186.9	53.28	98.05	57.67	58.21	58.76
680.7	130.9	476.0	21.51	224.3	64.77	117.7	69.47	69.85	70.65
935.9	142.3	571.2	32.16	299.1	88.67	156.9	93.35	93.13	94.57
1191	122.8	716.6	61.75	373.8	113.9	196.1	117.6	104.8	106.6
1361	87.82	856.8	81.31	448.6	140.6	235.3	142.3	128.1	130.8
1532	30.99	1047	131.8	486 0	154.5	254.9	154.9	151.3	155.3
1702	−52.32	1142	163.5	560.8	183.6	294.1	180.2	174.6	179.9
1912	−200.8	1238	200.3	603.1	200.9	325.5	200.8	194.0	200.6

$b_d = 7.53 \cdot 10^{-4}$ m^{-1}; $v = 54.66$ m/s

30°		45°		60°		75°		85°	
y, m	x, m	y, m	x, m	y, m	x, m	y, m	x, m	y, m	x, m
4.313	7.695	8.404	8.651	12.30	7.303	15.03	4.141	15.90	1.430
18.90	38.48	36.93	43.26	54.18	36.51	66.35	20.70	70.23	7.151
29.07	69.26	56.89	77.86	83.70	65.72	102.8	37.27	108.9	12.87
34.59	100.0	67.75	112.5	100.1	94.94	123.4	53.83	130.9	18.59
35.54	123.1	69.62	138.4	103.3	116.8	128.0	66.25	117.0	31.46
23.95	184.7	56.25	190.3	90.65	153.4	114.6	86.96	81.77	37.18
−4.343	238.6	26.54	233.6	54.21	189.9	73.79	107.7	41.94	41.47
−51.28	292.4	−21.54	276.8	−7.970	226.4	20.09	124.2	10.79	44.33
−119.0	346.3	−89.91	320.1	−98.89	262.9	−76.11	144.9	−69.72	50.05
−200.1	394.8	−200.3	371.1	−200.8	293.6	−200.4	164.4	−200.4	57.01

−45°		−30°		−15°		0°		15°	
y, m	x, m	y, m	x, m	y, m	x, m	y, m	x, m	y, m	x, m
8.489	8.264	7.167	11.88	4.797	16.17	0.7276	20.94	1.179	4.534
26.85	24.79	14.96	23.75	17.25	48.51	2.942	41.89	5.160	22.67
47.10	41.32	32.50	47.50	33.68	80.85	18.99	104.7	7.942	40.81
69.32	57.85	42.28	59.38	43.45	97.02	27.65	125.7	9.492	58.95
93.60	74.37	63.91	83.13	79.39	145.5	50.29	167.5	9.825	72.55
106.5	82.64	75.80	95.00	109.2	177.9	64.40	188.5	1.616	136.0
134.1	99.17	88.43	106.9	126.0	194.0	80.46	209.4	−21.80	195.0
148.7	107.4	116.0	130.6	163.6	226.4	141.2	272.3	−61.70	253.9
179.8	124.0	163.4	166.3	184.4	242.6	165.9	293.2	−120.1	312.9
200.7	134.4	200.7	190.8	200.7	254.4	200.8	319.7	−200.1	372.3

$b_d = 7.530 \cdot 10^{-4} \text{ m}^{-1}; \ v = 76.56 \text{ m/s}$

85°		75°		60°		45°		30°	
x, m	y, m	x, m	y, m	x, m	y, m	x, m	y, m	x, m	y, m
2.572	28.66	7.479	27.20	13.37	22.55	16.16	15.71	14.70	8.244
12.86	127.4	29.92	99.81	53.47	82.52	64.63	57.31	58.80	29.95
23.15	197.9	59.83	173.5	106.9	142.8	113.1	89.94	117.6	51.38
33.44	237.1	82.27	209.6	160.4	178.1	161.6	112.8	132.3	55.30
43.73	242.8	112.2	230.0	200.5	186.6	226.2	126.6	191.1	64.87
54.02	214.2	142.1	217.3	254.0	171.9	290.8	119.5	249.9	63.94
61.74	169.3	187.0	131.3	320.8	108.1	371.6	77.99	323.4	46.52
72.03	75.38	216.9	23.96	360.9	42.65	468.5	−27.20	396.9	8.982
82.32	−63.22	239.3	−87.74	414.4	−81.50	517.0	−106.5	485.1	−66.46
89.95	−200.8	257.5	−201.7	452.7	−201.2	562.8	−200.9	585.5	−200.7

15°		0°		−15°		−30°		−45°	
x, m	y, m	x, m	y, m	x, m	y, m	x, m	y, m	x, m	y, m
8.823	2.294	29.32	0.7307	20.75	5.950	14.24	8.452	9.488	9.639
35.29	8.318	58.64	2.967	41.50	12.70	28.49	17.37	19.98	19.59
70.59	14.27	117.3	12.23	62.25	20.27	42.73	26.76	37.95	40.43
105.9	17.74	175.9	28.38	103.7	38.01	71.22	47.03	56.93	62.58
202.9	13.26	234.6	52.08	145.2	59.40	99.71	69.36	75.90	86.11
291.2	−10.73	293.2	84.09	207.5	98.93	128.2	93.86	85.39	98.40
379.4	−56.63	351.9	125.3	249.0	130.7	170.9	135.0	104.3	124.1
441.2	−103.7	381.2	149.7	290.5	167.2	213.7	181.8	123.3	151.3
502.9	−164.7	410.5	176.7	311.2	187.3	227.9	198.7	142.3	180.2
533.2	−200.3	434.0	200.4	323.7	200.1	229.8	201.1	155.0	200.4

$b_d = 7.530 \cdot 10^{-4}$ m^{-1}; $v = 107.1$ m/s

85°		75°		60°		45°		30°	
x, m	y, m	x, m	y, m	x, m	y, m	x, m	y, m	x, m	y, m
4.381	46.96	12.81	46.74	23.34	39.47	29.09	28.35	27.48	15.43
13.14	139.7	38.43	133.3	116.7	176.0	116.4	103.8	109.9	56.12
21.91	219.8	64.06	209.5	210.1	271.6	203.6	162.8	192.3	87.48
43.81	364.6	89.68	274.1	303.4	317.1	261.8	191.5	274.8	108.2
61.34	414.1	140.9	363.3	373.4	313.2	378.1	219.8	357.2	116.8
74.49	408.9	192.2	392.4	443.5	272.5	494.5	202.6	439.6	111.6
87.63	365.9	243.4	355.1	513.5	190.0	581.8	154.5	549.5	80.18
109.5	201.9	294.7	244.0	560.2	108.3	669.0	70.86	686.9	−6.220
127.1	−32.38	358.7	−23.78	630.2	−62.95	756.3	−55.65	796.8	−122.1
136.1	−202.0	386.9	−202.9	672.2	−200.3	829.0	−201.8	852.7	−200.7

15°		0°		−15°		−30°		−45°	
x, m	y, m	x, m	y, m	x, m	y, m	x, m	y, m	x, m	y, m
17.05	4.435	41.05	0.7351	25.81	7.225	16.38	9.608	10.45	10.54
68.22	16.07	82.10	3.002	51.62	15.08	32.75	19.53	31.34	32.30
153.5	29.43	164.2	12.53	103.2	32.82	65.50	40.35	52.23	54.65
238.8	34.35	205.2	20.01	154.9	53.43	98.25	62.55	73.12	77.94
341.1	27.47	287.3	41.00	206.5	77.16	114.6	74.19	83.57	89.91
443.4	4.447	369.4	70.93	258.1	104.3	147.4	98.61	104.5	114.5
528.7	−29.06	451.5	111.1	309.7	135.2	180.1	124.6	125.4	140.1
614.0	−77.66	533.6	162.9	361.3	170.1	212.9	152.3	146.2	166.7
699.2	−143.8	574.7	193.8	387.1	189.2	245.6	181.9	156.7	180.3
757.2	−200.2	582.9	200.4	402.6	201.2	265.3	200.5	172.0	200.8

$b_d = 7.530 \cdot 10^{-4}$ m^{-1}; $v = 150$ m/s

85°		75°		60°		45°		30°	
x, m	y, m	x, m	y, m	x, m	y, m	x, m	y, m	x, m	y, m
6.962	78.11	20.50	75.06	38.22	64.88	49.60	48.49	49.53	27.86
27.85	292.1	81.98	280.3	152.9	241.0	148.8	138.1	148.6	78.79
55.70	514.7	143.5	447.2	267.6	381.8	297.6	249.8	297.2	141.0
76.58	619.7	205.0	564.3	382.2	476.2	446.4	326.7	445.8	182.0
97.47	661.0	286.9	624.4	496.9	512.6	595.2	358.8	594.4	196.3
125.3	608.4	327.9	607.9	573.4	498.8	694.4	349.6	693.5	187.8
146.2	480.6	389.4	518.1	649.8	450.1	793.6	311.3	842.1	141.7
160.1	345.0	430.4	408.8	726.3	361.3	892.8	238.8	941.1	84.44
174.1	158.7	491.9	150.5	840.9	134.0	1042	49.21	1080	−51.88
192.6	−200.1	545.2	−205.0	942.9	−202.4	1162	201.9	1199	−201.1

15°		0°		−15°		−30°		−45°	
x, m	y, m	x, m	y, m	x, m	y, m	x, m	y, m	x, m	y, m
32.55	8.471	57.47	0.7412	30.90	8.507	18.05	10.51	11.10	11.16
97.66	23.82	114.9	3.054	61.81	17.48	36.10	21.23	22.21	22.43
195.3	42.45	229.9	12.97	123.6	36.94	72.20	43.27	44.42	45.30
293.0	55.04	344.8	31.06	185.4	58.56	108.3	66.16	66.63	68.66
423.2	60.76	402.3	43.65	247.2	82.57	144.4	89.98	88.83	92.50
520.9	55.27	459.8	58.89	278.1	95.56	180.5	114.8	111.0	116.9
683.6	23.40	574.7	98.36	309.0	109.2	198.6	127.6	133.3	141.8
846	−43.85	689.6	151.8	370.9	138.8	234.7	154.0	155.5	167.3
976.6	−130.3	747.1	184.7	432.7	171.7	270.8	181.5	177.7	193.4
1056	−200.6	773.9	201.6	482.1	200.5	294.8	200.5	183.6	200.5

$b_d = 7.530 \cdot 10^{-4}$ m^{-1}; $v = 210$ m/s

85°		75°		60°		45°		30°	
x, m	y, m	x, m	y, m	x, m	y, m	x, m	y, m	x, m	y, m
10.27	115.7	30.43	111.9	58.14	99.11	78.93	77.46	84.62	47.74
41.07	437.8	121.7	422.9	174.4	285.9	236.8	222.0	253.9	135.5
71.87	705.6	213.0	679.6	348.9	526.4	394.6	348.2	423.1	210.0
92.40	840.9	304.3	855.3	523.3	696.2	552.5	449.6	592.4	267.1
112.9	932.3	365.2	913.7	639.6	755.0	710.3	518.9	677.0	287.5
143.7	973.2	426.0	917.0	755.9	759.0	868.2	545.6	846.2	308.0
174.5	888.7	486.9	859.5	930.3	637.3	1026	518.3	1015	294.9
205.	657.9	547.7	732.5	1047	445.3	1184	422.0	1185	238.4
225.9	393.8	639.0	369.6	1163	122.3	1421	92.77	1439	40.47
253.6	−202.9	716.1	−209.6	1240	−208.2	1539	−202.1	1608	−202.9

15°		0°		−15°		−30°		−45°	
x, m	y, m	x, m	y, m	x, m	y, m	x, m	y, m	x, m	y, m
60.77	15.83	80.46	0.7500	35.46	9.654	19.21	11.14	11.51	11.54
182.3	44.49	160.9	3.127	70.92	19.63	38.41	32.40	23.02	23.14
303.9	68.46	321.8	13.63	141.8	40.59	76.82	45.27	46.04	46.52
446.2	93.48	402.3	22.26	177.3	51.63	96.03	56.89	69.06	70.17
668.5	101.7	563.2	47.81	248.2	74.91	134.4	80.52	92.07	94.09
790.1	95.23	643.7	65.46	319.1	99.96	172.9	104.7	115.1	118.3
972.4	62.73	804.6	112.8	425.5	141.3	230.5	142.1	138.1	142.8
1155	−5.009	885.0	143.5	461.0	156.3	268.9	167.9	161.1	167.6
1337	−120.1	965.5	179.9	531.9	183.0	307.3	194.3	184.1	192.8
1428	−200.8	1008	201.8	557.9	200.4	316.3	200.6	191.1	200.4

$b_d = 1.054.10^{-3}\ m^{-1}$; $v = 46.2$ m/s

x, m	y, m	x, m	y, m	x, m	y, m	x, m	y, m	x, m	y, m
85°		75°		60°		45°		30°	
1.022	11.35	2.958	10.73	5.216	8.782	6.179	6.003	5.497	3.080
4.086	41.48	11.83	39.20	20.86	32.02	24.72	21.84	21.99	11.19
8.172	71.95	23.66	67.94	41.73	55.37	49.44	37.66	43.97	19.25
12.26	90.69	35.49	85.53	62.59	69.47	74.15	47.08	65.96	24.04
16.34	97.20	47.32	91.44	83.46	73.79	98.87	49.73	87.95	25.39
23.49	78.50	65.07	77.63	114.8	60.94	135.9	40.18	115.4	21.99
29.62	30.67	88.73	14.35	146.1	23.26	191.6	—7.409	148.4	9.994
35.75	—49.80	103.5	—54.37	177.4	—42.02	234.8	—75.88	225.4	—55.44
40.86	—146.1	118.3	—149.7	203.4	—120.8	271.9	—160.7	291.3	—161.2
43.18	—200.2	124.7	—200.7	223.8	—200.4	285.9	—200.0	309.5	—200.1
15°		0°		—15°		—30°		—45°	
3.239	0.8419	17.70	0.7290	14.14	4.287	10.69	6.522	7.589	7.855
12.96	3.054	35.40	2.953	28.29	9.591	21.37	13.76	15.18	16.25
22.67	4.788	53.10	6.730	42.43	15.94	42.74	30.48	30.36	34.73
38.87	6.587	88.50	19.19	70.71	31.94	64.11	50.33	45.53	55.53
51.82	7.018	123.9	38.63	99.00	52.59	74.80	61.48	60.71	48.81
113.4	—3.995	159.3	65.69	127.3	78.24	96.17	86.41	75.89	104.7
165.2	—31.63	194.7	101.1	155.6	109.3	117.5	115.0	91.07	133.4
262.3	—137.5	247.8	172.1	212.1	189.9	160.3	184.5	121.4	199.9
223.5	—85.77	230.1	145.8	183.9	146.3	138.9	147.6	106.2	165.0
299.3	—200.1	265.5	201.1	218.7	201.1	168.8	200.5	121.9	201.1

b_d 1.054·10⁻³ m⁻¹; v = 64.68 m/s

85°		75°		60°		45°		30°	
x, m	y, m	x, m	y, m	x, m	y, m	x, m	y, m	x, m	y, m
1.837	20.47	5.342	19.43	9.548	16.10	11.54	11.22	10.50	5.88
9.187	91.01	21.37	71.29	38.19	58.94	57.70	49.44	42.00	21.40
16.54	141.4	48.08	133.9	85.93	110.1	103.9	75.93	84.00	36.70
23.89	169.3	69.45	159.9	124.1	130.5	150.0	89.29	126.0	45.30
29.40	144.7	80.13	164.3	143.2	133.3	173.1	90.60	157.5	46.98
36.75	160.5	96.16	159.8	181.4	122.8	219.3	81.60	199.5	42.31
45.94	106.9	122.9	121.0	229.2	77.21	288.5	35.82	262.5	19.98
55.12	8.703	154.9	17.11	267.3	11.29	346.2	-36.76	325.5	-26.65
62.47	-108.8	176.3	-94.16	315.1	-115.7	392.4	-122.1	420.0	-146.2
66.88	-200.1	191.6	-200.2	338.6	-201.0	425.1	-200.5	450.1	-200.2

15°		0°		-15°		-30°		-45°	
x, m	y, m	x, m	y, m	x, m	y, m	x, m	y, m	x, m	y, m
6.302	1.639	24.78	0.7327	18.38	5.355	13.07	7.817	8.903	9.090
21.21	5.941	49.56	2.983	36.76	11.59	26.15	16.19	17.81	18.56
50.42	10.20	74.34	6.833	73.52	26.88	39.22	25.13	35.61	38.69
75.63	12.67	123.9	19.68	110.3	46.17	65.36	44.79	53.42	60.49
94.53	13.30	173.5	40.06	128.7	57.43	78.44	55.56	71.22	84.05
163.9	5.846	223.0	68.86	165.4	83.44	91.51	66.99	89.03	109.5
252.1	-28.83	272.6	107.2	202.2	114.5	117.7	91.90	206.8	136.9
327.7	-85.43	322.2	156.4	238.9	151.1	143.8	119.8	124.6	166.5
384.4	-147.6	346.9	185.5	275.7	194.0	169.9	150.8	142.4	198.4
422.3	-200.2	358.5	200.3	280.6	200.2	206.5	200.1	143.6	200.6

$b_d = 1.054 \cdot 10^{-3}$ m^{-1}; $v = 90.55$ m/s

85° x, m	85° y, m	75° x, m	75° y, m	60° x, m	60° y, m	45° x, m	45° y, m	30° x, m	30° y, m
3.130	34.97	9.151	33.38	16.67	28.19	20.78	20.25	19.63	11.02
12.52	129.4	36.60	123.4	66.69	103.8	83.11	74.14	78.50	40.08
25.04	222.64	73.21	215.5	133.4	180.0	145.4	116.3	157.0	68.31
37.56	283.8	109.8	268.9	183.4	215.1	228.5	150.6	215.9	80.33
46.94	297.5	137.3	280.3	233.4	228.3	290.9	157.2	255.1	83.40
59.46	274.7	164.7	265.0	266.7	223.7	353.2	144.7	353.3	71.18
71.98	201.8	192.2	221.1	333.4	178.5	436.3	93.48	431.8	37.50
81.37	109.4	228.8	111.8	383.5	108.6	498.6	24.21	510.3	−22.04
90.76	−23.13	256.2	−16.98	450.1	−44.96	561.0	−77.95	588.8	−113.5
99.83	−201.8	283.7	−200.3	495.1	−200.7	615.0	−200.4	643.1	−200.2

15° x, m	15° y, m	0° x, m	0° y, m	−15° x, m	−15° y, m	−30° x, m	−30° y, m	−45° x, m	−45° y, m
12.18	3.168	34.69	0.7379	23.24	6.580	15.36	9.056	10.00	10.13
48.73	11.48	69.39	3.026	46.48	13.89	30.71	18.50	20.01	20.50
97.45	19.57	104.1	6.982	92.97	30.85	61.42	38.63	40.02	42.01
134.0	23.20	173.5	20.42	139.5	51.22	92.13	60.51	60.03	64.61
170.5	24.54	208.2	30.19	185.9	75.39	122.8	84.29	80.04	88.36
243.6	19.62	277.5	56.65	232.4	103.8	153.6	110.1	110.0	126.3
328.9	−0.8200	346.9	93.66	278.9	137.0	184.3	138.2	130.1	153.3
414.2	−40.19	416.3	143.1	325.4	175.5	215.0	168.8	150.1	181.8
511.6	−113.8	451.0	173.3	348.6	197.0	230.3	185.0	160.1	196.6
589.6	−200.4	478.8	200.3	353.3	201.5	244.7	200.8	162.7	200.6

$b_d = 1.054 \cdot 10^{-3}$ m^{-1}; $v = 126.8$ m/s

30° x, m	30° y, m	45° x, m	45° y, m	60° x, m	60° y, m	75° x, m	75° y, m	85° x, m	85° y, m
35.38	19.90	35.43	34.63	27.30	46.34	14.64	53.62	4.973	55.79
106.1	52.28	141.7	127.6	109.2	172.2	56.56	200.2	19.89	208.7
212.3	100.7	248.0	199.8	191.1	272.7	117.1	351.7	34.81	339.6
318.4	130.0	354.3	244.9	273.0	340.1	161.0	421.6	54.70	442.7
424.6	140.2	425.1	256.3	354.9	366.1	205.0	446.0	69.62	472.1
530.7	126.6	531.4	238.9	436.9	342.2	248.9	419.3	79.54	464.7
636.9	82.89	637.7	170.5	518.8	258.0	292.8	334.9	99.46	380.3
743.0	0.8380	708.6	89.17	573.4	160.2	336.7	179.5	114.4	246.5
849.2	−132.5	779.4	−29.45	628.0	18.67	380.6	−77.48	129.3	29.71
889.3	−200.4	851.8	−201.6	687.1	−204.1	395.8	−201.1	139.9	−200.6

−45° x, m	−45° y, m	−30° x, m	−30° y, m	−15° x, m	−15° y, m	0° x, m	0° y, m	15° x, m	15° y, m
10.81	10.88	17.28	10.10	28.39	7.876	48.57	0.7452	23.25	6.051
21.62	21.91	34.55	20.45	56.78	16.31	97.14	3.087	93.01	21.89
43.24	44.44	69.11	41.96	113.6	35.02	145.7	7.198	162.8	33.83
64.85	67.61	103.7	64.62	170.3	56.43	242.9	21.52	232.5	41.24
86.49	91.48	138.2	88.54	227.1	80.94	340.0	45.51	302.3	43.40
108.1	116.1	172.8	113.8	283.9	109.0	388.6	61.82	372.1	39.48
129.7	141.5	224.6	154.6	340.7	141.0	437.1	81.46	488.3	16.71
151.4	167.8	259.2	184.0	397.4	177.6	534.3	132.3	624.8	−44.59
173.0	195.0	276.4	199.4	425.8	197.9	582.8	164.5	744.1	−133.5
177.3	200.5	277.6	200.4	429.6	200.7	631.4	201.9	806.9	−200.4

$b_d = 1.054 \cdot 10^{-3}$ m^{-1}; $v = 177.5$ m/s

30° x, m	30° y, m	45° x, m	45° y, m	60° x, m	60° y, m	75° x, m	75° y, m	85° x, m	85° y, m
60.45	34.10	56.38	55.33	41.53	70.49	21.74	79.95	7.334	82.63
181.3	96.75	169.1	158.5	124.6	204.2	65.21	231.5	29.33	312.7
362.7	172.2	338.3	287.6	249.2	376.6	130.4	429.7	51.34	504.0
483.6	205.14	507.4	370.7	415.3	522.7	194.4	610.9	80.64	665.9
604.5	220.0	620.1	389.7	498.4	546.1	282.6	659.0	102.7	695.1
725.4	210.7	732.9	370.2	581.4	526.5	347.8	613.9	124.7	634.8
846.2	170.3	789.3	342.8	664.5	455.2	381.2	532.2	139.3	538.3
967.1	89.14	902.0	243.7	747.5	318.1	434.7	370.6	154.0	385.3
1088	−47.84	1015	66.26	830.6	87.33	478.2	130.4	168.7	153.0
1179	−202.8	1118	−204.2	895.7	−200.1	516.6	−203.5	183.1	−204.6

−45° x, m	−45° y, m	−30° x, m	−30° y, m	−15° x, m	−15° y, m	0° x, m	0° y, m	15° x, m	15° y, m
11.33	11.37	18.69	10.86	33.28	9.108	68.00	0.7557	43.41	11.31
22.67	22.83	37.38	21.88	66.57	18.61	136.0	3.176	130.2	31.78
45.33	46.00	74.76	44.40	133.1	38.94	204.0	7.518	260.5	56.00
68.00	69.54	112.1	67.61	199.7	61.26	272.0	14.08	347.3	66.77
90.67	93.48	149.5	91.59	266.3	85.87	340.0	23.20	434.1	72.22
113.3	117.8	186.9	116.4	332.8	113.2	408.0	35.29	520.9	71.28
136.0	142.6	224.3	142.2	399.4	143.6	544.0	70.28	652.2	54.99
158.7	167.9	261.7	168.9	466.0	177.6	680.0	123.8	824.8	−3.578
181.3	193.7	299.1	196.8	499.3	196.2	748.0	159.5	998.4	−122.8
187.4	200.7	304.0	200.6	505.9	200.0	816.0	202.7	1074	−201.2

$b_d = 1.054 \cdot 10^{-3}$ m^{-1}; $v = 248.5$ m/s

85°		75°		60°		45°		30°	
x, m	y, m	x, m	y, m	x, m	y, m	x, m	y, m	x, m	y, m
10.09	114.2	30.09	111.1	58.75	100.6	83.25	82.05	96.40	54.59
40.37	437.3	90.24	324.4	176.2	292.5	249.7	236.9	289.2	155.6
70.64	710.8	210.6	688.3	325.5	543.0	416.2	373.2	482.0	240.7
111.0	933.1	300.9	863.0	528.7	714.8	666.0	518.8	674.8	300.8
131.2	957.3	361.1	908.2	846.2	760.2	832.5	553.0	867.6	322.9
151.4	914.8	421.2	883.5	705.0	755.6	915.7	543.2	964.0	313.9
171.6	797.5	481.4	777.2	822.5	678.8	999.0	510.7	1157	239.5
191.7	583.0	541.6	563.5	940.0	483.0	1165	356.6	1253	164.4
211.9	210.5	601.8	171.8	998.7	319.3	1332	14.84	1446	−104.6
225.1	−202.7	635.9	−210.6	1108	−208.9	1396	−206.8	1491	−201.2

15°		0°		−15°		−30°		−45°	
x, m	y, m	x, m	y, m	x, m	y, m	x, m	y, m	x, m	y, m
78.11	20.38	95.20	0.7707	37.38	10.14	19.61	11.36	11.64	11.66
234.3	57.20	190.4	3.307	74.77	20.54	39.22	22.81	23.28	23.37
390.5	87.05	285.6	8.003	149.5	42.20	78.44	45.98	46.57	46.93
546.8	107.2	300.8	15.34	224.3	65.18	117.7	69.54	69.85	70.68
703.0	113.7	476.0	25.90	299.1	89.73	156.9	93.54	93.13	94.64
781.1	110.2	571.2	40.41	373.8	116.1	196.1	118.0	116.4	118.8
937.3	85.76	761.6	85.06	448.6	144.7	235.3	143.0	139.7	143.3
1172	−13.30	856.8	117.7	523.4	175.8	274.5	168.6	163.0	168.0
1328	−143.0	952.0	159.3	560.8	192.5	313.7	194.9	186.3	193.0
1380	−202.6	1028	200.6	578.2	200.6	321.6	200.2	193.2	200.5

$b_d = 1.476 \cdot 10^{-3}$ m^{-1}; $v = 39.05$ m/s

85°		75°		60°		45°		30°	
x, m	y, m	x, m	y, m	x, m	y, m	x, m	y, m	x, m	y, m
0.7296	8.110	2.113	7.668	3.726	6.273	4.414	4.288	3.926	2.200
2.919	29.63	10.56	33.85	14.80	22.87	17.66	15.60	15.70	7.990
5.837	51.40	19.01	52.43	29.81	39.55	30.90	24.50	27.48	12.53
9.485	66.77	27.46	62.94	44.71	49.62	48.55	32.39	39.26	15.76
11.67	69.43	33.80	65.31	59.61	52.71	57.38	34.57	51.04	17.65
18.24	47.10	44.37	58.48	81.97	43.53	88.28	32.31	78.52	16.52
24.08	−13.60	61.27	18.04	115.5	−4.006	132.4	0.0026	117.8	0.4068
28.46	−88.90	76.06	−50.82	137.9	−61.68	163.3	−45.87	168.8	−49.69
31.37	−156.6	88.73	−140.4	160.2	−144.8	203.0	−139.4	208.1	−115.1
32.96	−200.3	95.28	−200.5	171.9	−200.6	221.7	−200.2	243.7	−200.2

15°		0°		−15°		−30°		−45°	
x, m	y, m	x, m	y, m	x, m	y, m	x, m	y, m	x, m	y, m
2.313	0.6014	14.96	0.7307	12.31	3.826	9.533	5.898	6.898	7.207
9.254	2.181	29.92	2.967	24.61	8.737	19.07	12.60	13.80	15.04
18.51	3.755	59.84	12.24	49.22	21.99	38.54	28.54	27.59	32.69
30.07	4.843	89.76	28.42	61.53	30.43	57.20	48.08	34.49	42.55
37.02	5.013	104.6	39.33	86.14	51.21	76.26	71.51	48.28	64.44
71.72	0.1403	134.6	67.32	110.8	77.64	95.33	99.20	62.08	89.45
113.4	−19.41	164.4	104.4	135.4	110.3	114.4	131.6	75.88	117.8
161.9	−63.98	194.5	152.1	160.0	150.1	123.9	149.7	89.67	150.0
208.2	−132.7	209.4	180.4	184.6	198.1	143.0	190.3	103.5	186.2
240.8	−200.2	219.4	201.2	186.2	201.6	147.4	200.6	108.5	200.6

$b_d = 1.476 \cdot 10^{-3}$ m^{-1}; $v = 54.66$ m/s

85° x, m	85° y, m	75° x, m	75° y, m	60° x, m	60° y, m	45° x, m	45° y, m	30° x, m	30° y, m
1.312	14.62	3.816	13.88	6.820	11.50	8.243	8.017	7.500	4.206
5.250	53.69	15.26	50.92	27.28	42.10	32.97	29.24	30.00	15.28
11.81	101.0	30.53	88.51	54.56	72.88	65.95	50.35	60.00	26.21
18.37	123.3	45.79	111.1	81.84	90.87	98.92	62.33	90.00	32.36
21.00	124.8	57.24	117.4	102.3	95.18	123.6	64.72	112.5	33.56
24.97	118.8	76.32	106.5	122.8	91.06	173.1	50.63	150.0	28.45
32.81	76.35	95.40	67.00	163.7	55.15	214.3	17.15	202.5	4.582
40.69	−12.20	114.5	−5.128	197.8	−7.009	255.5	−39.73	255.0	−42.16
47.25	−131.2	133.6	−118.2	238.7	−131.1	296.8	−125.5	307.5	−117.2
50.14	−200.8	143.7	−200.6	255.1	−200.6	323.4	−200.6	348.2	−200.4

15° x, m	15° y, m	0° x, m	0° y, m	−15° x, m	−15° y, m	−30° x, m	−30° y, m	−45° x, m	−45° y, m
4.502	1.170	20.94	0.7351	16.17	4.801	11.88	7.169	8.264	8.490
18.01	4.244	41.89	3.003	32.34	10.57	23.75	15.98	16.53	17.45
36.01	7.282	83.77	12.54	64.68	25.21	47.50	32.65	33.06	36.82
54.02	9.049	125.7	29.49	97.02	44.38	71.25	53.24	49.58	58.27
67.52	9.498	167.5	54.92	129.4	68.61	95.00	77.06	66.11	81.97
99.04	7.435	188.5	71.22	145.5	82.81	106.9	90.27	82.64	108.1
130.5	0.6744	230.4	112.0	177.9	115.9	130.6	119.6	99.17	137.0
171.1	−15.68	272.3	165.4	210.2	155.9	154.4	153.0	115.7	168.8
270.1	−99.81	293.2	197.7	226.4	178.9	178.1	191.2	124.0	185.9
335.2	−200.3	296.0	202.3	240.4	200.6	183.7	200.9	130.6	200.2

$b_d = 1.476 \cdot 10^{-3}$ m^{-1}; $v = 76.53$ m/s

85°		75°		60°		45°		30°	
x, m	y, m	x, m	y, m	x, m	y, m	x, m	y, m	x, m	y, m
2.235	24.98	6.537	23.85	11.91	20.14	14.85	14.46	14.02	7.871
8.842	92.45	26.15	88.15	47.63	74.17	59.36	52.95	56.07	28.63
17.88	161.7	52.29	153.9	83.36	117.1	118.7	90.95	98.13	44.63
26.83	202.7	78.44	192.1	131.0	153.7	163.2	107.6	140.2	55.19
33.53	212.5	98.05	200.2	166.7	163.1	207.8	112.3	182.2	59.57
40.24	203.5	124.2	181.2	202.4	155.1	267.1	97.02	252.3	50.84
49.18	160.8	150.3	124.5	262.0	96.94	326.5	52.59	322.4	18.03
58.12	78.14	176.5	22.26	309.6	0.7047	371.0	-4.149	378.5	-29.73
67.06	-55.89	196.1	-94.20	345.3	-111.0	430.4	-120.6	434.6	-101.7
73.62	-201.0	209.4	-200.2	366.8	-200.4	459.6	-200.6	488.3	-200.4

15°		0°		-15°		-30°		-45°	
x, m	y, m	x, m	y, m	x, m	y, m	x, m	y, m	x, m	y, m
8.701	2.263	29.32	0.7412	20.75	5.954	14.24	8.454	9.488	9.640
34.80	8.199	58.64	3.054	41.50	12.73	28.49	17.38	19.98	19.59
60.91	12.77	117.3	12.98	83.00	28.99	56.98	36.77	37.95	40.50
95.71	16.51	175.9	31.09	124.5	49.25	85.47	58.37	56.93	62.82
121.8	17.53	234.6	59.03	166.0	74.13	99.71	70.08	66.41	74.56
165.3	15.14	263.9	77.32	186.7	88.50	128.2	95.49	85.39	99.26
226.2	2.269	293.2	98.89	228.2	121.7	156.7	123.8	140.4	125.7
287.1	-23.76	351.9	153.5	269.7	161.5	185.2	155.4	123.3	154.2
400.3	-117.6	381.2	187.5	190.5	184.3	213.7	190.7	142.3	184.8
460.6	-200.5	392.9	202.5	304.3	200.7	221.3	200.8	151.8	201.0

$b_d = 1.476 \cdot 10^{-3}$ m^{-1}; $v = 107.1$ m/s

85° x, m	85° y, m	75° x, m	75° y, m	60° x, m	60° y, m	45° x, m	45° y, m	30° x, m	30° y, m
3.552	39.85	10.46	38.30	19.50	33.10	24.31	24.74	25.27	14.22
14.21	149.0	41.83	143.0	78.01	123.0	101.2	91.13	101.1	51.82
28.42	262.6	83.66	251.2	136.5	194.8	177.1	142.7	176.9	80.23
39.07	316.2	115.0	301.2	195.0	243.0	227.8	166.7	227.5	92.85
49.73	337.2	146.4	318.6	253.5	261.5	303.7	183.0	303.3	100.2
60.39	393.2	177.8	299.5	312.0	244.4	379.6	170.6	379.0	90.40
71.04	271.7	209.1	239.2	351.0	209.8	430.2	142.7	454.9	59.21
81.70	176.0	230.1	171.8	409.6	114.4	506.1	63.69	53.07	0.5985
92.36	21.22	261.4	16.20	468.1	−52.37	556.7	−21.04	606.5	−94.62
102.1	−201.6	289.0	−203.2	502.5	−202.6	626.8	−200.6	663.8	−200.4

15° x, m	15° y, m	0° x, m	0° y, m	−15° x, m	−15° y, m	−30° x, m	−30° y, m	−45° x, m	−45° y, m
16.61	4.322	41.05	0.7500	25.81	7.229	16.38	9.610	10.45	10.54
66.44	15.64	82.10	3.128	51.62	15.12	32.75	19.54	20.89	21.28
116.3	24.17	164.2	13.63	77.43	23.72	65.50	40.45	31.34	32.21
166.1	29.46	205.2	22.27	129.0	43.33	98.25	62.90	52.23	54.74
215.9	31.00	246.3	33.56	180.7	66.59	131.0	27.08	73.12	78.20
265.8	28.20	328.4	65.60	323.3	94.17	163.8	113.2	104.5	115.3
382.0	0.6475	369.4	87.24	283.9	126.9	196.5	141.5	135.8	155.2
465.1	−42.32	451.5	144.7	335.5	165.8	229.3	172.4	156.7	183.4
548.1	−111.8	492.6	182.0	361.3	187.9	245.6	188.8	167.1	198.1
619.5	−200.4	511.8	201.8	375.1	200.6	256.5	200.2	168.5	200.0

$b_d = 1.476 \cdot 10^{-3}$ m^{-1}; $v = 150$ m/s

85°		75°		60°		45°		30°	
x, m	y, m	x, m	y, m	x, m	y, m	x, m	y, m	x, m	y, m
5.238	59.02	15.53	57.10	29.66	50.56	40.27	39.52	43.18	24.36
20.95	223.4	62.10	215.8	118.7	189.9	161.1	146.8	172.7	89.08
36.67	360.0	108.7	346.7	207.7	302.2	241.6	205.4	259.1	123.0
57.62	475.6	155.3	436.4	287.0	355.2	362.4	264.8	345.4	146.7
73.34	496.5	201.8	470.7	356.0	390.1	442.9	278.4	431.8	157.2
83.81	457.3	232.9	457.3	415.3	376.1	523.5	264.4	518.1	150.5
99.53	384.5	279.5	373.7	474.6	325.1	604.1	215.3	604.5	121.6
115.2	200.9	326.0	188.6	534.0	227.2	724.8	47.33	734.0	20.65
125.7	−5.113	357.1	−28.08	593.3	62.38	765.1	−46.09	820.3	−103.5
132.5	−204.3	374.2	−205.2	649.7	−200.2	814.8	−203.0	867.8	−201.5

15°		0°		−15°		−30°		−45°	
x, m	y, m	x, m	y, m	x, m	y, m	x, m	y, m	x, m	y, m
31.01	8.076	57.47	0.7625	30.90	8.511	18.05	10.52	11.10	11.16
124.0	29.14	144.9	3.235	61.81	17.51	36.10	21.24	22.21	22.43
217.1	44.28	229.9	16.64	92.71	27.06	54.15	32.17	33.31	33.82
279.1	50.16	287.3	24.39	154.5	47.98	90.25	54.74	55.52	56.97
341.1	51.87	344.8	37.53	216.3	71.80	126.4	78.34	77.73	80.68
403.1	48.58	459.8	76.69	278.1	99.12	162.5	103.1	99.94	105.0
496.1	32.00	517.2	104.4	339.9	130.7	198.6	129.2	133.3	142.7
589.1	−2.556	574.7	139.2	401.8	167.6	234.7	156.9	155.5	168.8
744.2	−118.3	632.2	182.4	432.7	188.4	270.8	186.3	177.7	195.7
810.3	−200.3	655.2	202.5	449.1	200.2	287.6	200.8	181.4	200.3

$b_d = 1.476 \cdot 10^{-3}$ m^{-1} $v = 210$ m/s

85°		75°		60°		45°		30°	
x, m	y, m	x, m	y, m	x, m	y, m	x, m	y, m	x, m	y, m
7.209	81.54	21.49	79.37	41.96	71.83	59.46	58.61	68.85	38.99
28.83	312.3	85.97	303.5	167.9	272.9	237.9	219.9	206.6	111.2
50.46	507.7	150.4	491.7	251.8	387.9	356.8	308.1	344.3	171.9
72.09	541.0	193.4	583.9	355.7	478.0	475.7	370.6	482.0	214.9
93.71	683.8	257.9	648.7	416.6	543.0	594.6	395.0	619.7	230.6
108.1	653.4	300.1	631.1	503.6	539.7	713.6	364.8	757.4	205.4
112.5	569.6	343.9	555.2	587.5	484.9	832.5	254.7	826.3	171.1
137.0	416.4	386.9	402.5	671.4	345.0	892.0	155.4	964.0	38.63
151.4	150.4	429.8	122.7	755.3	61.39	951.4	10.60	1033	—74.69
162.2	—206.6	457.8	—203.9	798.7	—200.7	1011	—208	1088	—201.2

15°		0°		—15°		—30°		—45°	
x, m	y, m	x, m	y, m	x, m	y, m	x, m	y, m	x, m	y, m
55.79	14.56	80.46	0.7805	35.46	9.657	19.21	11.14	11.51	11.54
167.4	40.86	160.9	3.396	70.92	19.65	38.41	22.41	23.02	23.14
279.0	62.18	241.4	8.338	106.4	30.01	76.82	45.04	46.04	46.54
390.5	76.54	321.8	16.23	177.3	52.04	115.2	68.79	69.06	70.22
502.1	81.18	402.3	27.87	248.2	76.14	153.6	92.91	92.07	94.23
613.7	72.33	56.32	66.76	319.1	102.8	191.1	117.8	115.1	118.6
725.3	44.66	643.7	97.02	390.1	132.8	230.5	143.5	138.1	143.3
836.9	—9.503	724.1	137.3	461.0	166.8	268.9	170.2	161.1	168.5
948.5	—102.1	804.6	190.7	496.5	185.6	307.3	198.0	184.1	194.1
1027	—200.5	820.7	203.3	522.5	200.4	311.1	200.9	189.5	200.1

$b_d = 2.066 \cdot 10^{-3}$ m^{-1}; $v = 33$ m/s

85°		75°		60°		45°		30°	
x, m	y, m	x, m	y, m	x, m	y, m	x, m	y, m	x, m	y, m
0.5212	5.793	1.509	5.477	2.661	4.481	3.153	3.063	2.804	1.572
2.084	21.16	6.036	20.00	10.65	16.34	15.76	13.46	11.22	5.707
4.691	39.67	13.53	37.45	21.29	28.25	31.53	22.04	22.44	9.822
6.775	47.69	19.62	44.96	34.60	36.47	40.99	24.69	36.46	12.60
8.339	45.59	24.14	46.65	42.58	37.65	50.44	25.37	44.87	12.95
12.51	37.06	31.69	41.77	62.21	28.80	72.51	18.84	67.70	10.51
16.16	9.931	39.23	26.89	79.84	2.424	94.58	0.0018	84.13	0.2906
17.72	−17.30	48.29	−5.215	98.47	−44.06	119.8	−38.71	115.0	−28.29
22.41	−111.9	61.84	−87.97	117.1	−115.4	195.0	−99.59	143.0	−71.36
25.29	−200.2	73.19	−200.1	132.6	−200.2	172.7	−200.4	192.3	−200.3

15°		0°		−15°		−30°		−45°	
x, m	y, m	x, m	y, m	x, m	y, m	x, m	y, m	x, m	y, m
1.652	0.4296	12.64	0.7327	10.66	3.414	8.439	5.306	6.210	6.562
6.610	1.558	25.29	2.983	21.32	7.977	16.88	11.50	12.42	13.84
13.22	2.682	37.93	6.635	31.98	13.75	33.76	26.73	18.63	21.88
19.83	3.361	63.22	19.71	42.64	20.78	50.63	46.06	31.05	40.33
26.44	3.581	88.50	40.21	63.96	38.95	67.51	69.95	43.47	62.22
51.23	0.1002	101.1	53.65	85.28	63.13	84.39	98.98	55.89	87.92
87.58	−18.56	126.4	87.79	117.3	112.7	101.3	133.9	68.31	117.9
112.4	−41.87	151.7	133.2	138.6	156.4	118.1	175.8	80.73	152.8
153.7	−104.0	177.0	192.6	149.2	182.1	126.6	199.8	93.15	193.5
193.4	−200.1	180.4	201.8	156.4	200.9	127.1	201.5	95.22	200.9

$b_d = 2.066 \cdot 10^{-3}$ m^{-1}; $v = 46.2$ m/s

85° x, m	85° y, m	75° x, m	75° y, m	60° x, m	60° y, m	45° x, m	45° y, m	30° x, m	30° y, m
0.9375	10.44	2.726	9.913	4.871	8.216	5.888	5.727	5.357	3.004
3.750	38.35	13.63	44.03	12.49	30.07	29.44	26.23	21.43	10.92
0.562	60.65	21.81	65.22	38.97	52.06	52.99	38.74	42.86	18.72
10.31	80.75	32.71	79.34	58.46	64.91	70.66	44.52	64.29	23.11
14.06	88.99	40.88	83.84	73.07	67.99	88.32	46.23	80.36	23.97
18.75	81.88	54.51	76.05	97.43	59.53	117.8	39.17	117.9	16.95
23.44	54.54	70.87	39.55	121.8	32.32	159.0	5.560	144.6	3.273
29.06	—8.713	81.77	—3.663	141.3	—5.006	188.4	—38.81	187.5	—36.42
34.69	—115.2	98.12	—105.0	175.4	—113.1	212.0	—89.66	230.4	—103.5
37.78	—200.2	108.5	—200.8	193.2	—200.2	247.3	—200.6	270.2	—200.0

15° x, m	15° y, m	0° x, m	0° y, m	—15° x, m	—15° y, m	—30° x, m	—30° y, m	—45° x, m	—45° y, m
3.215	0.8360	17.70	0.7379	14.14	4.292	10.69	6.525	7.589	7.857
12.86	3.031	35.40	3.026	28.29	9.632	21.37	13.79	15.18	16.27
25.72	5.202	53.10	6.983	42.43	16.09	32.06	21.83	22.77	25.26
35.37	6.236	88.50	20.43	70.71	32.65	53.43	40.42	37.94	45.10
48.23	6.784	106.2	30.24	99.00	54.67	74.80	67.72	53.12	67.65
70.74	5.311	141.6	56.93	113.1	67.99	85.48	75.41	68.30	93.24
94.46	—0.5010	177.0	94.87	141.4	99.93	106.9	104.2	83.48	122.2
160.8	—37.78	212.4	146.1	169.7	140.1	128.2	138.3	98.66	155.2
221.9	—112.2	230.1	178.2	198.0	190.3	149.6	178.6	113.8	192.7
266.0	—200.2	241.9	202.4	203.7	201.8	160.3	201.4	116.9	200.8

$b_d = 2.066 \cdot 10^{-3}$ m^{-1}; $v = 64.68$ m/s

85° x, m	85° y, m	75° x, m	75° y, m	60° x, m	60° y, m	45° x, m	45° y, m	30° x, m	30° y, m
1.597	17.84	4.669	17.03	8.506	14.38	10.60	10.33	10.01	5.622
6.387	66.04	18.68	62.97	34.02	52.98	42.40	37.82	40.05	20.45
12.77	116.5	32.68	99.89	59.54	83.64	84.80	64.96	70.09	31.88
17.56	139.5	51.36	132.4	93.57	109.8	116.6	76.83	110.1	40.98
23.95	151.8	70.03	143.0	119.1	116.5	148.4	80.19	130.2	42.55
30.34	140.1	88.71	129.4	153.1	105.8	180.8	69.30	170.2	38.82
38.32	89.25	107.4	88.94	187.1	69.24	233.2	37.57	210.3	24.57
44.71	13.34	126.1	15.90	221.2	0.5034	265.0	-2.963	250.3	-2.267
51.10	-106.7	144.7	-102.2	255.2	-113.0	318.0	-113.6	320.4	-88.76
54.61	-201.0	155.5	-273.3	273.3	-200.8	345.2	-200.5	372.5	-200.5

15° x, m	15° y, m	0° x, m	0° y, m	-15° x, m	-15° y, m	-30° x, m	-30° y, m	-45° x, m	-45° y, m
6.215	1.616	24.78	0.7452	18.38	5.361	13.07	7.820	8.903	9.092
24.86	5.856	49.56	3.087	36.76	11.64	26.15	16.21	17.81	18.58
49.72	9.885	73.74	7.199	55.14	18.92	39.22	26.21	26.71	28.47
68.37	11.83	123.9	21.53	91.90	36.81	65.36	45.19	44.51	49.60
87.01	15.52	148.7	32.22	128.7	59.83	91.51	68.15	53.42	60.89
124.3	10.01	173.5	45.63	165.4	89.01	104.6	80.88	80.12	98.00
161.6	1.620	223.0	82.02	202.2	125.7	130.7	109.2	97.93	125.8
217.5	-24.28	272.6	134.3	238.9	171.6	156.9	141.9	115.7	156.5
304.5	-107.2	297.4	168.0	257.3	198.8	183.0	179.7	133.5	190.6
360.1	-201.0	318.8	202.3	258.5	200.7	196.1	200.9	138.3	200.3

$b_d = 2.066 \cdot 10^{-3}$ m^{-1}; $v = 90.55$ m/s

85°		75°		60°		45°		30°	
x, m	y, m	x, m	y, m	x, m	y, m	x, m	y, m	x, m	y, m
2.537	28.47	7.469	27.36	13.93	23.64	18.08	17.67	18.05	10.15
10.15	106.5	29.88	102.2	55.72	87.84	72.30	65.09	72.21	37.02
17.76	170.2	59.75	179.4	111.4	152.7	126.5	101.9	108.3	51.37
25.37	215.5	82.16	215.1	153.2	180.5	180.8	125.0	162.5	66.37
35.52	240.9	104.6	227.6	181.1	186.8	216.9	130.7	216.6	71.54
43.13	230.9	127.0	213.9	222.9	174.6	253.2	127.4	270.8	64.57
53.28	175.1	149.4	170.9	264.4	131.7	307.3	101.9	324.9	42.29
63.43	57.82	171.18	91.57	306.5	48.84	361.5	45.49	379.1	0.4275
68.51	−34.79	194.2	−39.53	334.3	−37.41	397.7	−15.03	451.3	−97.84
74.76	−201.9	211.6	−201.7	369.2	−201.6	464.0	−201.2	498.2	−201.0

15°		0°		−15°		−30°		−45°	
x, m	y, m	x, m	y, m	x, m	y, m	x, m	y, m	x, m	y, m
11.86	3.087	34.69	0.7557	23.24	6.586	15.36	9.058	10.00	10.13
47.46	11.17	69.39	3.176	46.48	13.94	30.71	18.52	20.01	20.51
83.05	17.26	104.1	7.519	69.73	22.14	46.07	28.42	40.02	42.09
118.6	21.04	173.5	23.22	116.2	41.45	76.78	49.67	60.03	64.90
142.4	22.10	208.2	35.34	162.7	65.40	92.13	61.10	80.04	89.09
237.3	11.78	242.9	50.93	185.9	79.46	138.2	99.16	90.04	101.8
284.7	−4.408	277.5	70.58	232.4	112.6	168.9	128.2	110.0	128.4
344.1	−38.46	346.9	125.2	278.9	153.9	199.6	161.0	130.1	157.0
403.4	−92.63	381.6	162.3	302.1	178.3	230.3	198.1	150.1	187.8
476.9	−200.5	411.7	201.3	322.3	201.9	232.4	200.7	158.1	200.8

$b_d = 2.066 \cdot 10^{-3}$ m^{-1}; $v = 126.8$ m/s

85°		75°		60°		45°		30°	
x, m	y, m	x, m	y, m	x, m	y, m	x, m	y, m	x, m	y, m
3.742	42.16	11.09	40.79	21.19	36.12	28.76	28.23	30.84	17.40
14.97	159.6	44.36	154.1	84.76	135.6	115.1	104.9	123.4	63.63
26.19	257.1	77.63	247.7	148.3	215.8	201.3	164.0	185.0	87.84
41.16	339.7	110.9	311.7	210.9	266.7	258.9	189.1	246.7	104.8
52.38	354.7	144.2	336.2	254.3	278.6	316.4	198.8	339.2	111.6
63.61	323.9	166.3	326.6	317.8	254.1	373.9	188.9	400.9	99.43
71.09	274.7	199.6	266.9	360.2	202.1	431.4	153.8	462.6	69.17
82.32	143.7	232.9	134.7	402.6	110.8	489.0	85.11	524.3	14.75
89.80	−3.652	255.1	−20.06	445.0	−41.45	546.5	−32.92	586.0	−73.95
96.16	−201.8	271.7	−204.5	273.2	−203.6	596.3	−203.7	642.5	−202.4

15°		0°		−15°		−30°		−45°	
x, m	y, m	x, m	y, m	x, m	y, m	x, m	y, m	x, m	y, m
22.15	5.769	48.57	0.7707	28.39	7.881	17.28	10.10	10.81	10.88
88.59	20.81	97.14	3.307	56.78	16.36	34.56	20.46	21.62	21.92
155.0	31.63	145.7	8.004	85.16	25.51	51.83	31.12	32.43	33.12
199.3	35.83	242.9	25.93	141.9	46.20	86.39	53.38	54.06	56.04
243.6	37.05	291.4	40.47	198.7	70.82	120.9	77.13	75.68	79.76
332.2	28.06	340.0	59.93	227.1	84.95	155.5	102.6	97.3	104.4
398.7	8.068	388.6	85.48	283.9	117.7	190.1	130.3	118.9	130.0
465.1	−24.52	437.1	118.7	340.7	158.1	224.6	160.4	151.4	170.8
553.7	−109.7	485.7	161.7	369.0	182.0	259.2	193.6	173.0	199.7
614.2	−200.5	521.3	201.1	389.9	201.3	266.1	200.7	173.7	200.7

266

$b_d = 2.066 \cdot 10^{-3}$ m^{-1}; $v = 177.5$ m/s

30°		45°		60°		75°		85°	
x, m	y, m	x, m	y, m	x, m	y, m	x, m	y, m	x, m	y, m
49.18	27.85	42.47	41.86	29.97	51.31	15.35	56.59	5.149	58.25
98.36	54.43	169.9	157.1	119.9	195.0	61.40	216.8	20.60	223.1
196.7	102.4	254.8	220.1	209.8	311.9	107.5	351.2	41.19	400.4
344.3	153.5	339.8	264.7	268.9	364.4	153.5	440.3	56.64	476.1
442.6	164.7	424.7	282.2	329.7	387.9	184.2	463.3	66.94	488.4
541.0	146.7	509.4	260.6	389.7	372.2	214.9	450.8	77.23	466.7
590.2	122.2	594.6	181.9	479.6	246.4	261.0	350.3	92.68	359.6
688.5	27.59	679.6	7.569	539.5	43.85	291.7	203.1	103.0	216.0
737.7	−53.35	722.1	−148.5	569.5	−135.7	322.4	−79.17	113.3	−44.58
796.7	−201.1	733.4	−204.3	577.5	−200.9	330.6	−205.2	117.1	−205.2

−80°		−30°		−15°		0°		15°	
x, m	y, m	x, m	y, m	x, m	y, m	x, m	y, m	x, m	y, m
11.33	11.37	18.69	10.84	33.28	9.112	68.00	0.7925	39.85	10.40
22.67	22.83	37.38	21.89	66.57	18.65	136.0	3.505	159.4	37.32
45.33	46.03	56.07	33.09	99.85	28.68	204.0	8.761	199.3	44.41
68.00	69.65	93.46	56.07	166.4	50.55	272.0	17.39	279.0	54.67
90.67	93.75	130.8	79.97	233.0	75.46	340.0	30.49	358.7	57.99
113.3	118.4	168.2	105.0	266.3	89.38	408.0	49.56	438.4	51.66
147.3	156.6	205.6	131.3	332.8	121.0	476.0	76.62	557.9	15.36
159.2	170.0	243.0	243.0	399.4	159.0	544.0	114.5	597.8	−6.788
280.4	196.5	432.7	181.2	432.7	181.2	612.0	167.4	717.3	−120.6
184.4	200.2	294.1	200.5	459.3	200.8	648.3	203.8	767.8	−200.9

$b_d = 2.066 \cdot 10^{-3} \ m^{-1}; \ v = 248.5 \ m/s$

85° x, m	85° y, m	75° x, m	75° y, m	60° x, m	60° y, m	45° x, m	45° y, m	30° x, m	30° y, m
6.690	75.92	20.4	74.25	39.82	68.41	58.52	57.91	72.79	41.39
26.76	293.9	80.16	287.0	159.3	262.9	234.1	249.7	218.4	118.7
40.14	424.3	140.3	468.8	278.8	423.1	292.6	266.8	291.2	153.1
60.21	574.8	200.4	584.8	358.4	491.4	409.6	342.1	436.7	206.9
80.27	634.7	240.5	603.4	438.1	508.8	526.7	376.9	582.3	226.5
93.65	609.5	280.5	561.4	477.9	492.0	585.2	371.1	655.1	215.2
113.7	448.0	320.6	440.0	557.5	387.8	702.2	289.2	727.9	183.0
127.1	191.6	360.7	174.4	637.0	118.8	760.8	194.9	873.5	15.79
133.8	−53.17	380.7	−94.35	667.0	−179.7	819.3	36.76	946.2	−167.1
136.5	−207.1	386.1	−205.6	679.7	−209.3	870.0	−205.1	956.0	−201.3

15° x, m	15° y, m	0° x, m	0° y, m	−15° x, m	−15° y, m	−30° x, m	−30° y, m	−45° x, m	−45° y, m
67.90	17.76	63.47	0.3499	37.38	10.14	19.61	11.36	11.64	11.66
203.7	49.76	190.4	3.811	112.2	31.32	39.22	22.82	46.57	46.94
271.6	63.17	253.9	7.499	149.5	42.45	58.83	34.37	69.85	70.74
339.5	74.08	380.9	20.95	224.3	66.17	98.05	57.80	93.13	94.80
475.3	84.84	444.3	31.98	281.7	78.92	137.3	81.76	116.4	119.2
543.2	82.18	507.7	47.09	336.5	106.8	196.1	118.9	139.7	143.9
679.0	51.09	634.7	94.99	411.2	138.7	235.3	144.7	163.0	168.9
746.9	16.58	698.1	131.8	486.0	176.5	274.5	171.6	174.6	181.7
814.8	−36.93	761.6	181.4	523.4	198.3	313.7	199.6	186.3	194.5
932.5	−203.0	787.0	205.8	528.4	201.4	315.1	200.6	191.7	200.5

268

$b_d = 2.893 \cdot 10^{-3}$ m^{-1}; $v = 27.89$ m/s

85° x, m	85° y, m	75° x, m	75° y, m	60° x, m	60° y, m	45° x, m	45° y, m	30° x, m	30° y, m
0.3723	4.138	1.078	3.912	1.901	3.201	2.252	2.188	2.003	1.123
1.489	15.12	4.312	14.29	7.604	11.67	11.26	9.612	10.02	4.920
3.350	28.34	8.623	24.76	15.21	20.18	27.02	17.16	20.03	8.041
4.839	34.07	12.93	31.17	22.81	25.32	36.03	18.12	26.04	9.003
5.956	35.42	17.25	33.32	30.42	26.89	45.04	16.48	32.05	9.252
8.190	30.45	23.71	28.29	39.92	23.60	67.56	0.0013	52.08	4.598
10.42	14.87	30.18	12.85	51.33	11.43	87.83	−32.15	62.10	−1.131
11.91	−1.889	34.49	−3.725	58.93	−2.075	110.3	−92.4	90.14	−30.98
16.01	−79.89	48.50	−101.2	83.64	−82.43	126.1	−155.9	128.2	−113.3
19.48	−200.1	56.44	−200.5	102.7	−200.4	134.6	−200.4	151.5	−200.0

15° x, m	15° y, m	0° x, m	0° y, m	-15° x, m	-15° y, m	-30° x, m	-30° y, m	-45° x, m	-45° y, m
1.180	0.3068	10.69	0.7351	9.200	3.048	7.420	4.755	5.543	5.936
5.902	1.343	21.37	3.003	18.40	7.307	14.84	10.48	11.09	12.68
10.62	2.067	32.06	6.904	27.60	12.85	22.26	17.24	16.63	20.29
15.34	2.471	53.43	20.07	46.00	28.13	37.10	34.11	27.71	38.26
18.89	2.558	74.80	41.33	64.40	49.71	51.94	55.95	38.80	60.32
33.77	−0.2749	85.48	55.47	82.80	78.73	66.78	83.61	44.34	73.07
70.82	−20.20	106.9	92.06	101.2	116.8	81.63	118.3	55.43	102.5
94.43	−48.34	128.2	142.4	119.6	166.6	96.47	161.6	66.51	138.1
120.4	−96.48	138.9	174.3	128.8	197.1	103.9	187.3	77.60	181.4
154.7	−200.1	146.7	201.2	130.0	201.5	107.3	200.4	82.03	201.3

$b_d = 2.893 \cdot 10^{-3}$ m^{-1}; $v = 39.05$ m/s

85°		75°		60°		45°		30°	
x, m	y, m	x, m	y, m	x, m	y, m	x, m	y, m	x, m	y, m
0.6696	7.459	1.947	7.081	3.480	5.869	4.206	4.090	3.827	2.146
3.348	33.17	9.734	31.45	13.92	21.48	16.82	14.92	15.31	7.797
6.027	51.53	17.52	48.78	27.84	37.18	33.65	25.69	30.61	13.37
8.705	61.71	23.36	56.67	41.76	46.36	50.47	31.80	45.92	16.51
10.71	63.66	29.20	59.88	52.19	48.56	63.09	33.02	57.40	17.12
14.73	52.47	35.04	58.22	69.69	42.52	84.11	27.98	80.36	13.44
18.75	18.62	46.73	39.45	90.47	17.42	113.6	3.971	103.3	2.338
20.76	−6.223	56.46	−6.236	100.9	−3.576	143.0	−44.51	137.8	−30.82
24.78	−82.29	68.14	−60.31	125.3	−80.81	180.8	−160.3	176.0	−98.39
28.62	−200.5	82.16	−200.1	147.1	−200.6	189.5	−200.0	209.9	−200.2

15°		0°		−15°		−30°		−45°	
x, m	y, m	x, m	y, m	x, m	y, m	x, m	y, m	x, m	y, m
2.297	0.5971	41.96	0.7412	12.91	3.833	9.533	5.902	6.898	7.210
6.890	1.680	29.92	3.054	24.61	8.791	19.07	12.64	13.80	15.07
16.08	3.387	44.88	7.082	36.92	14.97	28.60	20.26	20.69	23.62
25.26	4.454	74.80	20.95	61.53	31.38	47.66	38.45	34.49	42.97
34.45	44.846	104.7	43.96	86.14	54.07	66.73	61.11	48.28	65.70
66.61	0.3441	119.7	59.55	98.45	68.17	76.26	74.36	55.18	78.51
89.57	−9.224	149.6	100.9	123.1	103.0	95.33	105.4	68.98	107.4
130.9	−42.87	179.5	159.7	147.7	148.7	114.14	143.6	82.77	141.5
174.6	−109.0	194.5	198.0	160.0	176.8	133.5	191.2	96.57	182.0
210.5	−200.2	195.5	200.2	169.0	200.2	136.6	200.2	102.1	200.4

$b_d = 2.893 \cdot 10^{-3}$ m^{-1}; $v = 54.66$ m/s

85° x, m	85° y, m	75° x, m	75° y, m	60° x, m	60° y, m	45° x, m	45° y, m	30° x, m	30° y, m
1.141	12.74	3.335	12.17	6.076	10.27	7.572	7.379	7.452	4.016
4.562	47.17	13.34	44.98	24.30	37.84	30.29	27.02	28.61	14.61
9.124	82.52	23.34	71.35	42.53	59.74	53.00	42.37	50.07	22.77
13.69	103.4	36.68	94.58	66.83	78.40	83.29	54.89	71.52	28.16
17.11	108.4	50.02	102.1	85.06	83.21	106.0	57.28	92.98	30.39
22.81	95.25	63.36	92.44	115.4	70.93	136.3	49.50	128.7	25.94
28.51	52.55	80.04	52.90	145.8	28.19	166.6	26.83	157.4	13.67
33.08	−8.429	90.04	11.35	164.0	−16.39	189.3	−2.117	178.8	−1.619
37.64	−104.6	103.4	−72.99	188.3	−108.0	234.7	−103.1	236.0	−76.00
40.68	−200.6	115.9	−201.3	204.5	−200.1	260.5	−201.1	284.7	−200.2

15° x, m	15° y, m	0° x, m	0° y, m	−15° x, m	−15° y, m	−30° x, m	−30° y, m	−45° x, m	−45° y, m
4.439	1.155	20.94	0.7500	16.17	4.808	11.88	7.173	8.264	8.493
17.76	4.183	41.89	3.128	32.34	10.63	23.76	15.02	16.53	17.47
35.51	7.132	62.83	7.345	48.51	17.58	35.63	23.58	24.79	26.96
48.83	8.453	104.7	22.30	80.85	35.33	59.38	43.17	41.32	47.64
62.15	8.942	146.6	48.11	97.02	46.44	83.13	66.55	57.85	70.92
93.23	6.462	167.5	66.15	113.2	59.26	106.9	94.53	74.37	97.25
119.9	−0.2988	188.5	88.40	145.5	91.01	130.6	128.2	90.90	127.3
208.6	−65.24	230.4	149.1	177.9	132.9	154.4	169.2	107.4	161.7
244.2	−117.9	251.3	190.3	210.2	188.8	166.3	193.0	115.7	181.0
281.3	−200.3	256.9	202.9	216.7	202.1	170.2	201.6	123.4	200.4

$b_d = 2.893 \cdot 10^{-3}$ m^{-1}; $v = 76.53$ m/s

85°		75°		60°		45°		30°	
x, m	y, m	x, m	y, m	x, m	y, m	x, m	y, m	x, m	y, m
1.812	20.33	5.335	19.54	9.950	16.89	12.91	12.62	12.89	7.253
7.249	76.04	21.34	72.97	39.80	62.74	51.65	46.49	51.58	26.44
14.50	134.0	37.35	116.4	69.65	99.39	90.38	72.82	90.26	40.93
19.94	161.3	58.69	153.7	99.50	124.0	129.1	89.25	128.9	49.47
25.37	172.1	74.69	162.5	129.4	133.4	154.9	93.39	154.7	51.10
32.62	158.4	96.03	145.1	159.2	124.7	206.6	81.04	219.2	36.89
39.87	108.9	117.5	87.66	199.0	77.95	245.3	48.81	270.8	0.3054
47.12	10.83	133.4	8.266	228.9	6.803	284.1	−10.73	309.5	−48.28
52.56	−117.2	149.4	−124.1	248.8	−66.92	322.8	−112.4	348.1	−123.9
54.97	−202.1	155.6	−201.0	272.3	−201.5	344.7	−201.6	374.8	−200.5

15°		0°		−15°		−30°		−45°	
x, m	y, m	x, m	y, m	x, m	y, m	x, m	y, m	x, m	y, m
8.477	2.205	29.32	0.7625	20.75	5.963	14.24	8.458	9.488	9.642
33.90	7.978	56.84	3.236	41.50	12.80	28.49	17.41	18.98	19.61
59.32	12.33	87.96	7.737	62.25	20.64	42.73	26.92	28.46	29.93
84.74	15.03	146.7	24.42	83.00	29.60	71.22	47.82	47.44	51.76
110.2	15.82	175.9	37.63	124.5	51.59	99.71	71.72	56.93	63.33
169.5	8.413	205.2	54.98	145.2	65.01	128.2	99.32	85.39	101.2
203.4	−3.149	234.6	77.36	186.7	98.04	156.7	131.6	104.4	129.6
254.2	−33.92	293.2	142.6	228.2	141.8	185.2	169.9	123.3	161.2
305.1	−86.85	322.5	189.6	249.0	169.0	199.4	191.9	142.3	196.8
367.2	−200.6	328.4	200.6	269.7	200.9	205.1	201.3	144.2	200.6

$b_d = 2.893 \cdot 10^{-3}$ m^{-1}; $v = 107.1$ m/s

85°		75°		60°		45°		30°	
x, m	y, m	x, m	y, m	x, m	y, m	x, m	y, m	x, m	y, m
2.673	30.11	7.921	29.14	15.13	25.80	20.54	20.16	22.03	12.43
13.36	139.1	31.68	110.1	60.54	96.89	82.18	74.90	88.11	45.45
21.38	202.6	55.45	176.9	105.9	154.2	143.8	117.1	132.2	62.74
29.40	242.7	79.21	222.6	151.3	190.5	184.9	135.1	176.2	74.84
37.42	253.3	103.0	240.1	181.6	199.0	226.0	142.0	220.3	80.18
45.43	231.3	126.7	223.7	211.9	191.9	287.6	124.9	308.4	62.06
50.78	196.2	150.5	166.1	257.3	144.4	328.7	88.83	374.5	10.54
58.80	102.5	174.3	47.52	302.7	31.82	369.8	24.15	396.5	−17.44
64.14	−2.608	182.2	−14.33	317.8	−29.61	410.9	−86.14	440.6	−97.67
70.02	−202.5	197.8	−202.6	345.1	−200.4	437.6	−202.5	447.6	−200.6

15°		0°		−15°		−30°		−45°	
x, m	y, m	x, m	y, m	x, m	y, m	x, m	y, m	x, m	y, m
15.82	4.121	41.05	0.7805	25.81	7.237	16.38	9.612	10.45	10.54
63.28	14.87	82.10	3.396	51.62	15.19	32.75	19.57	20.89	21.29
110.7	22.59	123.1	8.339	77.43	23.97	49.13	29.90	31.34	32.25
142.4	25.59	164.2	16.24	129.0	44.65	81.88	51.90	52.23	54.94
174.0	26.47	205.2	27.91	154.9	56.91	114.6	76.06	73.12	78.78
221.5	22.83	287.3	67.15	206.5	86.49	147.1	102.9	94.02	104.0
268.9	11.60	328.4	98.09	258.1	125.2	180.1	133.3	114.9	130.9
300.6	−1.304	369.4	140.1	283.9	149.1	212.9	168.2	135.8	159.9
395.5	−78.38	410.5	197.8	309.7	177.1	229.3	187.8	156.7	191.4
466.2	−200.5	413.2	202.4	328.6	200.7	239.1	200.3	162.3	200.3

$b_d = 2.893 \cdot 10^{-3}$ m^{-1}; $v = 150$ m/s

85° x, m	85° y, m	75° x, m	75° y, m	60° x, m	60° y, m	45° x, m	45° y, m	30° x, m	30° y, m
3.678	41.60	10.97	40.49	21.41	36.65	30.34	29.90	35.13	19.89
14.71	159.4	43.86	154.8	85.64	139.3	121.4	112.2	140.5	73.11
25.74	259.0	76.76	250.8	149.9	222.8	182.0	157.2	175.6	87.71
36.78	327.1	109.7	314.5	192.7	260.5	242.7	189.1	245.9	109.6
47.81	348.9	131.6	331.0	235.5	277.0	303.4	201.5	316.2	117.7
55.17	333.4	153.5	322.2	299.7	247.4	364.1	186.1	351.3	144.4
66.20	256.8	186.4	250.2	342.6	176.0	424.7	130.0	421.6	87.29
77.23	76.73	219.3	62.62	385.4	31.32	485.4	5.407	491.8	19.71
80.91	-31.84	230.3	-56.55	406.8	-96.95	515.8	-106.1	526.9	-38.11
84.59	-201.0	239.0	-203.3	418.9	-205.8	534.0	-205.0	585.5	-201.7

15° x, m	15° y, m	0° x, m	0° y, m	-15° x, m	-15° y, m	-30° x, m	-30° y, m	-45° x, m	-45° y, m
28.46	7.426	38.31	0.3450	30.90	8.518	18.05	10.52	11.10	11.16
113.9	96.66	76.63	1.492	61.81	17.58	36.10	21.25	22.21	22.44
142.3	31.72	114.9	3.640	92.71	27.28	54.15	32.23	33.31	33.84
199.3	39.05	153.3	7.039	154.5	49.21	90.25	55.04	55.52	57.09
256.2	41.42	191.6	12.00	185.4	61.79	126.4	79.25	77.73	81.03
341.6	31.25	268.2	28.27	216.3	75.76	162.5	105.3	99.94	105.8
427.0	-4.849	344.8	57.03	278.1	109.1	198.6	133.6	121.1	131.6
483.9	-52.10	383.1	78.30	339.9	152.8	234.7	165.0	144.4	158.5
540.8	-129.8	459.8	142.0	370.9	180.2	252.7	182.2	166.6	187.0
575.0	-200.1	505.7	200.8	391.5	201.1	270.8	200.5	176.9	200.9

$b_d = 2.893 \cdot 10^{-3}$ m⁻¹; $v = 210$ m/s

85°		75°		60°		45°		30°	
x, m	y, m	x, m	y, m	x, m	y, m	x, m	y, m	x, m	y, m
4.778	54.23	14.31	53.03	28.24	48.86	41.80	41.36	51.99	29.57
19.11	209.9	57.25	205.0	113.8	187.8	125.4	120.3	156.0	84.80
33.45	344.2	100.2	334.9	199.1	302.2	209.0	190.0	208.0	109.3
43.00	410.6	143.1	417.7	256.0	551.0	292.6	244.3	312.0	147.8
57.34	453.4	171.8	431.0	312.9	363.4	376.2	269.2	415.9	161.8
71.67	410.5	200.4	401.0	341.3	351.4	418.0	265.1	467.9	153.7
81.23	320.0	229.0	314.3	398.2	277.0	501.6	206.6	519.9	130.7
90.79	136.9	257.6	124.6	455.1	84.88	585.2	26.26	571.9	86.72
95.56	−37.98	272.0	−67.39	483.6	−128.3	627.0	−185.1	623.9	11.28
98.27	−207.1	278.2	−208.3	490.2	−209.4	629.8	−206.4	696.7	−200.3

15°		0°		−15°		−30°		−45°	
x, m	y, m	x, m	y, m	x, m	y, m	x, m	y, m	x, m	y, m
48.50	12.69	53.64	0.3558	35.46	9.663	19.21	11.15	11.51	11.54
145.5	35.54	107.3	1.592	70.92	19.70	38.41	22.42	23.02	23.14
194.0	45.12	160.9	4.027	106.4	30.19	57.62	33.83	34.53	34.82
242.5	52.91	268.2	14.41	177.3	53.03	96.03	57.16	57.55	58.41
339.5	60.60	321.8	23.77	212.8	65.69	134.4	81.33	80.56	82.39
436.5	51.27	429.1	58.66	248.2	79.43	192.1	119.7	115.0	119.2
533.5	11.85	482.7	84.03	319.1	111.3	230.5	247.3	183.1	144.6
582.0	−26.38	536.4	122.8	390.1	151.9	268.9	177.0	161.1	170.6
679.0	−172.6	590.0	178.5	425.5	177.0	288.1	192.9	184.1	197.6
691.9	−200.4	611.5	207.5	453.9	200.2	297.1	200.5	186.4	200.4

$b_d = 4.05 \cdot 10^{-3}$ m^{-1}; $v = 23.54$ m/s

30°		45°		60°		75°		85°	
x, m	y, m	x, m	y, m	x, m	y, m	x, m	y, m	x, m	y, m
1.431	0.8019	1.609	1.563	1.358	2.286	0.7699	2.794	0.2659	2.956
5.423	2.912	6.434	5.685	5.431	8.335	3.080	10.20	1.064	10.80
11.45	5.011	11.26	8.927	10.86	14.41	6.159	17.69	2.127	18.73
17.17	6.257	17.69	11.80	14.65	18.61	9.239	22.26	3.191	23.61
22.90	6.608	24.13	12.94	21.43	19.21	12.32	23.80	4.254	25.30
34.34	4.452	35.39	10.46	29.84	15.86	20.02	13.42	6.116	20.43
44.36	−0.8076	48.26	0.0010	39.38	37.47	24.64	−2.661	8.243	2.006
70.11	−31.26	57.91	−13.85	52.96	−31.44	30.03	−33.38	10.64	−38.04
97.30	−99.37	85.25	−90.05	67.89	−103.6	36.96	−97.44	13.30	−116.7
118.9	−200.1	104.7	−200.4	79.34	−200.4	43.50	−200.5	15.01	−200.3

−45°		−30°		−15°		0°		15°	
x, m	y, m	x, m	y, m	x, m	y, m	x, m	y, m	x, m	y, m
4.909	5.343	6.487	4.251	7.915	2.727	9.031	0.7379	0.8431	0.2192
9.818	11.59	12.97	9.557	15.83	6.724	18.06	3.027	3.372	0.7950
14.73	18.80	19.46	15.99	23.74	12.08	27.09	6.988	6.745	1.369
24.54	36.40	32.43	32.62	39.57	27.36	45.15	20.51	10.96	1.765
29.45	46.98	45.41	55.05	47.49	37.57	63.22	42.81	13.49	1.827
39.27	72.30	51.89	68.85	63.32	64.10	72.25	57.92	26.14	0.051
49.09	104.3	64.87	102.8	79.15	100.8	90.31	98.26	59.86	−24.33
58.91	145.3	77.84	147.7	94.98	151.7	108.4	157.2	88.52	−74.82
68.73	198.7	84.33	175.8	102.9	184.7	117.4	197.1	107.9	−133.0
69.05	200.8	89.52	201.8	106.6	202.4	118.0	200.1	122.9	−200.1

$b_d = 4.05\cdot10^{-3}$ m^{-1}; $v = 33$ m/s

85°		75°		60°		45°		30°	
x, m	y, m	x, m	y, m	x, m	y, m	x, m	y, m	x, m	y, m
0.4783	5.328	1.391	5.058	2.485	4.192	3.004	2.922	2.733	1.533
1.913	19.57	5.563	18.56	9.942	15.34	12.02	10.66	10.93	5.569
4.305	36.80	11.13	32.26	19.88	26.56	24.03	18.35	21.87	9.553
6.218	44.08	15.30	38.97	29.83	33.12	36.05	22.71	32.80	11.79
7.653	45.47	20.86	42.78	37.28	34.69	45.06	23.59	41.00	12.23
10.52	37.48	27.81	38.80	47.22	34.69	63.09	18.45	57.40	9.582
13.39	14.02	34.74	24.42	62.14	16.49	81.11	2.837	73.80	1.670
15.78	−19.70	41.72	−1.869	72.08	−2.554	102.1	−31.79	90.20	−12.36
19.61	−113.0	52.84	−77.34	94.45	−80.26	123.2	−91.06	133.9	−90.90
21.70	−200.3	62.35	−200.0	111.9	−200.2	145.3	−200.1	162.7	−200.4

15°		0°		−15°		−30°		−45°	
x, m	y, m	x, m	y, m	x, m	y, m	x, m	y, m	x, m	y, m
1.641	0.4265	12.64	0.7452	10.66	3.422	8.439	5.311	6.210	6.566
8.203	1.866	25.29	3.088	21.32	8.046	16.88	11.55	12.42	13.88
14.76	2.859	37.93	7.203	31.98	13.99	25.32	18.80	18.63	22.00
16.69	3.298	63.22	21.60	53.30	30.42	42.19	36.73	31.05	40.97
24.61	3.461	88.50	46.09	63.96	41.27	50.63	47.66	43.47	64.17
47.58	0.2458	101.1	63.10	74.62	54.17	67.51	74.24	55.89	92.64
62.34	−5.716	126.4	109.9	95.94	87.42	84.39	108.7	68.31	127.9
95.15	−32.48	13.91	141.9	117.3	133.8	101.3	154.0	80.73	172.4
134.5	−99.27	151.7	182.0	127.9	164.0	109.7	182.4	86.94	199.3
165.6	−200.0	156.8	201.0	138.6	200.5	114.8	201.4	87.36	201.2

$b_d = 4.05 \cdot 10^{-3}$ m^{-1}; $v = 46.2$ m/s

85° x, m	85° y, m	75° x, m	75° y, m	60° x, m	60° y, m	45° x, m	45° y, m	30° x, m	30° y, m
0.8147	9.103	2.382	8.690	4.340	7.339	5.408	5.244	5.109	2.868
3.259	33.69	32.13	14.36	24.03	21.63	19.30	20.44	20.44	10.43
6.517	58.95	19.06	56.08	30.38	42.67	37.86	30.26	35.46	16.26
9.776	73.86	28.59	70.01	47.74	56.00	59.49	39.20	51.09	20.11
12.22	77.44	35.73	72.96	60.46	59.44	75.72	40.91	66.41	2171
17.11	63.74	47.64	62.23	78.12	54.00	113.6	23.33	102.2	14.91
21.18	28.48	59.55	29.10	99.82	28.27	135.2	−1.512	127.7	−1.157
23.63	−6.021	66.70	−4.421	117.2	−11.70	156.8	−43.95	153.3	−29.55
27.70	−97.61	78.61	−95.06	138.9	−99.26	183.9	−133.8	178.8	−74.90
30.39	−200.8	86.63	−201.1	155.3	−201.2	196.5	−204.4	217.5	−200.2

15° x, m	15° y, m	0° x, m	0° y, m	−15° x, m	−15° y, m	−30° x, m	−30° y, m	−45° x, m	−45° y, m
3.171	0.8247	17.70	0.7557	14.14	4.302	10.69	6.531	7.589	7.861
12.68	2.988	35.40	3.177	28.29	9.717	21.37	13.84	15.18	16.30
25.37	5.094	53.10	7.523	42.43	16.39	32.06	22.00	22.77	25.38
34.88	6.038	88.50	23.28	70.71	34.22	53.43	41.34	37.94	45.70
44.39	6.387	106.2	35.52	99.00	59.60	74.80	65.55	53.12	69.46
66.59	4.616	123.9	51.41	113.1	75.93	85.48	79.92	68.30	97.57
85.62	−0.2135	141.6	71.73	127.3	95.30	106.9	114.5	83.48	131.3
123.7	−21.39	177.0	130.5	155.6	146.1	128.2	159.3	98.66	172.7
190.3	−116.4	194.7	173.2	169.7	179.7	138.9	187.0	106.2	197.2
218.7	−200.1	204.1	201.4	177.3	200.8	143.9	201.6	107.3	200.7

$b_d = 4.05 \cdot 10^{-3}$ m^{-1}; $v = 64.68$ m/s

85°		75°		60°		45°		30°	
x, m	y, m	x, m	y, m	x, m	y, m	x, m	y, m	x, m	y, m
1.295	14.52	3.811	13.96	7.107	12.06	9.222	9.016	9.210	5.181
5.178	54.31	15.24	52.12	28.43	44.82	46.11	40.15	36.84	18.89
9.062	86.85	26.68	83.16	49.75	70.99	73.78	56.80	55.27	26.21
12.95	110.0	41.92	109.8	78.18	92.10	92.22	63.75	82.89	33.84
18.12	122.9	53.35	116.1	92.39	95.30	110.7	66.71	110.5	36.50
24.60	106.9	72.41	96.33	120.8	83.69	147.6	57.88	147.4	30.12
31.07	48.16	91.46	27.98	149.3	41.71	184.4	23.21	175.0	15.70
34.95	−17.75	99.08	−20.17	170.6	−19.09	202.9	−7.667	202.6	−8.647
38.84	−127.5	110.5	−134.8	191.9	−125.5	239.9	−115.0	248.7	−88.52
40.52	−202.1	114.7	−200.6	201.4	−201.9	256.4	−200.5	282.1	−200.0

15°		0°		−15°		−30°		−45°	
x, m	y, m	x, m	y, m	x, m	y, m	x, m	y, m	x, m	y, m
6.053	1.575	24.78	0.7707	18.38	5.372	13.07	7.826	8.903	9.095
24.21	5.699	45.56	3.308	36.76	11.74	26.15	16.26	17.81	18.60
42.37	8.807	73.34	8.007	55.15	19.27	39.22	25.38	26.71	28.57
60.53	10.74	99.12	15.36	91.90	38.67	65.36	46.07	44.51	50.09
78.63	11.30	123.9	25.98	110.3	51.10	91.51	80.88	62.32	74.19
109.0	8.592	173.5	60.54	128.7	65.82	104.6	85.26	80.12	101.6
139.2	0.2360	198.2	87.06	165.4	104.2	130.7	119.1	97.93	133.2
169.5	−15.42	223.0	122.6	202.2	160.2	156.9	162.2	115.7	170.7
248.2	−111.7	247.8	171.1	220.6	198.3	169.9	188.5	124.6	192.3
281.9	−200.4	261.0	204.5	221.8	201.2	175.2	200.2	128.2	201.5

$b_d = 4.05 \cdot 10^{-3}\ m^{-1}$; $v = 90.55$ m/s

x, m	*y*, m	*x*, m	*y*, m	*x*, m	*y*, m	*x*, m	*y*, m	*x*, m	*y*, m
85°		75°		60°		45°		30°	
1.909	21.51	5.658	20.81	10.81	18.43	14.67	14.40	15.73	88.77
7.636	81.41	22.63	78.63	43.24	69.21	58.70	53.50	62.94	32.47
15.27	144.7	39.61	126.4	64.86	97.87	102.7	83.65	78.67	39.04
21.00	173.3	56.58	159.0	97.30	129.4	132.1	96.48	110.1	49.66
26.73	180.9	73.55	171.5	129.7	142.2	161.4	101.4	157.3	57.27
32.45	165.2	90.53	159.8	183.8	103.1	220.1	78.46	220.3	44.33
38.18	122.3	107.5	118.6	216.2	22.73	264.1	17.25	267.5	7.525
43.91	39.83	124.5	33.95	227.0	—21.15	278.8	—16.80	283.2	—12.45
47.73	—55.28	130.1	—10.23	248.6	—162.5	308.2	—122.1	330.3	—110.9
51.10	—203.3	144.3	—201.9	252.6	—202.8	321.9	—202.3	354.6	—201.7
15°		0°		—15°		—30°		—45°	
11.30	2.943	34.69	0.7925	23.24	6.598	15.36	9.063	10.00	10.13
45.20	10.62	69.39	3.505	46.48	14.04	30.71	18.56	20.01	20.53
67.80	14.58	104.1	8.764	69.73	22.51	46.07	28.56	30.01	31.23
101.7	18.28	138.8	17.41	116.2	43.52	76.78	50.42	50.02	53.69
124.3	18.90	173.5	30.57	56.70	92.13	62.48	70.03	70.03	77.91
169.5	14.32	208.2	49.87	162.7	72.25	122.8	89.56	90.04	104.4
214.7	—0.9313	242.9	77.63	184.9	90.77	153.6	121.9	110.0	133.8
259.9	—31.97	277.5	117.6	232.4	140.3	184.3	161.7	130.1	167.2
316.4	—107.9	312.2	176.6	255.7	174.1	199.6	185.6	140.1	185.8
353.3	—200.3	323.8	202.6	271.2	201.6	208.8	201.5	147.4	200.4

$b_d = 4.05 \cdot 10^{-3}$ m^{-1}; $v = 126.8$ m/s

85°		75°		60°		45°		30°	
x, m	y, m	x, m	y, m	x, m	y, m	x, m	y, m	x, m	y, m
2.627	29.72	7.832	28.92	15.29	26.18	21.67	21.36	25.09	14.21
13.14	139.4	23.50	84.43	61.17	99.47	86.68	80.16	75.28	40.51
21.02	204.3	46.99	158.3	107.1	159.1	130.0	112.3	125.5	62.65
28.90	242.9	70.49	212.8	137.6	186.1	173.4	135.1	175.6	78.31
34.15	249.2	93.99	236.4	168.2	197.9	216.7	144.0	225.8	84.05
42.03	225.5	117.5	219.1	198.8	189.9	260.0	132.9	301.1	62.35
49.91	151.8	141.0	146.7	229.4	155.9	303.4	92.82	351.3	14.08
55.17	54.81	156.6	44.73	275.3	22.37	346.7	3.862	376.4	−27.22
57.79	−22.74	164.5	−40.39	290.6	−69.25	368.4	−75.78	401.5	−87.34
61.30	−203.3	173.1	−201.1	303.8	−202.2	389.3	−204.7	430.8	−201.0

15°		0°		−15°		−30°		−45°	
x, m	y, m	x, m	y, m	x, m	y, m	x, m	y, m	x, m	y, m
20.33	5.305	32.38	0.3499	28.39	7.892	17.28	10.10	10.81	10.89
61.00	14.89	64.76	1.537	56.78	16.46	34.56	20.49	21.62	21.93
101.7	22.66	129.5	7.501	85.16	25.88	51.83	31.32	32.43	33.16
142.3	27.89	161.9	13.03	141.9	48.29	86.39	53.94	54.06	56.28
183.0	29.59	194.3	20.97	170.3	61.98	103.7	66.09	75.68	80.47
264.3	16.27	226.7	32.06	198.7	77.98	138.2	92.50	97.30	106.1
305.0	−3.463	259.0	47.32	227.1	96.99	172.8	122.7	118.9	133.5
366.0	−61.53	291.4	68.15	283.9	148.3	207.3	158.4	140.5	163.4
406.6	−133.3	356.2	135.1	312.3	184.1	224.6	179.1	162.2	196.6
431.0	−200.7	395.0	206.0	323.6	201.2	240.7	200.7	164.3	200.2

$b_d = 4.05 \cdot 10^{-3}$ m^{-1}; $v = 177.5$ m/s

85°		75°		60°		45°		30°	
x, m	y, m	x, m	y, m	x, m	y, m	x, m	y, m	x, m	y, m
3.413	38.73	10.22	37.88	20.32	34.90	29.86	29.54	37.14	21.12
10.24	114.0	30.67	111.4	60.95	102.3	89.57	85.96	111.4	60.57
20.48	216.5	61.34	210.9	101.6	164.1	149.3	136.1	148.5	78.10
30.72	293.3	92.02	283.8	162.5	236.0	209.0	174.5	222.8	105.6
40.96	323.8	122.7	307.8	223.5	259.6	268.7	192.3	297.1	115.6
47.78	310.9	143.1	286.5	243.8	251.0	298.6	189.4	334.2	109.8
58.02	228.6	163.6	224.5	284.5	197.8	358.3	147.6	371.4	93.35
64.85	97.77	184.0	88.97	325.1	60.63	418.0	18.75	445.6	8.055
68.26	−27.13	194.3	−48.14	345.4	−91.68	447.9	−132.2	482.8	−85.28
70.88	−210.9	200.4	−202.5	353.5	−201.9	455.8	−200.3	508.8	−200.5

15°		0°		−15°		−30°		−45°	
x, m	y, m	x, m	y, m	x, m	y, m	x, m	y, m	x, m	y, m
34.64	9.061	45.33	0.3830	33.28	9.122	18.69	10.87	11.33	11.38
103.9	25.39	90.67	1.660	66.57	18.74	37.38	21.91	22.67	22.84
138.6	32.23	136.0	4.307	99.85	29.00	56.07	33.16	45.33	46.10
173.2	37.80	181.3	8.903	166.4	52.41	93.46	56.44	68.00	69.90
242.5	43.29	272.0	27.83	199.7	66.22	130.8	81.16	90.67	94.42
311.8	36.62	317.3	45.33	233.0	82.08	168.2	107.9	113.3	119.9
381.1	8.461	362.7	71.73	299.6	123.1	205.6	137.6	136.0	146.5
415.7	−18.84	408.0	111.8	332.8	150.8	243.0	171.7	158.7	174.8
485.0	−123.3	453.3	174.7	366.1	185.9	261.7	191.1	170.0	189.8
512.7	−201.3	466.9	200.5	379.4	202.7	270.4	200.7	177.6	200.0

$b_d = 4.05 \cdot 10^{-3}$ m^{-1}; $v = 248.5$ m/s

85°		75°		60°		45°		30°	
x, m	y, m	x, m	y, m	x, m	y, m	x, m	y, m	x, m	y, m
4.239	48.22	12.74	47.34	25.68	44.23	38.83	38.55	51.27	29.27
16.96	188.3	38.23	140.0	77.03	130.4	116.5	113.0	153.8	84.51
29.67	311.4	76.47	267.6	154.1	246.5	194.1	179.8	256.3	129.9
38.15	371.4	114.7	260.3	205.4	303.3	271.8	229.5	307.6	144.9
50.87	401.1	140.2	383.6	256.8	326.8	349.4	243.6	358.9	150.9
59.35	369.4	152.9	379.3	282.4	320.1	388.3	227.9	410.1	143.4
67.83	280.6	178.4	331.0	308.1	297.0	427.1	188.4	461.4	115.0
72.06	197.9	203.9	197.1	359.5	172.0	465.9	110.5	512.7	52.25
76.30	54.78	216.7	49.17	385.1	19.03	504.8	−51.21	563.9	−81.26
79.69	−212.1	226.0	−202.7	401.4	−203.4	521.6	−200.4	587.8	−205.5

15°		0°		−15°		−30°		−45°	
x, m	y, m	x, m	y, m	x, m	y, m	x, m	y, m	x, m	y, m
55.07	14.45	63.47	0.3827	37.38	10.15	19.61	11.36	11.64	11.66
165.2	40.37	95.20	0.9470	74.77	20.63	39.22	22.83	23.28	23.38
220.3	50.59	126.9	1.859	112.2	31.55	58.83	34.41	46.57	46.98
275.4	57.53	190.4	5.165	186.9	55.49	98.05	58.02	69.85	70.88
330.4	59.33	253.9	11.53	224.3	69.04	137.3	82.47	93.13	95.17
385.5	53.10	317.3	23.04	261.7	84.21	196.1	121.6	116.4	120.0
440.6	34.24	380.8	43.22	299.1	101.6	235.3	150.2	139.7	145.4
495.6	−5.029	444.3	78.36	373.8	147.0	274.5	181.9	163.0	171.8
550.7	−80.54	507.7	140.9	411.2	178.3	294.1	199.5	186.3	199.3
596.6	−202.6	545.8	202.9	433.6	201.4	295.4	200.7	187.0	200.3

$b_d = 5.669 \cdot 10^{-3}$ m^{-1}; $v = 19.92$ m/s

85°		75°		60°		45°		30°	
x, m	y, m	x, m	y, m	x, m	y, m	x, m	y, m	x, m	y, m
0.1899	2.111	1.5499	1.996	0.9699	1.633	1.149	1.116	1.022	0.5728
0.7597	7.713	2.200	7.288	3.879	5.954	4.596	4.061	4.088	2.080
1.709	14.46	4.949	13.65	8.729	11.12	10.34	7.555	8.176	3.580
2.469	17.38	7.149	16.38	12.61	13.29	14.94	8.998	13.29	4.593
3.229	18.01	8.799	17.00	15.52	13.72	18.72	18.38	16.35	4.720
4.748	12.26	13.75	11.19	22.31	10.49	28.72	5.385	25.55	2.786
6.078	−0.9638	17.60	−1.901	29.10	0.8832	35.62	−1.378	31.68	−0.5768
7.977	−35.97	22.00	−27.83	36.86	−19.17	50.55	−29.88	49.06	−20.60
9.687	−91.08	26.95	−76.33	51.40	−94.55	66.64	−91.05	76.65	−99.13
11.50	−200.5	33.34	−200.1	60.97	−200.1	80.81	−200.2	92.49	−200.2

15°		0°		−15°		−30°		−45°	
x, m	y, m	x, m	y, m	x, m	y, m	x, m	y, m	x, m	y, m
0.6022	0.1565	7.633	0.7413	6.791	2.447	5.642	3.795	4.318	4.790
2.409	0.5678	15.27	3.055	13.58	6.222	11.28	8.725	8.636	10.57
4.818	0.9776	22.90	7.091	20.37	11.15	16.93	14.89	12.95	17.43
7.829	1.261	38.16	21.09	33.95	26.88	28.21	31.45	21.59	34.83
9.635	1.305	45.80	31.56	47.53	50.48	33.85	42.20	30.23	58.18
18.67	0.03652	53.43	44.82	54.32	66.30	39.49	54.90	34.55	72.65
32.52	−7.227	68.69	82.10	61.12	85.29	50.78	87.67	43.18	109.1
45.77	−21.25	83.96	141.3	74.70	137.5	62.06	134.4	51.72	160.2
62.03	−50.60	91.59	185.2	81.49	174.3	67.70	165.9	56.14	194.5
96.47	−200.1	94.13	203.3	85.56	201.8	72.59	200.0	57.00	202.4

$b_d = 5.669 \cdot 10^{-3}$ m^{-1}; $v = 27.89$ m/s

30°		45°		60°		75°		85°	
x, m	y, m	x, m	y, m	x, m	y, m	x, m	y, m	x, m	y, m
1.952	1.095	2.146	2.087	1.775	2.994	0.9933	3.613	0.3417	3.806
7.809	3.978	10.73	9.193	8.877	13.25	3.973	13.26	1.367	13.98
15.62	6.823	17.17	13.11	15.98	20.47	7.946	23.04	3.075	26.29
23.43	8.423	25.75	16.22	21.30	23.65	10.93	27.83	4.441	31.48
29.28	8.735	32.19	16.85	26.63	24.78	14.90	30.55	5.466	32.48
44.90	5.405	45.06	13.18	40.83	16.63	22.85	22.49	8.200	22.49
54.66	−0.1546	57.94	2.026	51.48	−1.824	28.81	3.182	10.25	1.618
72.24	−18.34	70.81	−18.26	63.91	−41.23	31.79	−11.65	11.62	−20.23
101.5	−82.37	92.27	−81.77	72.79	−87.16	39.73	−75.35	14.69	−105.9
125.2	−200.2	110.9	−200.7	84.98	−200.7	47.22	−200.6	16.42	−200.6

−45°		−30°		−15°		0°		15°	
x, m	y, m	x, m	y, m	x, m	y, m	x, m	y, m	x, m	y, m
5.543	5.942	7.420	4.763	9.200	3.059	10.69	0.7500	1.172	0.3047
11.09	12.73	14.84	10.55	18.40	7.396	21.37	3.128	4.687	1.105
16.63	20.47	22.26	17.48	27.60	13.17	32.07	7.352	9.375	1.896
27.71	39.21	29.68	25.69	46.00	29.83	42.74	13.68	14.06	2.356
38.80	63.36	44.52	46.67	64.40	55.21	53.43	22.43	17.58	2.472
44.34	78.02	59.36	75.54	73.60	72.26	74.80	49.00	26.95	1.762
55.43	114.4	74.20	115.9	92.00	119.1	96.17	92.75	35.15	−0.1826
66.51	164.4	81.63	142.5	101.2	151.9	106.9	124.7	80.86	−40.90
72.06	197.7	89.05	175.4	110.4	194.8	117.5	167.5	113.7	−124.0
72.43	200.2	93.99	202.2	111.6	201.7	124.0	201.2	129.0	−200.0

$b_d = 5.669 \cdot 10^{-3}$ m^{-1}; $v = 39.05$ m/s

30°		45°		60°		75°		85°	
x, m	y, m	x, m	y, m	x, m	y, m	x, m	y, m	x, m	y, m
3.649	2.049	3.863	3.765	3.100	5.242	1.702	6.207	0.5819	6.502
14.60	7.453	15.45	13.78	12.40	19.31	6.806	22.95	2.328	24.07
25.54	11.62	30.91	23.67	21.70	30.48	13.61	40.06	4.655	42.10
36.49	14.37	42.49	28.00	34.10	40.00	20.42	50.01	6.983	52.76
47.44	15.51	54.08	29.22	43.40	42.45	25.52	52.11	8.729	55.31
65.68	13.24	81.13	17.38	65.10	29.55	35.73	41.10	12.80	41.85
87.58	2.098	96.58	—1.080	80.60	0.1834	45.94	5.793	16.29	4.862
94.88	—4.099	115.9	—41.40	96.10	—55.08	51.05	—24.52	17.46	—14.55
127.7	—53.50	135.2	—114.0	108.5	—134.5	59.55	—108.0	20.95	—109.9
165.4	—200.6	148.0	—200.2	114.8	—201.3	64.66	—200.6	22.67	—200.4

—45°		—30°		—15°		0°		15°	
x, m	y, m	x, m	y, m	x, m	y, m	x, m	y, m	x, m	y, m
6.898	7.216	9.533	5.910	12.31	3.846	14.96	0.7625	2.265	0.5590
13.80	15.12	19.07	12.71	24.61	8.903	29.92	3.236	9.060	2.134
20.69	23.79	28.60	20.52	36.92	15.37	44.88	7.743	31.71	4.562
34.49	43.90	28.13	29.49	49.22	23.49	59.84	14.68	47.56	3.297
48.28	68.68	57.97	51.74	73.84	45.97	74.80	24.55	61.15	—0.1525
55.18	83.35	76.26	81.63	86.14	61.23	104.7	56.01	92.86	—19.09
68.98	118.9	95.33	123.1	110.8	103.5	119.7	79.93	140.4	—94.04
82.77	166.6	104.9	150.4	123.1	133.3	134.6	112.1	151.8	—127.1
89.67	197.6	114.4	184.4	135.4	172.1	149.6	156.6	163.1	—171.6
90.59	202.3	118.2	200.5	142.7	201.9	160.6	202.2	168.7	—200.0

285

$b_d = 5.669 \cdot 10^{-3}$ m^{-1}; $v = 54.66$ m/s

30°		45°		60°		75°		85°	
x, m	y, m	x, m	y, m	x, m	y, m	x, m	y, m	x, m	y, m
6.579	3.701	6.587	6.440	5.077	8.617	2.722	9.969	0.9247	10.37
26.31	13.49	26.35	23.72	20.31	32.01	10.89	37.23	3.699	38.80
52.63	22.71	46.11	37.15	35.54	50.71	19.05	59.40	7.397	68.36
65.79	25.24	59.29	43.38	50.77	63.24	29.94	78.40	10.17	82.31
78.94	26.07	79.05	47.65	66.00	68.07	38.11	82.93	12.95	87.79
105.3	21.51	105.4	41.35	91.38	54.60	54.44	62.27	13.87	87.60
138.2	0.1558	131.7	16.58	111.7	17.80	65.33	19.98	17.57	76.35
164.5	−35.66	144.9	−5.476	121.8	−13.63	70.77	−14.41	22.19	34.40
197.4	−125.0	171.3	−82.15	137.1	−89.65	78.94	−96.29	24.97	−12.68
212.1	−201.0	190.6	−200.2	148.7	−200.4	84.57	−200.2	29.87	−201.9

−45°		−30°		−15°		0°		15°	
x, m	y, m	x, m	y, m	x, m	y, m	x, m	y, m	x, m	y, m
8.264	8.498	11.88	7.181	16.17	4.824	20.94	0.7806	4.324	1.125
16.53	17.51	23.75	15.09	32.34	10.76	41.89	3.396	17.29	4.070
24.79	27.11	35.63	23.84	48.51	18.07	62.83	8.345	30.27	6.290
41.32	48.47	47.50	33.61	64.68	27.03	83.77	16.27	43.24	7.668
57.85	73.57	71.25	56.98	80.85	38.05	104.7	28.05	56.21	8.070
66.11	87.97	95.00	87.46	113.2	68.48	125.7	44.89	90.80	3.108
82.64	121.9	118.8	129.1	129.4	89.59	146.6	68.63	103.8	−1.606
99.17	165.6	130.6	156.3	145.5	116.5	167.5	102.3	138.4	−24.83
107.4	193.2	142.5	190.3	161.7	151.7	188.5	151.9	194.6	−121.7
109.6	201.4	145.7	200.8	177.9	200.0	202.5	200.0	215.0	−200.4

$b_d = 5.669 \cdot 10^{-3}$ m^{-1}; $v = 76.53$ m/s

85°		75°		60°		45°		30°	
x, m	y, m	x, m	y, m	x, m	y, m	x, m	y, m	x, m	y, m
1.364	15.36	4.041	14.86	7.722	13.16	10.48	20.29	11.24	6.341
5.454	58.15	16.17	56.16	30.89	49.43	31.45	29.48	44.96	23.19
10.91	103.4	28.29	90.26	54.05	78.66	62.89	53.47	56.20	27.89
15.00	123.8	40.41	113.6	77.22	97.20	94.34	68.92	89.91	38.18
19.09	129.2	52.54	122.5	92.66	101.5	115.3	72.46	112.4	40.91
24.54	110.2	68.70	107.0	123.6	84.64	146.7	63.74	157.3	31.66
30.00	52.30	84.87	49.09	146.7	40.38	178.2	31.02	191.1	5.375
32.73	−1.331	92.95	−7.309	162.2	−15.11	199.2	−12.00	213.5	−26.95
35.45	−90.73	101.0	−105.7	169.9	−56.93	220.1	−87.19	247.3	−117.9
37.27	−201.2	105.3	−202.3	184.8	−202.6	236.5	−200.3	263.4	−202.2

15°		0°		−15°		−30°		−45°	
x, m	y, m	x, m	y, m	x, m	y, m	x, m	y, m	x, m	y, m
8.072	2.102	19.55	0.3450	20.75	5.979	14.24	8.465	9.488	9.646
24.21	5.908	39.09	1.492	41.50	12.95	28.49	17.48	18.98	19.64
48.43	10.41	58.64	3.641	62.25	21.20	42.73	27.16	28.46	30.05
72.65	16.06	78.19	7.042	83.00	31.08	56.98	37.64	47.44	52.41
88.79	13.50	117.3	18.95	124.5	57.91	85.47	61.81	66.41	77.49
137.2	5.919	136.8	28.39	145.2	76.47	114.0	92.10	75.90	91.39
153.4	−0.6652	156.4	41.03	166.0	100.2	142.4	132.3	94.88	123.0
185.6	−22.83	195.5	80.26	186.7	131.3	156.7	158.3	113.9	161.8
234.1	−93.61	215.0	110.6	207.5	173.8	170.9	190.2	123.3	123.3
266.6	−200.3	250.2	201.1	218.6	203.8	174.7	200.1	129.0	201.0

$b_d = 5.669 \cdot 10^{-3}$ m^{-1}; $v = 107.1$ m/s

30° x, m	30° y, m	45° x, m	45° y, m	60° x, m	60° y, m	75° x, m	75° y, m	85° x, m	85° y, m
17.92	10.15	15.48	15.26	10.92	18.70	5.594	20.66	1.876	21.23
53.77	28.94	46.44	44.04	32.77	54.38	16.78	60.31	5.629	62.03
89.62	44.75	92.87	80.20	65.54	101.0	39.16	128.0	13.44	132.2
125.5	55.93	123.8	96.47	98.31	132.9	55.94	160.5	20.63	173.5
161.3	60.04	154.8	102.8	120.2	141.3	67.13	168.9	24.39	178.0
197.2	53.46	201.2	83.81	142.0	135.7	83.92	156.5	28.15	170.1
250.9	10.06	247.7	2.759	174.8	89.80	100.7	104.8	33.78	131.0
268.9	−19.44	263.1	−54.13	196.6	15.98	111.9	31.95	39.41	39.15
286.8	−62.38	278.6	−150.6	207.5	−49.47	117.5	−28.85	41.28	−16.25
317.2	−203.6	283.8	−202.3	220.7	−220.7	125.5	−202.8	44.41	−202.1

−45° x, m	−45° y, m	−30° x, m	−30° y, m	−15° x, m	−15° y, m	0° x, m	0° y, m	15° x, m	15° y, m
10.45	10.54	16.38	9.618	25.81	7.254	27.37	0.3558	14.52	3.789
20.89	21.31	32.75	19.62	51.62	15.34	82.10	4.028	43.57	10.64
31.34	32.34	49.13	30.09	77.43	24.56	109.5	8.104	62.61	16.19
52.23	55.39	65.50	41.15	103.2	35.33	136.8	14.43	101.7	19.92
73.12	80.23	98.25	65.68	129.0	48.22	191.6	37.54	130.7	21.13
94.02	107.7	131.0	94.98	180.7	84.08	218.9	57.35	159.8	18.83
114.9	138.9	163.8	132.2	206.5	110.2	246.3	86.17	203.3	5.597
135.8	176.1	180.1	155.5	232.3	145.5	273.7	129.3	217.8	−2.474
146.2	198.0	196.5	183.5	258.1	196.3	301.0	198.8	305.0	−134.0
147.6	201.1	205.2	201.1	259.8	200.5	303.8	208.2	321.9	−200.1

$b_d = 5.669 \cdot 10^{-3}$ m^{-1}; $v = 150$ m/s

85°		75°		60°		45°		30°	
x, m	y, m	x, m	y, m	x, m	y, m	x, m	y, m	x, m	y, m
2.438	27.67	7.303	27.06	14.51	24.93	21.33	21.10	26.53	15.08
7.314	81.42	29.21	104.6	58.05	95.79	63.98	61.40	79.58	43.26
14.63	154.6	51.12	170.9	101.6	154.2	106.6	97.21	106.1	55.79
21.94	209.5	73.02	213.1	130.6	179.1	149.3	124.7	159.2	75.40
29.25	231.3	87.63	219.9	159.6	185.4	191.9	137.7	212.2	82.54
36.57	209.4	102.2	204.6	174.2	179.3	234.6	125.4	265.3	66.68
41.44	163.3	116.8	160.3	203.2	141.3	255.9	105.4	318.3	5.754
46.32	69.84	13.15	63.55	232.2	43.30	298.6	13.40	344.8	−60.91
48.76	−19.38	138.8	−34.38	246.7	−65.48	319.9	−94.46	37.14	−198.8
51.11	−211.5	144.6	−205.9	255.4	−203.4	330.6	−201.8	372	−206.3

15°		0°		−15°		−30°		−45°	
x, m	y, m	x, m	y, m	x, m	y, m	x, m	y, m	x, m	y, m
24.74	6.472	38.31	0.3718	30.90	8.534	18.05	10.52	11.10	11.16
74.23	18.13	76.63	1.748	61.81	17.72	36.10	21.29	22.21	22.45
98.98	23.02	114.9	4.674	927.1	27.84	54.15	32.36	33.31	33.90
148.5	29.76	153.3	9.999	123.6	39.31	72.20	43.82	55.52	57.38
173.2	30.92	191.6	19.05	154.5	52.75	108.3	68.37	66.63	69.49
272.2	6.044	229.9	33.92	185.4	69.08	144.4	96.23	99.94	108.1
296.9	−13.46	268.2	58.13	216.3	89.71	180.5	129.7	122.1	136.5
321.7	−42.93	306.5	98.17	247.2	116.9	198.6	149.7	144.4	168.4
371.2	−162.3	344.8	169.0	278.1	154.8	216.6	173.0	155.5	186.3
380.2	−203.5	356.3	201.7	304.9	202.2	234.7	200.8	163.6	200.4

$b_d = 5.669 \cdot 10^{-3}$ m⁻¹; $v = 210$ m/s

85°		75°		60°		45°		30°	
x, m	y, m	x, m	y, m	x, m	y, m	x, m	y, m	x, m	y, m
3.028	34.45	9.104	33.81	18.34	31.59	27.73	27.53	36.62	20.91
9.084	101.9	27.31	100.0	55.02	93.17	83.20	80.68	73.24	41.16
18.17	195.4	54.62	191.2	110.0	176.1	138.7	128.4	109.9	60.36
27.25	265.3	81.93	257.7	146.7	216.6	192.1	164.0	183.1	92.76
36.33	28.65	100.1	274.0	183.4	233.4	249.6	174.0	256.3	107.8
39.36	279.2	109.2	271.0	201.7	228.6	277.3	162.8	292.9	102.4
45.42	238.7	127.4	236.4	320.1	212.1	305.1	134.5	329.6	82.14
51.47	141.4	145.7	140.8	256.8	122.9	332.8	78.91	366.2	37.32
54.50	39.13	154.8	35.12	275.1	13.60	360.5	-37.29	402.8	-58.05
57.23	-201.4	162.6	-214.7	289.2	-211.8	377.2	-112.4	427.2	-208.2

15°		0°		-15°		-30°		-45°	
x, m	y, m	x, m	y, m	x, m	y, m	x, m	y, m	x, m	y, m
39.34	10.32	80.46	0.9996	35.46	9.675	19.21	11.15	11.51	11.54
78.67	20.05	53.64	0.3961	70.92	19.81	38.41	22.44	23.02	23.15
118.0	28.84	107.3	2.005	106.4	30.65	57.62	33.91	34.53	34.85
157.3	36.14	160.9	5.843	141.8	42.55	76.82	45.62	57.55	58.57
196.7	41.09	214.6	13.78	177.3	56.09	115.2	70.03	80.56	82.90
236.0	42.36	268.2	29.33	212.8	72.15	172.9	110.9	115.1	121.2
314.7	24.46	321.8	59.33	248.2	92.12	211.3	143.6	138.1	148.6
354.0	-3.592	375.5	119.2	283.7	118.3	230.5	162.6	161.1	178.4
393.4	-57.43	402.6	172.8	319.1	154.9	249.7	184.5	172.6	194.5
439.3	-204.6	413.0	202.7	349.9	201.3	262.5	201.0	176.5	200.2

$b_d = 7.937 \cdot 10^{-3}$ m^{-1}; $v = 16.84$ m/s

85°		75°		60°		45°		30°	
x, m	y, m	x, m	y, m	x, m	y, m	x, m	y, m	x, m	y, m
0.1357	1.508	0.3928	1.426	0.6928	1.166	0.8207	0.7972	0.7300	0.4091
0.5427	5.509	1.571	5.206	2.771	4.253	3.283	2.901	2.920	1.486
1.085	9.556	3.143	9.024	6.235	7.940	7.386	5.396	6.570	2.757
1.628	12.04	5.107	11.70	9.006	9.494	10.67	6.427	9.490	3.281
2.177	12.91	6.285	12.14	11.08	9.800	13.13	6.604	11.68	3.372
3.256	9.646	9.820	7.990	15.93	7.496	20.52	3.846	18.25	1.990
4.206	1.023	12.57	−1.358	21.48	−0.7561	25.44	−0.9840	22.63	−0.4120
5.562	−22.46	16.89	−29.49	27.02	−16.04	36.93	−23.58	38.85	−11.29
7.462	−90.98	21.60	−90.68	34.64	−52.86	43.50	−45.95	55.48	−74.14
8.732	−200.6	25.32	−200.3	46.39	−200.5	61.69	−200.3	70.98	−200.4

15°		0°		−15°		−30°		−45°	
x, m	y, m	x, m	y, m	x, m	y, m	x, m	y, m	x, m	y, m
0.4301	0.1118	6.451	0.7453	5.813	2.205	4.886	3.389	3.776	4.284
1.721	0.4056	12.90	3.090	11.63	5.795	9.772	7.991	7.552	9.651
3.871	0.7534	19.35	7.218	17.44	10.93	14.66	13.94	11.33	16.22
5.592	0.9005	32.25	21.84	29.06	26.72	24.43	30.66	18.88	33.63
6.882	0.9321	38.70	3.308	34.88	38.03	29.32	41.98	22.66	44.96
13.33	−0.0261	45.15	47.71	40.69	52.27	34.20	55.84	26.43	58.40
26.67	−8.179	58.06	91.88	46.50	70.27	39.09	72.91	33.99	94.82
36.56	−20.97	64.51	126.3	58.13	124.0	48.86	121.7	41.54	152.6
49.04	−48.29	70.96	177.5	63.94	167.3	53.75	158.7	45.31	198.5
74.54	−200.1	73.11	201.3	67.43	204.1	57.65	200.5	45.54	202.3

$b_d = 7.937 \cdot 10^{-3}$ m⁻¹; $v = 23.57$ m/s

85°		75°		60°		45°		30°	
x, m	y, m	x, m	y, m	x, m	y, m	x, m	y, m	x, m	y, m
0.2440	2.718	0.7095	2.580	1.268	2.139	1.533	1.491	1.395	0.7820
0.9762	9.983	2.838	9.468	5.072	7.828	6.131	5.436	5.578	2.842
2.196	18.78	5.676	16.46	10.14	13.55	12.26	9.361	11.16	4.874
3.172	22.49	8.514	20.65	15.22	16.90	18.39	11.59	16.73	6.017
3.905	23.20	10.64	21.82	19.02	17.70	22.99	12.03	20.92	6.240
5.369	19.12	16.32	16.07	27.90	13.29	32.19	9.413	29.28	4.889
6.833	7.151	20.58	2.273	35.51	1.500	41.38	1.447	37.65	0.8520
7.565	—2.268	22.70	—8.323	39.31	—7.732	44.45	—2.580	52.99	—15.08
10.01	—57.68	30.51	—80.93	50.72	—54.53	64.37	—52.21	79.49	—85.23
12.33	—200.0	35.48	—200.4	63.95	—200.1	83.84	—200.2	95.29	—200.3

15°		0°		—15°		—30°		—45°	
x, m	y, m	x, m	y, m	x, m	y, m	x, m	y, m	x, m	y, m
0.8370	0.2176	9.031	0.7557	7.915	2.740	6.487	4.261	4.909	5.351
3.348	0.7891	18.06	3.178	15.83	6.838	12.97	9.645	9.818	11.66
6.696	1.354	27.09	7.536	23.74	12.50	19.46	16.31	14.73	19.05
10.04	1.683	36.12	14.17	39.57	29.66	25.95	24.47	24.54	37.82
12.56	1.766	45.15	23.51	47.49	41.98	38.92	46.44	29.45	49.69
25.11	—0.1304	63.22	53.22	55.40	51.67	45.41	61.13	34.36	63.71
39.34	—7.768	72.25	76.10	63.32	77.82	51.89	79.26	39.27	80.48
59.43	—32.00	81.28	107.8	79.15	140.8	64.87	131.7	49.09	126.3
70.31	—54.74	90.31	154.7	87.07	195.9	71.36	172.7	54.00	159.6
99.05	—200.2	96.33	202.5	87.59	200.7	74.81	202.6	58.58	202.6

$b_d = 7.937 \cdot 10^{-3}$ m^{-1}; $v = 33$ m/s

85°		75°		60°		45°		30°	
x, m	y, m	x, m	y, m	x, m	y, m	x, m	y, m	x, m	y, m
0.4156	4.645	1.215	4.434	2.214	3.745	2.759	2.689	2.607	1.463
1.663	17.19	4.861	16.39	8.857	13.79	11.04	9.846	10.43	5.324
3.325	30.07	9.723	28.61	17.71	23.91	22.08	16.91	18.25	8.299
4.988	37.69	14.58	35.72	24.36	28.57	30.35	20.00	26.07	10.16
6.235	39.51	18.23	37.22	31.00	30.32	38.63	20.87	33.88	11.08
8.729	32.52	25.52	29.36	46.50	21.11	55.19	14.66	49.52	8.623
11.22	9.319	32.82	4.138	57.57	0.1310	66.23	3.216	62.56	1.499
12.05	−3.072	36.46	−17.51	64.21	−20.63	82.78	−29.57	80.80	−18.91
14.96	−78.48	43.75	−95.18	77.50	−96.08	99.34	−96.61	104.3	−74.78
16.85	−200.6	48.05	−200.7	85.47	−201.3	110.7	−201.1	124.7	−200.7

15°		0°		−15°		−30°		−45°	
x, m	y, m	x, m	y, m	x, m	y, m	x, m	y, m	x, m	y, m
1.618	0.4207	12.64	0.7707	10.66	3.439	8.439	5.323	6.210	6.574
6.471	1.524	25.29	3.309	21.32	8.193	16.88	11.65	12.42	13.95
12.94	2.599	37.93	8.018	31.98	14.53	25.32	19.17	18.63	22.27
17.80	3.081	50.57	15.42	53.30	33.47	33.76	28.11	31.05	42.44
22.65	3.259	63.22	26.23	63.96	47.18	50.63	51.62	37.26	54.80
35.59	2.060	75.86	41.49	74.62	64.93	59.07	67.19	43.47	69.18
43.68	−0.1089	88.50	62.89	85.28	88.32	67.51	86.43	49.68	86.14
69.57	−16.67	101.1	93.50	95.94	120.3	84.39	143.0	62.10	131.9
105.2	−80.54	113.8	140.1	106.6	167.2	92.83	188.8	68.31	164.6
128.5	−200.1	123.9	201.7	112.3	203.8	94.52	200.6	73.69	203.1

$b_d = 7.937 \cdot 10^{-3}$ m^{-1}; $v = 46.2$ m/s

30°		45°		60°		75°		85°	
x, m	y, m	x, m	y, m	x, m	y, m	x, m	y, m	x, m	y, m
4.699	2.643	4.705	4.600	3.626	6.155	1.944	7.121	0.6605	7.410
18.80	9.635	18.82	16.94	14.50	22.87	7.777	26.59	3.302	33.71
32.89	14.92	32.94	26.54	25.38	36.22	15.56	46.71	5.284	48.83
36.99	18.03	47.05	32.53	36.26	45.17	21.39	56.00	7.265	58.79
56.38	18.62	61.17	33.92	47.14	48.62	27.22	59.24	9.247	62.70
79.88	13.44	84.70	22.65	61.64	42.70	35.00	52.86	12.55	54.53
98.68	0.1113	98.81	4.669	76.15	21.28	44.72	23.84	15.85	24.57
117.5	−25.47	117.6	−40.95	87.03	−9.738	50.55	−10.29	17.83	−9.056
141.0	−89.27	131.7	−108.5	101.5	−91.88	56.39	−68.78	20.47	−93.13
158.2	−200.8	141.2	−201.2	109.6	−201.4	62.22	−201.9	21.95	−201.5

−45°		−30°		−15°		0°		15°	
x, m	y, m	x, m	y, m	x, m	y, m	x, m	y, m	x, m	y, m
7.589	7.869	10.69	6.543	14.14	4.323	17.70	0.7925	3.088	0.8036
15.18	16.37	21.37	13.94	28.29	9.898	35.40	3.506	12.35	2.907
22.77	25.63	32.06	22.40	42.43	17.07	53.10	8.775	21.62	4.493
30.36	36.80	53.43	43.58	56.57	26.29	70.80	16.48	30.88	5.764
45.53	59.79	64.11	57.11	70.71	38.21	88.50	30.89	40.15	5.785
53.12	72.24	74.80	73.43	84.86	53.75	106.2	51.05	58.68	3.785
60.71	90.99	85.48	94.57	99.00	74.37	123.9	81.67	74.12	−1.147
75.89	135.0	96.17	119.2	113.1	102.6	141.6	131.2	105.0	−24.11
83.48	165.6	106.9	153.6	127.3	143.7	147.2	180.3	139.0	−86.96
90.05	200.8	117.5	203.7	140.5	205.7	155.8	202.0	162.4	−200.8

$b_d = 7.937 \cdot 10^{-3}$ m^{-1}; $v = 64.68$ m/s

	85°		75°		60°		45°		30°
x, m	y, m	x, m	y, m	x, m	y, m	x, m	y, m	x, m	y, m
0.9740	10.97	2.887	10.62	5.516	9.402	7.487	7.348	8.028	4.529
3.896	41.53	11.55	40.12	22.06	35.32	29.95	27.30	24.08	12.85
7.792	73.83	20.21	64.47	38.61	56.18	52.41	42.68	48.17	22.86
10.71	88.44	28.87	81.14	55.16	69.43	67.38	49.23	64.22	27.27
13.64	92.32	37.53	87.52	66.19	72.53	82.36	51.76	80.28	29.22
17.53	78.73	46.19	81.54	88.25	60.46	104.8	45.53	104.4	25.88
21.43	37.36	54.85	60.53	104.8	28.84	127.3	22.16	136.5	3.839
23.38	—0.9506	63.51	17.32	115.8	—10.79	142.3	—8.570	152.5	—19.25
25.32	—64.81	69.28	—34.70	126.9	—82.89	157.2	—62.28	176.6	—84.19
27.17	—203.3	76.79	—202.9	134.9	—203.0	173.7	—203.6	194.5	—200.8

	15°		0°		—15°		—30°		—45°
x, m	y, m	x, m	y, m	x, m	y, m	x, m	y, m	x, m	y, m
5.765	1.502	16.52	0.3499	18.38	5.396	13.07	7.837	8.903	9.101
23.06	5.418	33.04	1.537	36.76	11.95	26.15	16.36	17.81	18.66
34.59	7.437	49.56	3.812	55.14	20.09	39.22	25.76	26.71	28.78
51.89	9.326	82.60	13.06	73.52	30.41	52.29	36.29	35.61	39.60
63.42	9.645	115.6	32.36	91.90	43.78	65.36	48.30	53.42	64.07
92.25	—5.950	132.2	48.18	110.3	61.56	91.51	78.93	71.22	94.31
109.5	—0.4752	148.7	70.60	128.7	85.98	104.6	99.33	89.03	134.8
144.1	—28.51	165.2	103.4	147.0	121.4	117.7	125.3	97.93	161.8
178.7	—97.27	181.7	155.3	165.4	178.2	130.7	160.1	106.8	196.7
199.7	—200.8	191.6	205.9	170.3	200.4	142.1	203.2	108.0	202.3

$b_d = 7.937 \cdot 10^{-3}$ m^{-1}; $v = 90.55$ m/s

85° x, m	85° y, m	75° x, m	75° y, m	60° x, m	60° y, m	45° x, m	45° y, m	30° x, m	30° y, m
1.340	15.16	3.996	14.76	7.803	13.36	11.06	10.90	12.80	7.250
6.702	71.12	11.99	43.08	23.41	38.84	33.17	31.46	38.41	20.67
10.72	104.2	27.97	91.42	46.82	72.12	66.34	57.28	64.01	31.96
14.74	123.9	39.96	114.6	70.22	94.93	88.45	68.90	89.62	39.95
17.42	127.1	47.95	120.6	86.83	101.0	110.6	73.45	115.2	42.88
20.10	121.5	55.94	117.3	101.4	96.90	132.7	67.83	140.8	38.18
24.13	93.60	63.94	103.2	124.8	64.15	154.8	47.36	179.2	7.183
26.81	56.24	75.92	52.88	140.4	11.42	176.9	1.970	192.0	−13.89
29.49	−11.60	83.92	−20.61	148.2	−35.33	188.0	−38.66	217.6	−91.96
32.17	−205.2	90.84	−201.2	160.0	−200.7	206.8	−204.9	232.6	−202.9

15° x, m	15° y, m	0° x, m	0° y, m	−15° x, m	−15° y, m	−30° x, m	−30° y, m	−45° x, m	−45° y, m
10.37	2.706	23.13	0.3630	15.49	4.320	15.36	9.073	10.00	10.13
31.12	7.597	46.26	1.661	46.48	14.28	30.71	18.65	20.01	30.57
51.87	11.56	69.39	4.308	77.48	26.96	46.07	28.89	30.01	31.38
72.61	14.23	92.52	8.910	8.910	124.0	55.75	61.42	40.04	40.02
93.36	15.09	115.6	16.36	139.5	69.92	92.13	66.43	60.03	67.27
114.1	13.45	138.8	28.00	154.9	87.93	107.5	82.77	70.03	80.96
145.2	3.998	161.0	46.01	170.4	111.6	122.8	102.4	90.04	112.8
155.6	−1.767	185.0	74.28	185.9	144.1	138.2	127.1	110.0	154.6
207.5	−68.03	208.2	121.4	201.4	192.9	153.6	159.7	120.1	182.3
239.3	−201.7	228.1	202.8	204.5	206.0	167.9	202.9	125.4	200.1

$b_d = 7.937 \cdot 10^{-3}$ m^{-1}; $v = 126.8$ m/s

85°		75°		60°		45°		30°	
x, m	y, m	x, m	y, m	x, m	y, m	x, m	y, m	x, m	y, m
1.741	19.76	5.216	19.33	10.37	17.81	15.23	15.07	18.95	10.77
5.224	58.16	15.65	56.83	31.10	52.18	45.70	43.86	56.84	30.90
10.45	110.4	31.30	107.6	62.20	97.77	76.17	69.44	75.79	39.85
15.67	149.6	46.95	144.8	93.30	127.9	106.6	89.05	113.7	53.86
20.90	165.2	62.60	157.1	114.0	132.4	137.1	98.10	151.6	58.96
24.38	158.6	73.03	146.1	124.4	128.1	152.3	96.61	189.5	47.63
28.60	116.6	83.46	114.5	145.1	100.9	182.8	75.29	227.4	4.110
33.09	49.88	93.89	45.39	165.9	30.93	213.3	9.568	243.3	−43.51
34.83	−13.84	99.11	−24.56	176.2	−46.67	228.5	−67.47	265.3	−142.0
36.80	−204.3	104.2	−200.7	184.5	−210.0	239.7	−208.4	210.9	−202.2

15°		0°		−15°		−30°		−45°	
x, m	y, m	x, m	y, m	x, m	y, m	x, m	y, m	x, m	y, m
17.67	4.623	32.38	0.3827	18.93	5.201	17.28	10.11	10.81	10.89
53.02	12.95	48.57	0.9470	56.78	16.69	34.56	20.56	21.62	21.96
70.69	16.44	64.76	1.859	94.63	30.66	51.83	31.47	32.43	33.27
106.0	21.26	97.14	5.167	113.6	39.21	69.11	43.02	43.24	44.87
123.7	22.08	129.5	11.55	151.4	61.76	103.7	69.17	64.87	69.35
159.1	18.68	161.9	23.13	170.3	77.39	120.9	86.66	86.49	96.48
194.4	4.317	194.3	43.75	189.3	97.82	155.5	124.7	108.1	128.2
212.1	−9.612	226.7	81.41	208.2	125.8	172.8	152.6	129.8	168.6
247.4	−62.90	259.0	159.7	227.1	166.9	190.1	140.5	140.5	194.6
280.4	−204.4	268.8	203.8	238.5	203.2	193.5	200.6	142.7	200.5

$b_d = 7.934 \cdot 10^{-3}$ m^{-1}; $v = 177.5$ m/s

85°		75°		60°		45°		30°	
x, m	y, m	x, m	y, m	x, m	y, m	x, m	y, m	x, m	y, m
2.163	24.60	6.503	24.15	13.10	22.57	19.81	19.67	26.16	14.93
8.651	96.09	19.51	71.43	39.30	66.55	59.43	57.63	78.47	43.12
15.14	158.9	45.52	155.1	78.60	125.8	99.05	91.74	104.6	55.63
21.63	199.3	58.52	183.8	104.8	154.7	138.7	117.1	130.8	66.26
25.95	204.6	71.53	195.7	131.0	166.7	178.3	124.3	183.1	77.01
30.28	188.5	78.03	193.5	144.1	163.3	198.1	116.3	209.2	73.17
34.60	143.2	91.04	168.9	170.3	128.4	217.9	96.11	235.4	58.67
38.93	27.95	97.54	142.6	183.4	87.76	237.7	56.36	261.6	26.66
41.09	−193.8	110.5	25.09	196.5	9.712	257.5	−26.64	287.7	−41.46
41.17	−216.1	116.8	−210.7	207.9	−204.6	272.1	−213.6	309.5	−203.7

15°		0°		−15°		−30°		−45°	
x, m	y, m	x, m	y, m	x, m	y, m	x, m	y, m	x, m	y, m
28.10	7.374	22.67	0.0905	22.19	6.039	18.69	10.87	11.33	11.38
56.19	14.32	45.33	0.4128	44.38	12.32	37.38	21.95	22.67	22.86
84.29	20.60	68.00	1.067	66.57	18.94	56.07	33.32	34.00	34.47
140.5	29.35	90.67	2.198	88.76	26.06	74.76	45.11	56.67	58.27
168.8	30.28	136.0	6.804	110.9	33.91	112.1	70.68	68.00	70.58
224.8	17.47	181.3	17.24	133.1	42.80	130.8	85.10	102.0	110.3
252.9	−2.566	226.6	39.98	155.3	53.23	168.2	120.0	124.7	140.9
281.0	−41.09	272.0	91.46	177.5	65.93	186.9	142.8	136.0	158.4
309.1	−122.7	294.7	143.7	221.9	103.5	205.6	172.0	146.3	178.2
322.2	−206.4	312.8	222.9	275.2	207.2	219.3	260.1	158.7	201.2

$b_d = 7.937 \cdot 10^{-3}$ m^{-1}; $v = 248.5$ m/s

85°		75°		60°		45°		30°	
x, m	y, m	x, m	y, m	x, m	y, m	x, m	y, m	x, m	y, m
2.596	29.58	7.825	29.12	15.93	27.49	24.62	24.51	34.07	19.51
7.787	87.90	23.48	86.49	47.79	81.49	73.87	72.24	68.13	38.56
15.57	169.9	46.95	166.7	95.58	155.3	98.49	94.79	102.2	56.72
20.76	214.6	62.60	209.5	127.4	190.9	147.7	133.6	136.3	73.26
28.55	246.2	86.08	235.1	143.4	200.2	197.0	153.0	170.3	86.86
31.15	243.1	161.7	211.0	159.3	201.1	221.6	148.3	238.5	94.33
36.34	211.7	109.7	179.9	175.2	191.0	246.2	127.2	272.5	77.53
41.53	120.0	117.4	124.8	207.1	115.8	270.9	77.30	306.6	30.23
44.12	−9.248	125.2	0.6694	223.0	−2.146	295.5	−57.96	340.7	−114.3
45.25	−212.6	128.9	−214.5	231.0	−231.2	303.7	−213.2	348.6	−220.2

15°		0°		−15°		−30°		−45°	
x, m	y, m	x, m	y, m	x, m	y, m	x, m	y, m	x, m	y, m
27.63	7.328	31.73	0.0953	24.92	6.739	19.61	11.37	11.64	11.66
82.90	21.24	63.47	0.4611	49.84	13.64	39.22	22.85	23.18	23.39
110.5	27.52	95.20	1.274	74.77	20.78	58.83	34.51	34.92	35.18
138.2	32.99	158.7	5.610	99.69	28.29	78.44	46.40	46.57	47.07
193.4	39.45	190.4	10.43	124.6	36.36	117.7	71.41	69.85	71.26
221.1	38.64	222.1	18.63	149.5	45.27	156.9	99.45	104.8	109.3
276.3	20.09	285.6	56.54	174.5	55.47	196.1	134.1	128.1	137.0
304.0	−5.166	317.3	99.91	224.3	82.87	215.7	156.2	151.3	168.1
359.3	−166.5	349.1	194.1	274.1	131.4	235.3	184.3	163.0	185.9
363.4	−200.7	355.4	229.9	309.0	200.5	244.5	200.5	171.5	200.2

$b_d = 1.111 \cdot 10^{-2}$ m^{-1}; $v = 14.23$ m/s

85°		75°		60°		45°		30°	
x, m	y, m	x, m	y, m	x, m	y, m	x, m	y, m	x, m	y, m
0.0969	1.077	0.2806	1.018	0.4948	0.8331	0.5862	0.5694	0.5214	0.2922
0.3876	3.935	1.122	3.719	1.979	3.038	2.931	2.502	2.607	1.281
0.8721	6.377	2.245	6.445	4.454	5.672	5.276	3.855	4.693	1.970
1.260	8.867	3.648	8.359	6.433	6.781	7.621	4.591	6.779	2.344
1.550	9.221	4.489	8.674	7.917	7.000	9.379	4.717	8.343	2.408
2.511	5.541	7.295	5.000	11.88	4.862	14.07	3.149	11.99	1.800
3.101	−0.4917	8.979	−0.9697	15.34	−0.5402	18.17	−0.7029	16.16	−0.2943
4.264	−23.39	12.06	−21.06	21.28	−19.23	26.38	−16.84	29.20	−18.63
5.911	−106.1	16.83	−96.91	31.17	−103.2	37.52	−68.30	45.37	−86.94
6.531	−200.4	18.95	−200.4	34.75	−200.5	46.33	−200.6	53.48	−200.1

15°		0°		−15°		−30°		−45°	
x, m	y, m	x, m	y, m	x, m	y, m	x, m	y, m	x, m	y, m
0.3072	0.0799	5.452	0.7501	4.966	1.996	4.216	3.030	3.285	3.827
1.229	0.2897	10.90	3.132	9.931	5.439	8.431	7.352	6.571	8.828
2.458	0.4988	16.36	7.379	14.90	10.54	12.65	13.15	9.856	15.17
3.687	0.6249	21.81	13.80	19.86	17.58	16.86	20.69	13.14	23.09
4.916	0.6658	27.26	22.87	29.79	38.38	25.29	42.67	16.43	32.91
9.832	−0.0716	32.71	35.28	34.76	55.84	29.51	58.58	23.00	60.59
19.36	−6.146	43.61	76.05	39.72	78.31	33.72	79.72	26.28	80.62
27.96	−18.18	49.07	112.0	44.69	111.0	37.94	109.4	29.57	108.0
42.40	−61.11	54.52	177.9	49.66	166.7	42.16	156.7	32.85	149.3
56.45	−200.1	55.61	200.3	51.64	205.5	44.68	206.4	35.26	200.9

$b_d = 1.111 \cdot 10^{-2}$ m^{-1}; $v = 19.92$ m/s

85°		75°		60°		45°		30°	
x, m	y, m	x, m	y, m	x, m	y, m	x, m	y, m	x, m	y, m
0.1743	1.942	0.5068	1.843	0.9058	1.528	1.095	1.065	0.9961	0.5586
0.6973	7.130	2.027	6.763	3.623	5.591	4.379	3.883	3.984	2.030
1.394	12.41	4.054	11.76	6.340	8.810	8.758	6.686	6.973	3.176
2.092	15.61	6.081	14.75	9.964	11.64	12.04	8.001	10.90	4.156
2.789	16.57	7.602	15.59	13.59	12.64	16.42	8.595	14.94	4.457
3.835	13.66	11.15	12.52	20.83	8.483	27.37	3.398	23.91	2.307
5.055	3.065	14.70	1.623	28.98	−8.135	36.13	−9.318	32.87	−4.503
6.275	−17.42	18.24	−19.52	37.14	−44.43	45.98	−37.29	44.82	−23.37
8.018	−78.97	22.81	−74.44	43.48	−104.1	54.74	−85.97	59.76	−75.52
9.163	−201.2	26.35	−200.9	47.58	−201.0	62.55	−200.8	71.39	−200.1

15°		0°		−15°		−30°		−45°	
x, m	y, m	x, m	y, m	x, m	y, m	x, m	y, m	x, m	y, m
0.5979	0.1654	7.633	0.7626	6.791	2.464	5.642	3.809	4.318	4.801
2.391	0.5636	15.27	3.239	13.58	6.367	11.28	8.842	8.636	10.67
4.185	0.8818	22.90	7.766	20.37	11.99	16.93	15.33	12.95	17.80
6.577	1.160	30.53	14.80	27.16	19.70	22.57	23.57	17.27	26.44
8.968	1.261	38.16	25.00	33.95	30.04	33.85	47.34	21.59	36.96
21.52	−1.495	45.80	39.42	40.74	43.87	39.49	64.65	30.23	66.17
37.67	−15.56	53.43	49.99	47.53	62.65	45.14	88.01	34.55	87.27
50.22	−39.10	61.06	90.76	54.32	89.32	50.78	112.19	38.86	116.3
62.78	−84.50	68.69	143.5	61.12	130.9	56.42	180.4	43.18	161.3
74.65	−200.1	73.27	205.9	67.45	207.8	57.92	206.0	45.49	200.4

$b_d = 1.111 \cdot 10^{-2}$ m^{-1}; $v = 27.89$ m/s

85°		75°		60°		45°		30°	
x, m	y, m	x, m	y, m	x, m	y, m	x, m	y, m	x, m	y, m
0.2969	3.318	0.8681	3.167	1.582	2.675	1.971	1.921	1.862	1.045
1.484	14.90	3.472	11.71	6.326	9.850	7.884	7.033	7.447	3.803
2.375	21.48	6.945	20.44	11.07	15.55	15.77	12.08	13.03	5.927
3.563	26.92	10.42	25.51	17.40	20.41	21.68	14.29	18.62	7.230
4.453	28.22	13.02	26.59	22.14	21.66	27.59	14.91	24.20	7.911
5.938	24.79	17.36	22.68	28.47	19.68	39.42	10.47	33.51	6.753
7.422	13.68	22.57	7.013	39.54	3.946	53.22	−7.394	48.41	−2.091
8.907	−7.423	26.91	−19.00	47.45	−21.01	63.07	−33.27	59.58	−16.50
10.69	−56.06	31.25	−67.99	55.35	−68.63	72.93	−81.73	78.20	−68.15
12.41	−201.0	35.39	−200.6	63.05	−201.8	81.93	−200.5	92.78	−201.2

15°		0°		−15°		−30°		−45°	
x, m	y, m	x, m	y, m	x, m	y, m	x, m	y, m	x, m	y, m
1.156	0.3005	7.124	0.3376	9.200	3.080	7.420	4.778	5.543	5.954
4.622	1.090	14.25	1.427	18.40	7.588	14.84	10.69	11.09	12.84
8.089	1.697	21.37	3.399	27.60	13.89	22.26	18.00	22.17	30.30
12.71	2.200	35.62	10.66	36.80	22.50	29.68	27.07	27.71	41.54
16.18	2.328	49.87	23.96	46.00	34.16	37.10	38.45	33.26	55.17
28.89	0.6449	64.11	46.75	55.20	50.11	44.52	52.95	38.80	72.14
46.22	−8.741	71.24	63.90	64.40	72.59	51.94	71.96	44.34	94.12
62.40	−28.67	78.36	87.54	73.60	106.6	59.36	98.19	49.88	124.5
79.74	−73.06	92.61	185.9	82.80	167.9	66.78	138.1	55.43	173.6
96.34	−200.3	94.03	206.8	85.87	204.8	73.22	201.0	57.64	202.9

$b_d = 1.111 \cdot 10^{-2}$ m^{-1}; $v = 39.05$ m/s

85° x, m	85° y, m	75° x, m	75° y, m	60° x, m	60° y, m	45° x, m	45° y, m	30° x, m	30° y, m
0.4818	5.293	1.389	5.086	2.590	4.396	3.361	3.286	3.356	1.888
1.887	19.79	5.555	18.99	10.36	16.33	13.44	12.10	10.07	5.339
3.302	31.65	11.11	33.36	18.13	25.87	23.53	18.96	23.50	10.66
5.189	41.98	15.28	40.00	25.90	32.27	33.61	23.23	33.56	12.88
6.605	44.79	19.44	42.31	33.67	34.73	40.33	24.31	40.28	13.80
8.492	41.22	25.00	37.76	44.03	30.50	57.14	18.95	57.06	9.604
10.85	23.88	30.53	22.82	56.98	9.081	73.94	−2.794	77.20	−7.706
12.27	2.818	36.11	−7.350	64.75	−17.42	84.02	−29.25	90.62	−32.26
14.15	−46.46	41.66	−70.62	72.52	−65.63	94.11	−77.48	104.0	−78.74
16.04	−202.2	45.46	−201.8	80.21	−200.6	103.6	−200.4	116.9	−202.3

15° x, m	15° y, m	0° x, m	0° y, m	−15° x, m	−15° y, m	−30° x, m	−30° y, m	−45° x, m	−45° y, m
2.206	0.5740	9.973	0.3750	8.204	2.446	9.533	5.927	6.898	7.227
8.824	2.077	19.95	1.492	16.41	5.455	19.07	12.86	13.80	15.22
15.44	3.209	29.92	3.643	24.61	9.148	28.60	21.11	20.69	24.19
22.06	3.912	39.89	7.053	32.82	13.68	38.13	31.09	27.59	34.41
28.68	4.117	49.81	12.06	49.22	26.08	47.66	43.44	34.49	46.29
46.32	1.586	69.79	28.85	57.43	34.56	57.20	59.15	48.28	77.73
61.77	−5.621	79.79	42.27	73.83	58.65	66.73	79.97	55.18	99.94
79.41	−22.61	89.76	61.09	82.04	76.25	76.26	109.5	62.08	130.5
99.27	−62.11	109.7	134.1	98.45	135.8	85.79	157.1	68.98	178.7
121.0	−200.3	117.7	207.4	107.5	209.1	90.88	200.8	71.28	203.3

$b_d = 1.111 \cdot 10^{-2}$ m^{-1}; $v = 54.66$ m/s

30°		45°		60°		75°		85°	
x, m	y, m	x, m	y, m	x, m	y, m	x, m	y, m	x, m	y, m
5.734	3.235	5.348	5.249	3.940	6.715	2.062	7.584	0.6957	7.839
17.20	9.178	21.39	19.50	15.76	25.22	8.248	28.66	2.783	29.67
34.41	16.33	37.44	30.49	27.58	40.13	16.50	50.73	4.870	47.81
45.87	19.48	48.13	35.16	35.46	47.17	22.68	60.37	7.653	63.17
57.34	20.87	58.83	36.97	47.28	51.80	26.81	62.51	9.740	65.94
68.81	19.99	74.87	32.52	55.16	49.95	23.99	58.24	11.83	60.22
86.01	12.86	85.57	23.12	66.98	37.58	39.18	43.24	13.91	44.58
103.2	−4.539	101.6	−6.122	74.86	20.60	45.36	12.37	16.00	14.52
120.4	−40.40	117.7	−76.29	86.67	−29.04	51.55	−53.90	18.09	−46.29
142.8	−202.4	126.6	−201.1	98.10	−204.2	55.74	−204.2	19.71	−201.4

−45°		−30°		−15°		0°		15°	
x, m	y, m	x, m	y, m	x, m	y, m	x, m	y, m	x, m	y, m
8.264	8.509	11.88	7.199	10.78	3.111	13.96	0.3558	4.118	1.073
16.53	17.60	23.75	15.25	21.56	6.750	27.92	1.592	16.47	3.870
33.06	38.37	35.63	24.46	32.34	11.07	41.89	4.030	24.71	5.312
41.31	50.67	47.50	35.31	43.12	16.27	55.85	8.117	32.94	6.334
49.58	64.92	59.38	48.50	64.68	30.48	69.81	14.49	45.30	6.889
57.85	81.98	71.25	65.19	86.24	53.17	83.77	24.07	65.89	4.250
66.11	103.4	83.13	87.44	97.02	70.09	97.74	38.39	90.60	−8.139
74.37	132.1	95.00	119.7	107.8	93.58	111.7	60.16	111.2	−32.07
82.64	175.7	106.9	174.8	129.4	194.7	125.7	95.62	131.8	−83.77
85.94	201.9	110.8	205.8	130.4	205.2	143.8	210.2	147.8	−201.7

$b_d = 1.111 \cdot 10^{-2}$ m^{-1}; $v = 76.53$ m/s

85°		75°		60°		45°		30°	
x, m	y, m	x, m	y, m	x, m	y, m	x, m	y, m	x, m	y, m
0.9574	10.83	2.854	10.54	5.573	9.540	7.897	7.784	9.145	5.178
3.829	41.48	11.42	40.30	22.29	36.25	23.69	22.47	27.43	14.76
6.702	67.43	19.98	65.30	38.01	57.99	47.38	40.92	45.72	22.83
9.574	85.13	28.54	81.86	50.16	67.81	63.18	49.22	64.01	28.54
12.44	90.81	34.25	86.15	61.31	72.12	78.67	52.46	82.30	30.63
14.36	86.78	39.96	83.82	72.45	69.21	86.87	51.53	100.6	27.27
17.23	66.86	48.52	65.13	83.60	56.82	102.7	42.76	109.7	22.72
19.15	40.17	54.23	37.77	94.74	30.29	118.5	20.64	128.0	5.130
22.02	−52.31	59.94	−14.72	105.9	−25.24	134.3	−27.62	155.5	−65.68
23.23	−208.5	65.65	−206.6	115.7	−205.4	149.8	−200.5	169.5	−200.8

15°		0°		−15°		−30°		−45°	
x, m	y, m	x, m	y, m	x, m	y, m	x, m	y, m	x, m	y, m
7.409	1.933	19.55	0.3718	13.83	3.898	14.24	8.481	9.488	9.654
29.64	6.940	29.32	0.9056	27.67	8.273	28.49	17.62	28.46	30.34
51.87	10.16	39.09	1.748	41.50	13.41	42.73	27.74	37.95	41.73
66.69	10.78	58.64	4.676	55.33	19.25	56.98	39.27	47.44	54.20
81.51	9.606	78.19	10.02	83.00	35.52	71.22	52.96	56.93	68.17
103.7	2.855	97.74	19.16	96.83	47.18	85.47	70.01	66.41	84.36
126.0	−13.56	117.3	34.48	110.7	62.79	99.71	92.61	75.90	103.9
148.2	−48.60	136.8•	60.77	138.3	118.3	114.0	125.5	85.39	129.1
163.0	−96.22	156.4	111.3	152.2	180.1	128.2	183.0	94.88	164.5
176.1	−202.6	172.0	218.1	154.9	201.0	131.0	201.4	101.2	200.0

$b_d = 1.111 \cdot 10^{-2}$ m⁻¹; $v = 107.1$ m/s

85°		75°		60°		45°		30°	
x, m	y, m	x, m	y, m	x, m	y, m	x, m	y, m	x, m	y, m
1.244	14.12	3.726	13.81	7.405	12.72	10.88	10.77	13.53	7.696
3.731	41.54	7.452	27.37	29.62	48.87	32.64	31.33	40.60	22.07
6.219	67.16	14.90	53.36	51.83	78.68	54.40	49.60	67.67	34.03
8.707	89.61	22.36	76.87	66.64	91.38	76.17	63.60	94.74	41.35
9.950	99.03	29.81	96.13	81.45	94.61	97.93	70.07	108.3	42.11
14.93	118.0	40.99	111.7	88.85	91.47	119.7	63.99	135.3	34.02
18.66	106.9	52.16	104.4	96.26	84.32	141.5	36.24	148.89	22.57
21.14	83.31	59.61	81.81	111.1	52.81	152.3	6.834	162.4	2.936
23.63	35.63	67.07	32.42	125.9	−33.41	163.2	−48.20	175.9	−31.08
26.49	−213.1	75.02	−211.8	132.8	−203.7	173.0	−207.5	196.7	−204.0

15°		0°		−15°		−30°		−45°	
x, m	y, m	x, m	y, m	x, m	y, m	x, m	y, m	x, m	y, m
12.62	3.302	13.68	0.0888	17.21	4.766	16.38	9.631	10.45	10.55
25.25	6.408	41.05	0.9996	34.41	9.940	32.75	19.73	31.34	32.54
50.50	11.75	68.42	3.555	68.83	22.42	49.13	30.56	41.78	44.23
75.75	15.18	82.10	5.845	86.03	30.50	65.50	42.52	52.23	56.65
88.37	15.77	109.5	13.82	103.2	40.68	81.88	56.25	73.12	85.13
113.6	13.35	136.8	29.66	120.4	54.16	98.25	72.83	83.51	102.4
138.9	3.084	150.5	42.80	137.7	73.04	114.6	94.25	94.02	123.4
164.1	−21.90	177.9	91.34	154.9	101.7	131.0	124.7	104.5	150.6
189.4	−82.71	191.6	143.3	172.1	152.0	147.4	175.6	114.9	190.0
205.4	−206.9	199.8	206.4	182.4	212.4	153.8	203.4	117.0	200.6

$b_d = 1.111 \cdot 10^{-2}$ m^{-1}; $v = 150$ m/s

85°		75°		60°		45°		30°	
x, m	y, m	x, m	y, m	x, m	y, m	x, m	y, m	x, m	y, m
1.545	17.57	4.645	17.25	9.357	16.12	14.15	14.05	18.68	10.67
6.179	68.64	18.58	67.29	18.71	32.00	42.45	41.16	56.05	30.80
10.81	113.5	32.51	110.8	28.07	47.53	70.75	65.53	93.41	47.33
15.45	142.4	41.80	131.3	37.43	62.54	99.05	83.65	112.1	52.80
18.54	146.2	51.09	139.8	56.14	89.85	127.4	88.79	130.8	55.01
20.08	142.5	60.38	132.1	74.86	110.5	141.5	83.06	149.5	52.26
23.17	121.8	69.67	101.9	93.57	119.1	155.7	68.65	168.1	41.91
26.26	72.13	78.96	17.92	112.3	108.2	169.8	40.26	186.8	19.04
29.35	−138.4	83.60	−169.7	131.0	62.68	184.0	−19.03	205.5	−29.62
29.59	−232.7	83.91	−227.9	149.4	−213.8	195.7	−211.0	224.2	−217.9

15°		0°		−15°		−30°		−45°	
x, m	y, m	x, m	y, m	x, m	y, m	x, m	y, m	x, m	y, m
20.07	5.267	19.16	0.0927	20.60	5.637	18.05	10.53	11.10	11.16
40.14	10.23	38.31	0.4340	41.21	11.60	36.10	21.37	22.21	22.48
60.21	14.71	57.47	1.156	82.41	25.42	54.15	32.70	33.31	34.02
80.28	18.44	95.78	4.659	103.0	34.13	72.20	44.82	55.52	58.17
120.4	21.62	114.9	8.229	123.6	45.11	90.25	58.22	77.73	85.03
160.6	12.48	134.1	13.92	144.2	59.93	108.3	73.73	88.83	100.3
180.6	−1.833	153.3	22.92	164.8	81.57	126.4	92.86	99.84	117.8
200.7	−29.35	191.6	60.82	185.4	116.9	118.6	111.0	138.0	138.8
220.8	−87.61	210.7	103.4	206.0	190.6	162.5	157.9	122.1	165.6
234.8	−208.1	229.9	214.9	208.1	203.6	174.5	202.3	132.5	200.9

$b_d = 1.111 \cdot 10^{-2}$ m^{-1}; $v = 210$ m/s

85°		75°		60°		45°		30°	
x, m	y, m	x, m	y, m	x, m	y, m	x, m	y, m	x, m	y, m
1.854	21.13	5.589	20.80	11.38	19.64	17.59	17.50	24.33	13.94
9.270	102.7	16.77	61.78	22.76	39.08	35.18	34.76	48.67	27.54
12.98	138.4	33.54	119.4	34.13	58.20	70.35	67.71	73.00	40.51
16.69	165.1	50.31	160.4	45.51	76.80	87.94	82.59	121.7	62.04
20.39	175.9	61.48	167.9	56.89	94.54	105.5	95.42	146.0	67.99
25.96	151.2	72.66	150.7	79.65	125.2	140.7	109.3	170.0	67.38
27.81	127.0	78.25	128.5	102.4	143.0	175.9	90.85	194.7	55.38
29.66	85.75	83.84	89.11	125.2	136.5	193.5	55.21	219.0	21.59
31.52	−6.606	89.43	0.4781	147.9	82.71	211.1	−41.40	243.3	−81.64
32.44	−239.0	92.41	−247.0	165.4	−203.6	218.1	−213.0	250.6	−200.4

15°		0°		−15°		−30°		−45°	
x, m	y, m	x, m	y, m	x, m	y, m	x, m	y, m	x, m	y, m
19.74	5.235	26.82	0.0986	11.82	3.185	19.21	11.15	11.51	11.54
39.48	10.32	53.64	0.4962	35.46	9.702	38.41	22.49	34.53	34.92
59.22	15.17	80.46	1.437	59.10	16.53	57.62	34.13	57.55	59.04
78.96	19.66	107.3	3.363	82.74	23.89	76.82	46.26	69.06	71.59
98.70	23.56	134.1	7.082	106.4	32.18	115.2	73.54	80.56	84.70
138.2	28.18	160.9	14.08	141.8	48.01	134.4	90.22	92.07	98.60
157.9	27.60	187.7	27.19	165.5	63.01	153.6	111.1	115.1	130.7
197.4	14.35	214.6	52.40	189.1	85.31	172.9	139.5	126.6	150.5
236.9	−38.19	241.4	107.4	212.8	123.2	192.1	186.8	138.1	175.3
264.5	−211.0	262.8	263.0	236.4	210.8	195.9	200.7	147.3	201.3

Bibliography

1. Kuleshov, N.A. and Yu.N. Anistratov. 1968. Tekhnologiya otkrytykh gornykh rabot (Open-pit Mining Technology). Nedra, Moscow, 400 p.
2. Rzhevskii, V.V., Yu.N. Anistratov and S.A. Il'in. 1964. Otkrytye gornye raboty v slozhnykh usloviyakh (Open-pit Mining in Difficult Conditions). Nedra, Moscow, 294 p.
3. Rzhevskii, V.V. 1966. Tekhnologiya, mekhanizatsiya i avtomatizatsiya protsessov na kar' erakh (Technology, Mechanization and Automation of Processes in Open-pit Mines). Nedra, Moscow, 652 p.
4. Dokuchaev, M.M. 1960. Primenenie massovykh vzryvov v narodnom khozyaistve SSSR (Application of Large-scale Explosions in the Economy of the USSR). In *Primenenie massovykh vzryvov v gornoi promyshlennosti i stroitel' stve*. izd. GOSINTI, Moscow, p. 5–11.
5. Vasil'ev, G.A. 1960. Vskrytie Altyn-Topkanskogo mestorozhdeniya poleznykh iskopaemykh massovymi vzryvami na sbros (Overburden Removal at the Altyn-Topkanskii Mines through Large-scale Blasting). In *Primenenie massovykh vzryvov v gornoi promyshlennosti i stroitel' stve*. izd. GOSINTI, Moscow, p. 37–39.
6. Doronicheva, L.A. 1960. Massovye vzryvy dlya vskrytiya Baiinchanskogo mestorozhdeniya v KNR (Large-scale Explosions for Overburden Removal at the Baiinchanskii Mine in the Korean People's Democratic Republic). In *Primenenie massovykh vzryvov v gornoi promyshlennosti i stroitel' stve*. izd. GOSINTI, Moscow, p. 50–56.
7. Kuznetsov, V.M., M.A. Lavrent'ev and E.N. Sher. 1960. O napravlennom vybrose grunta pri pomoshchi VV (Directional Blasting of Rock with the Help of Explosives). PMTF, No. 4, p. 49–50.
8. Chernigovskii, A.A. 1965. Raschet ploskikh zaryadov dlya vskrytiya poleznykh iskopaemykh (Designing Slab Charges for Stripping Minerals). Nedra, Moscow, 96 p.
9. Chernigovskii, A.A. 1971. Metod ploskikh sistem zaryadov v gornom dele i stroitel' stve (Method of Slab System of Charges in Mining and Construction Work). Nedra, Moscow, 244 p.
10. Kuznetsov, V.M. and E.N. Sher. 1963. Napravlennyi vzryv v grunte

(Directed Explosion in the Soil). In *Vzryvnye raboty v sovremennykh usloviyakh*. No. 51/8, Gosgortekhizdat, Moscow, p. 22–39.

11. Paporotskii, L.A. 1963. Opytnye vzryvy s odnostoronnim napravlennym vybrosom (Experimental Explosions with Directional Blasting). In *Vzryvnye raboty v sovremennykh usloviyakh*. No. 51/8, Gosgortekhizdat, Moscow, p. 39–50.

12. Chernigovskii, A.A. 1963. Napravlennyi vybros gornoi porody sistemoi skvazhinnykh zaryadov (Directional Blasting of Rock by Deep-hole Charges). In *Vzryvnye raboty v sovremennykh usloviyakh*, No. 51/8, Gosgortekhizdat, Moscow, p. 13–22.

13. Baum, F.A., S.S. Grigoryan and N.S. Sanasaryan. 1964. Opredelenie impul' sa vzryva vdol' obrazvyushchei skvazhiny i optimal' nykh parametrov skvazhinnogo zaryada (Determining the Blast Pulse along the Generatrix of the Blast Hole and Calculating the Optimum Parameters of Deep-hole Charges). In *Vzryvnoe delo*. No. 54/11, Gosgortekhizdat, Moscow, p. 53–102.

14. Davydov, S.A. and V.A. Kuznetsov. 1970. Vzryvanie na vybros transheinymi zaryadami VV (Excavating Explosion Using Chamber Charges). In *Vzryvnoe delo*. No. 69/26, Nedra, Moscow, p. 168–174.

15. Tekhnicheskie pravila vedeniya vzryvnykh rabot v e'nergeticheskom stroitel' stve (Technical Rules for Blasting Operations in Construction of Power Plants). E'nergiya, Moscow, 1972, 208 p.

16. Andreev, K.K. and A.F. Belyaev. 1970. Teoriya vzryvchatykh veshchestv (Theory of Explosives). Oborongiz, Moscow, 596 p.

17. Baum, F.A., K.P. Stanyukovich and B.I. Shekhter. 1959. Fizika vzryva (Physics of Explosion). Fizmatgiz, Moscow, 800 p.

18. Pokrovskii, G.I. and A.A. Chernigovskii. 1963. Raschet zaryadov pri massovykh vzryvakh na vybros (Designing Charges for Large-scale Excavation Blast). Gosgortekhizdat, Moscow, 88 p.

19. Rodionov, V.N. and A.N. Romashov. 1959. O mekhanizme vybrosa grunta (Mechanism of Blasting Rock). In *Doklady soveshchaniya po narodnokhozyaistvennomu ispol'zovaniyu vzryva*. izd. SO Akademiya Nauk SSSR, Novosibirsk, p. 11–40.

20. Rodionov, V.N., A.N. Romashov and A.P. Sukhotin. 1959. Vzryv v grunte (Blasting in Rock). In *Doklady soveshchaniya po narodnokhozyaistvennomu ispol'zovaniyu vzryva*. Izd. SO Akademiya Nauk SSSR, Novosibirsk, p. 3–39.

21. Dokuchaev, M.M., A.N. Rodionov and A.N. Romashov. 1963. Vzryvy na vybroc (Excavation Blast). Izd. AN SSSR, Moscow, 108 p.

22. Pokrovskii, G.I., I.S. Fedorov and M.M. Dokuchaev. 1963. Primenenie napravlennogo vzryva v gidrotekhnicheskom stroitel' stve (Application of Directional Blasting in Hydraulic Engineering). Gosstroiizdat, Moscow, 224 p.

23. Lyakhov, G.M. and G.I. Pokrovskii. 1962. Vzryvnye volny v gruntakh (Blast Waves in Rock). Gosgortekhizdat, Moscow, 104 p.

24. Rodionov, V.N., I.A. Sizov and V.M. Tsvetkov. 1969. Issledovaniya polosti pri kamufletnom vzryve (Study of Hollow Space in Case of Implosion). In *Vzryvnoe delo*. No. 64/21, Nedra, Moscow, p. 5–24.

25. Khanukaev, A.N. 1962. E'nergiya voln napryazhenii pri razrushenii porod vzryvom (Energy of Stress Waves while Crushing Rock by Blasting). Gosgortekhizdat, Moscow, 200 p.

26. Pokrovski, G.I. and I.S. Fedorov. 1968. Tsentrobezhnoe modelirovanie v stroitel'nom dele (Centrifugal Modeling in Construction Work). Stroiizdat, Moscow, 248 p.

27. Assonov, V.A., P.A. Demchuk and D.S. Kuznetsov. 1964. Opredelenie optimal'noi dliny peschanoglinistoi zaboiki shpurov (Determining the Optimum Length of Sand and Loam Stemming in a Blast Hole). In *Vzryvnoe delo*. No. 55/12, Nedra, Moscow, p. 60–68.

28. Demchuk, P.A. and D.S. Kuznetsov. 1964. Opredelenie parametrov vodyanoi zaboiki shpurov c pomoshch'yu nomograficheskikh metodov (Determining the Parameters of Water Stemming of Blast Holes with the Help of Nomographic Methods). In *Vzryvnoe delo*. No. 55/12, Nedra, Moscow, p. 231–238.

29. Komir, V.M., I.A. Semenyuk and L.F. Petryashin. 1971. E'ksperimental'nye issledovaniya vliyaniya ukorochennoi zaboiki na rezul'taty vzryva (Experimental Analysis of Effect of Contracted Stemming on Results of Blasting). In *Vzryvnoe delo*. No. 70/27, Nedra, Moscow p. 279–285.

30. Antonyan, T.S. and E.P. Gorbacheva. 1968. Vliyanie zaboiki na e'ffektivnost' vzryva skvazhinnykh zaryadov (Influence of Stemming on Blast Effectiveness of Deep-hole Charges). In *Referativnaya informatsiya o peredovom opyte*. Series V, vyp. 5–6 (35–36): izd TsBTI, Moscow, p. 9–12.

31. Seinov, N.P., I.F. Zharikov, B.S. Vlasov and V.G. Udachin. 1972. Ob e'ffectivnosti primeneniya aktivnoi zaboiki (Effectiveness of Application of Active Stemming). In *Vzryvnoe delo*. No. 71/28, Nedra, Moscow, p. 134–139.

32. Kragel' skii, I.V. and N.E'. Vinogradova. 1962. Koe'ffitsienty treniya (Coefficients of Friction). Mashgiz, Moscow, 220 p.

33. Chernigovskii, A.A. 1970. Vneshnyaya ballistika i droblenie porody pri vzryve na vybros i sbros (External Ballistics and Crushing of Rock during Blasting and Projection of Muck). In *Vzryvnoe delo*. No. 69/26, Nedra, Moscow, p. 66–78.

34. Tekhnicheskie pravila vedeniya vzryvnykh rabot na dnevnoi poverkhnosti (Technical Rules for Conducting Explosions on an Open Surface). Nedra, Moscow, 1972, p. 240.

35. Baron, L.I. 1960. Kuskovatost' i metody ee izmereniya (Lumpiness and Methods of its Measurement). Izd. AN SSSR, Moscow, 124 p.
36. Okunev, B.N. 1943. Osnovy ballistiki (Fundamentals of Ballistics). Voenizdat, Moscow, 524 p.
37. Pokrovskii, G.I. 1959. Novaya forma napravlennogo deistviya vzryva (A New Technique of Directional Blasting). In *Doklady soveshchaniya po narodnokhozyaistvennomu ispol'zovaniyu vzryva*. Izd. AN SSSR, Novosibirsk, p. 11–17.
38. Kutuzov, B.N. and V.K Rubtsov. 1971. Fizika vzryvnogo razrusheniya gornykh porod (Physics of Rock Crushing by Blasting). Izd. MGI, Moscow, 178 p.
39. Rubtsov, V.K. 1976. Raschety zadannogo vykhoda krupnykh i melkykh kuskov porody na kar'erakh (Calculating the Throw of Large and Small Rock Fragments in Open-pit Mines). In *Vzryvnoe delo*. No. 62/19, Nedra, Moscow, p. 84–99.
40. Sukhanov, A.F. and B.N. Kutuzov. 1967. Razrushenie gornykh porod (Rock Fragmentation). Nedra, Moscow, 340 p.
41. Gaek, Yu.V., M.F. Drukovannyi and V.V. Mishin. 1963. O skorosti rasprostraneniya treshchin v gornykh porodakh i tverdykh telakh i metody ee izmereniya (Fissure Propagation Velocity in Rock and Solids and Methods of its Measurement). In *Vzryvnoe delo*. No. 51/8, Gosgortekhizdat, Moscow, p. 85–96.
42. Marchenko, L.N. 1964. Issledovaniya protsessa obrazovaniya i razvitiya treshchin v tverdykh sredakh v zavisimosti ot konstruktsii zaryada (Studying the Process of Formation and Development of Fissures in Hard Mediums Depending on the Design of the Charge). In *Vzryvnoe delo*. No. 54/11, Nedra, Moscow, 224 p.
43. Mosinets, V.N. 1967. Usloviya khrupkogo i plasticheskogo razrusheniya gornykh porod vzryvom (Condition for Brittle and Plastic Breaking of Rocks by Explosion). In *Problema razrusheniya gornykh porod vzryvom*. Nedra, Moscow, p. 77–100.
44. Maksimova, E.P. 1962. Laboratornye issledovaniya kharakteristik kuskovatosti pri droblenii vzryvom (Laboratory Analysis of Characteristics of Lumpiness in Case of Crushing through Blasting). In *Vzryvoe delo*. No. 50/7, Gosgortekhizdat, Moscow, p. 103–109.
45. Baum, F.A. 1963. Protsessy razrusheniya gornykh porod vzryvom (Processes of Crushing Rock through Blasting). In *Vzryvnoe delo*. No. 52/9, Gosgortekhizdat. Moscow, p. 263–285.
46. Filippov, V.K. 1961. Napravlenie rasprostraneniya treshchin, obrazovavshikhsya pri razrushenii krepkikh gornykh porod vzryvom (Direction of Spread of Fissures Formed while Crushing Rock through Blasting). In *Vzryvnoe delo*. No. 47/4, Gosgortekhizdat, Moscow, p. 172–177.
47. Drukovannyi, M.F. 1961. O mekhanizme razrusheniya gornykh porod

pri korotkozamedlennom vzryvanii (Rock-breaking Mechanism in Case of Short-delay Firing). In *Vzryvnoe delo*. No. 47/4, Gosgortekhizdat, Moscow, p. 184–196.

48. Zakalinskii, V.M. 1963. Issledovanie otboiki rudy gruppovymi kontsentratsionnymi skvazhinami metodom e' lektrogidtrodinamicheskikh analogii (Studying the Tamping of Ore with a Series of Converging Holes Using the Method of Electrohydrodynamic Analogies). Izd. AN SSSR, Series '*Metallurgiya i gornoe delo*', No. 5, p. 159–166.

49. Bud'ko, A.V. 1963. O novom metode massovoi otboiki rudy puchkami sblizhennykh skvazhin (New Method of Large-scale Tamping of Ore by a Series of Converging Holes). *Gornyi zhurnal*, No. 4, p. 40–41.

50. Bud'ko, A.V. and V.M. Zakalinskii. 1964. Otboika rudy puchkami sblizhennykh impul'snykh skvazhin s primeneniem mnogoshpindel'-nykh burovykh stankov (Tamping the Ore with a Series of Converging Sprung Holes by Using Multiple Drilling Machines). In *Referativnaya informatsiya o peredovom opyte*. Izd. TsBTI, Moscow, p. 10–12.

51. Sundukov, A.A. and V.P. Sas. 1970. Vzryvanie mergelei pri pomoshchi parnosblizhennykh skvazhinnykh zaryadov diametrom 400 mm (Blasting of Marl with the Help of Paired-converging Holes of 400 mm Diameter). In *Vzryvnoe delo*. No. 69/26, Nedra, Moscow, p. 211–216.

52. Korovin, V.D. 1970. O rezul'tatakh primeneniya parno-sblizhennykh skvazhinnykh zaryadov v karbonatnykh porodakh kar' erov Samarskoi Luki (Results of Application of Paired Converging-hole Charges in Calcareous Rocks in Open-pit Mines of Samarskoi Luki). In *Vzryvnoe delo*. No. 69/26, Nedra, Moscow, p. 216–220.

53. Gushchin, V.P., V.I. Lokhov, V.P. Sorokin and M.I. Shadov. 1971. Issledovaniye e'ffectivnosti primeneniya parallel'nykh parnosblizhennykh sploshnykh kolonkovykh zaryadov na Khramtsovskom kar'ere No. 1–2 tresta Cheremkhovugol' (Studying the Effectiveness of Application of Parallel Pair-converging Continuous Deep-hole Charges in the Khramtsov Open-pit Mines Nos. 1–2 of the Cheremkhovugol' Company). In *Vzryvnoe delo*. No. 70/27, Nedra, Moscow, p. 257–261.

54. Petryashin, L.F., V.D. Petrenko, V.S. Kravtsov and N.F. Borodin. 1971. Opredelenie optimal'nogo rasstoyaniya mezhdu parno-sblizhennymi skvazhinami dlya uslovii granitnykh kar'erov (Determining Optimum Distance between Pair-converging Holes for Granite Open-pit Mines). In *Vzryvnoe delo*. No. 70/27, Nedra, Moscow, p. 261–268.

55. Petryashin, L.F., V.S. Kravtsov, V.D. Petrenko and N.F. Borodin. 1971. Opredelenie ratsional'nogo diametra parno-sblizhennykh skvazhinnykh zaryadov (Determining Rational Diameter of Pair-converging Hole Charges). In *Vzryvnoe delo*. No. 70/27, Nedra, Moscow, p. 268–271.

56. Levchik, S.P. and N.P. Seinov. 1963. K voprosu o deistvii vzryva v tverdoi srede (Problem of Effect of Blasting in Hard Medium). Izv. AN

SSSR, Series 'Metallurgiya i gornoe delo', No. 3, p. 170–174.

57. Demidyuk, G.P. and V.S. Ivanov. 1963. Vliyanie formy odinochnogo zaryada na droblenie tverdoi sredy vzryvom (Effect of Shape of Single Charge on Crushing of a Hard Medium through Blasting). In *Vzryvoe delo*. No. 53/10, Gosgortekhizdat, Moscow, p. 47–58.

58. Toporkov, V.A. 1961. Vliyanie diametra i dliny zaryada na stepen' drobleniya porody (Effect of Diameter and Length of Charge on the Degree of Crushing of Rock). In *Vzryvnoe delo*. No. 47/4, Gosgortekhizdat, Moscow, p. 205–217.

59. Azarkovich, A.E., B.I. Chistosredov and S.A. Davydov. 1960. O vliyanii rasstoyaniya mezhdu skvazhinnymi i shpurovymi zaryadami na rezul'taty vzryva (Effect of the Distance between Deep-hole and Blast-hole Charges on the Explosion Result). In *Vzryvnoe delo*. No. 45/2, Gosgortekhizdat, Moscow, p. 63–75.

60. Danchev, P.S. 1962. O zavisimosti e'ffekta drobleniya sredy ot koe'ffitsienta zaryazhaniya (Dependence of Effect of Crushing on Charging Coefficient). In *Vzryvnoe delo*. No. 50/7, Gosgortekhizdat, Moscow, p. 71–83.

61. Demidyuk, G.P. 1964. K voprosu upravleniya deistviem vzryva skvazhinnykh zaryadov (Problem of Controlling the Effect of Deep-hole Charges). In *Vzryvnoe delo*. No. 54/11, Nedra, Moscow, p. 174–185.

62. Mel'nikov, N.V. 1964. Povyshenie poleznoi raboty vzryva pri otboike poleznykh iskopaemykh (Improving the Effectiveness of Blasting in Strip Mining). In *Vzryvnoe delo*. No. 54/11, Nedra, Moscow, p. 7–34.

63. Marchenko, L.N. 1965. Uvelichenie e'ffectivnosti vzryva pri dobyvanii poleznykh iskopaemykh (Increasing Blast Effectiveness while Crushing Mineral Ore). Nauka, Moscow, 224 p.

64. Pokrovskii, G.I. and I.S. Fedorov. 1967. Deistvie udara i vzryva v deformiruemykh sredakh (Effect of Impact and Explosions in Deformable Mediums). Stroiizdat, Moscow, 276 p.

65. Turuta, N.U., Yu.I. Blagodarenko, D.F. Panchenko, and A.V. Karpinskii. 1972. E'ksperimental'nye issledovaniya polya skorostei smeshcheniya pri vzryve v treshchinovatykh gornykh porodakh (Experimental Investigations of Field of Displacement Rates through Blasting in Fissured Rocks). In *Vzryvnoe delo*. No. 71/28, Nedra, Moscow, p. 118–133.

66. Mosinets, V.N. 1963. E'nergeticheskie i korrelyatsionnye svyazi protsessa razrusheniya porod vzryvom (Energy and other Correlations Describing the Process of Breaking Rock through Blasting). Izd-vo Kirgiz, AN SSSR, Frunze, 232 p.

67. Mosinets, V.N. and I.A. Tangaev. 1967. Nekotorye osobennosti protsessa razrusheniya gornykh porod vzryvom (Some Peculiarities of the Process of Crushing Rock through Blasting). In *Problema razrusheniya gornykh porod vzryvom*. Nedra, Moscow, p. 109–125.

68. Vlasov, O.E. and S.A. Smirnov. 1972. Osnovy rascheta drobleniya porod vzryvom (Fundamentals of Calculating the Crushing of Rock through Blasting). Izd-vo AN SSSR, Moscow, 104 p.

69. Drukovannyi, M.F., Yu.V. Gaek and V.V. Mishin. 1962. K voprosu o vliyanii treshchinovatosti na kharakter razrusheniya porodnogo massiva vzryvom (Effect of Fissures on the Nature of Breaking of a Massif through Blasting). In *Vzryvnoe delo*. No. 50/7, Gosgortekhizdat, Moscow, p. 31–44.

70. Panchenko, D.F. 1969. Skorosti treshchinoobrazovaniya i sdvizheniya pri vzryvanii skvazhinnykh zaryadov v treshchinovatom massive (Rate of Fissure Formation and Movement during Firing of Deep-hole Charges in a Fissured Massif). In *Vzryvnoe delo*. No. 67/24, Nedra, Moscow, p. 83–89.

71. Turuta, N.U. and A.V. Bruyakin. 1965. Razrushenie treshchinovatykh gornykh porod vzryvom pri razlichnykh parametrakh zaryadov vzryvnykh veshchestv (Breaking of Fissured Rock by Blasting with Different Parameters of Explosive Charges). In *Vzryvnoe delo*. No. 57/14, Nedra, Moscow, p. 82–90.

72. Rubtsov, V.K. 1965. Granulometricheskii sostav massivov gornykh porod pri vzryvnoi otboike (Granulometric Composition of Rock Massifs during Explosive Cutting). In *Vzryvnoe delo*. 59/16, Nedra, Moscow, p. 64–70.

73. Kutuzov, B.N. and V.K. Rubtsov. 1969. O zavisimosti fraktsionnogo sostava vzorvannoi massy ot srednego diametra kuska (Dependence of Fractional Composition of Blasted Mass on Average Diameter of Fragments). *Gornyi zhurnal*, No. 12, p. 33–35.

74. Turuta, N.U., A.T. Galimullin, D.F. Panchenko and A.V. Karpinskii. 1967. Issledovanie razrusheniya krepkikh porod vzryvom dlya dostizheniya bol'shei stepeni drobleniya porod (Study of Breaking of Large Rocks by Blasting for Achieving a High Degree of Crushing). In *Vzryvnoe delo*. No. 62/19, Nedra, Moscow, p. 104–111.

75. Rubtsov, V.K. 1962. Osobennosti razrusheniya treshchinovatykh massivov deistviem vzryva (Peculiarities of Breaking Fissured Massifs through Blasting). In *Vzryvnoe delo*. No. 50/7, Gosgortekhizdat, Moscow, p. 114–121.

76. Turuta, N.U., A.T. Galimullin, D.F. Panchenko and A.V. Karpinskii. 1966. Issledovanie vliyaniya estestvennoi treshchinovatosti izvestnyakov na e'ffectivnost' vzryvnykh rabot (Studying the Effect of Natural Fissures in Limestone on the Effectiveness of Blasting). In *Vzryvnoe delo*. No. 59/16, Nedra, Moscow, p. 181–193.

77. Novozhilov, M.G., M.F. Drukovannyi and L.M. Geiman. 1963. Vliyanie diametra zaryada na intensivnost' drobleniya khrupkikh tel vzryvom (Effect of Diameter of the Charge on Intensity of Crushing a Triable

Medium through Blasting). In *Vzryvnoe delo*. No. 53/10, Gosgortekhizdat, Moscow, p. 59–76.

78. Licheli, G.P. and L.I. Baron. 1963. Issledovanie vliyaniya treshchinovatosti na droblenie sredy vzryvom pri otsutstvii poverkhnosti obnazheniya (Studying the Effect of Fissures on Crushing of the Medium through Blasting in the Absence of a Free Surface). In *Vzryvnoe delo*. No. 53/10, Gosgortekhizdat, Moscow, p. 83–98.

79. Baron, L.M. and G.P. Licheli. 1962. Issledovanie drobyashchei sposobnosti vzryvchatykh veshchestv pri vzryvanii treshchinovatykh porod (Studying the Crushing Property of Explosives while Blasting Fissured Rocks). In *Vzryvnoe delo*. No. 50/7, Gosgortekhizdat, Moscow, p. 83–98.

80. Drukovannyi, M.F., V.M. Komir and N.I. Myachina. 1966. Ob otsenke ispol'zovaniya e'nergii vzryva pri razlichnykh parametrakh burovzryvnykh rabot (Utilization of Blast Energy with Different Parameters for Drilling and Blasting Operations). In *Vzryvnoe delo*. No. 59/16, Nedra, Moscow, p. 70–76.

81. Usik, I.N. 1971. O ratsional'noi vysote ustupa pri vzryvanii v zazhatoi srede (Rational Height of Ledge in Case of Blasting in a Compressed Medium). In *Vzryvnoe delo*. No. 70/27, Nedra, Moscow, p. 152–161.

82. Drukovannyi, M.F., E'.I. Efremov, I.N. Usik and V.A. Glyavin. 1971. Otsenka e'konomicheskoi e'ffectivnosti vzryvaniya na neubrannuyu gornuyu massu (Evaluating the Cost Effectiveness of Blasting by Directional Technique without Involving a Mode of Mechanical Transport). In *Vzryvnoe delo*. No. 70/27, Nedra, Moscow, p. 120–124.

83. Rodionov, V.N. 1963. K voprosu o povyshenii e'ffectivnosti vzryva v tverdoi srede (Problem of Improving the Effectiveness of Blasting in a Hard Medium). In *Vzryvnoe delo*. No. 51/8, Gosgortekhizdat, Moscow, p. 50–60.

84. Kozhevnikov, V.I. 1966. Parametry burovzryvnykh rabot s primeneniem konturnogo vzryvaniya v podzemnykh vyrabotkakh (Parameters of Drilling and Blasting in Mines). In *Vzryvnoe delo*. No. 61/18, Nedra, Moscow, p. 5–11.

85. Davydov, S.A. and N.A. Palivoda. 1966. Konturnoe vzryvanie na otkrytykh razrabotkakh (Peripheral Blasting in Open-pit Mines). In *Vzryvnoe delo*. No. 61/18, Nedra, Moscow, p. 11–21.

86. Feshchenko, A.A. 1966. Konturnoe vzryvanie na stroitel'stve Krasnoyarskoi GE'S (Peripheral Blasting for the Construction of the Krasnoyar Hydroelectric Power Plant). In *Vzryvnoe delo*. No. 61/18, Nedra, Moscow, p. 21–34.

87. Yuzhakov, S.V. and V.I. Sheiman. 1970. Primenenie konturnogo vzryvaniya pri massovom vzryve na rykhlenie kamernymi zaryadami (Application of Peripheral Charges in Large-scale Blasting for Loosening with

Chamber Charges). In *Vzryvnoe delo*. No. 69/26, Nedra, Moscow, p. 155–161.

88. Kamenka, B.I. 1970. Osobennosti tekhnologii i perspektivy razvitiya konturnogo vzryvaniya (Special Feature of Technology and Prospects for Peripheral Blasting). In *Vzryvnoe delo*. No. 69/26, Nedra, Moscow, p. 161–167.

89. Gabrie'lyan, S.S. 1969. Kombinirovannoe peremeshchenie porod pri bestransportnoi sisteme razrabotki (Combined Displacement of Rock in case of Mining without Using a Means of Transportation). *'Ugol'* No. 1, p. 39–40.

90. Osnovnye printsipy e'ffectivnosti razrabotki mestorozhdenii otkrytym sposobom (Basic Principles of Effective Mining Work by Open-pit Method). In *E'kspress-informatsiya, 'Ugol'naya promyshlennost'*. 1968, No. 27, p. 11–12.

91. Baron, V.L. and L.I. Baron. 1967. K otsenke otboiki naklonnymi skvazhinami v kar'erakh kak metoda potochnoi tekhnologii (Evaluation of Cutting by Inclined Blast Holes in Open-pit Mines as a Method of Mass Production Technology). In *Vzryvnoe delo*. No. 62/19, Nedra, Moscow, p. 31–35.

92. Pashkov, A.D., L.S. Kaonstantinov and V.S. Kurtov. 1967. Tekhniko-e'konomicheskaya otsenka metoda naklonnykh skvazhin (Techno-economic Evaluation of Inclined-hole Method). In *Vzryvnoe delo*. No. 62/19, Nedra, Moscow, p. 230–236.

93. Dokuchaev, M.M., A.T. Galimullin, N.U. Turuta and M.M. Zaitsev. 1971. Vzryvanie naklonnymi skvazhinnymi zaryadami na kar'erakh (Blasting by Inclined Deep-hole Charges in Open-pit Mines). Nedra, Moscow, 208 p.

94. Gluskin, L.I., P.F. Korsakov and A.A. Kozhevnikov. 1964. Issledovanie e'ffectivnosti vzryvaniya naklonnykh skvazhinnykh zaryadov malogo diametra v gneiso-granitakh (Studying the Effectiveness of Explosion of Inclined Deep-hole Charges of Small Diameter in Gneiso-granites). In *Vzryvnoe delo*. No. 54/11, Nedra, Moscow, p. 137–145.

95. Danchev, P.S., Ya.M. Puchkov and V.V. Vetluzhskikh. 1964. Vliyanie vremeni zamedleniya na kachestvo drobleniya gornoi massy, vzorvan-noi skvazhinnymi zaryadami (Effect of Delay on the Quality of Crushing of Rock Exploded through Deep-hole Charges). In *Vzryvnoe delo*. No. 55/12, Nedra, Moscow, p. 188–195.

96. Krivtsov, V.A. 1967. E'ksperimental'noe opredelenie radiusa tsilindri-cheskoi polosti v zavisimosti ot velichiny zaglubleniya zaryada (Experimental Calculation of Radius of Cylindrical Cavity, Depending on the Depth of the Charge). *'Fizika goreniya i vzryva'*, No. 3, p. 449–454.

97. Demidyuk, G.P. 1967. Primenenie e'nergeticheskogo printsipa k ras-chetu skvazhinnykh zaryadov na kar'erakh (Application of the Energy

Principle for Calculating Deep-hole Charges for Open-pit Mines). In *Vzryvnoe delo*. No. 62/19, Nedra, Moscow, p. 36–51.

98. Novozhilov, M.G., F.I. Kucheryavyi, V.S. Khokhryakov et al. 1971. Tekhnologiya otkrytoi razrabotki mestorozhdeniya poleznykh iskopaemykh (Technology of Open-pit Mining). Part I, Nedra, Moscow, 512 p.

99. Baron, L.I., B.D. Rossi and S.P. Levchik. 1960. Drobyashchaya sposobnost' vzryvchatykh veshchestv dlya gornykh porod (Crushing Property of Explosives used in Rock Blasting). Gosgortekhizdat, Moscow, 112 p.

100. Ivanov, N.S. 1963. Issledovanie diametra zaryada na e'ffectivnost' vzryva v sloistoi srede (Studying the Diameter of the Charge for Effectiveness of Blasting in a Stratified Medium). In *Vzryvnoe delo*. No. 53/10, Gosgortekhizdat, Moscow, p. 76–89.

101. Mosinets, V.N. 1971. Deformatsiya gornykh porod vzryvom (Deformation Caused in Rocks by Blasting). Izd-vo Ilim, Frunze, 188 p.

102. Zhunusov, K.K. 1953. Sovershenstvovanie burovzryvnykh rabot na kar'erakh Altaya (Improvement in Drilling and Blasting Work in Open-pit Mines of the Altai Mountains). In *Vzryvnoe delo*. No. 51/8, Gosgortekhizdat, Moscow, p. 175–186.

103. Burshtein, M.P. and N.P. Seinov. 1964. Opyt primeneniya skvazhinnykh zaryadov, rassredotochennykh vozdushnymi promezhutkami, na kar'erakh Altyn-Topkanskogo kombinata (Experience of Using Deep-hole Charges, Separated by Air Gaps, in the Open-pit Altyn-Topkan Mines). In *Vzryvnoe delo*. No. 54/11, Nedra, Moscow, p. 257–265.